"十三五"国家重点出版物出版规划项目
卓越工程能力培养与工程教育专业认证系列规划教材
（电气工程及其自动化、自动化专业）
普通高等教育"十一五"国家级规划教材

自动化专业英语教程
Specified English for Automation

第4版

主　编	王宏文		
副主编	李练兵	刘作军	
参　编	李　洁	王　萍	孙进生
	陈志军	林　燕	耿　昕
	薛忠辉	暴永辉	綦　建
	江春冬	孙　昊	岳大为
	孙曙光	梁　涛	雷兆明
主　审	杨　鹏	李彦平	

机械工业出版社

本书是"十三五"国家重点出版物出版规划项目 卓越工程能力培养与工程教育专业认证系列规划教材、普通高等教育"十一五"国家级规划教材,是针对高等工科院校自动化专业"科技英语阅读"课程的需要、在第 3 版的基础上修订而成的。本书包括电气与电子工程基础、控制理论、计算机控制技术、过程控制系统、网络化与信息化控制及自动化技术的综合应用 6 部分,内容涉及智能控制综述、DSP、嵌入式系统、电力系统自动化、智能电网、大数据应用、知识自动化、云计算、智慧城市和智慧企业,并新增了人工智能研究的前沿热点问题综述、人工智能在机器学习领域的应用等内容,涵盖了自动化专业各个发展方向,内容新颖、全面、系统、精炼。每篇文章后都附有词汇表和注解,并配有 30 篇英语翻译及应用文知识,专业、学科介绍,自动化学科相关的期刊、会议、科技前沿等诸多内容,使读者在学习并掌握专业词汇和翻译技能的同时开阔眼界。本书可作为自动化专业本科生及研究生专业英语课程的教材,也可供有关工程技术人员参考。

图书在版编目（CIP）数据

自动化专业英语教程/王宏文主编. —4 版. —北京：机械工业出版社，2018.12（2024.12 重印）

普通高等教育"十一五"国家级规划教材 "十三五"国家重点出版物出版规划项目 卓越工程能力培养与工程教育专业认证系列规划教材. 电气工程及其自动化、自动化专业

ISBN 978-7-111-61331-2

Ⅰ. ①自⋯ Ⅱ. ①王⋯ Ⅲ. ①自动化技术–英语–高等学校–教材 Ⅳ. ①TP1

中国版本图书馆 CIP 数据核字（2018）第 258770 号

机械工业出版社（北京市百万庄大街22号 邮政编码100037）
策划编辑：于苏华 责任编辑：于苏华 王 康 刘琴琴
责任校对：李 伟 封面设计：鞠 杨
责任印制：单爱军
北京虎彩文化传播有限公司印刷
2024 年 12 月第 4 版第 10 次印刷
184mm×260mm・18.75 印张・459 千字
标准书号：ISBN 978-7-111-61331-2
定价：59.80 元

电话服务 网络服务
客服电话：010-88361066 机 工 官 网：www.cmpbook.com
　　　　　010-88379833 机 工 官 博：weibo.com/cmp1952
　　　　　010-68326294 金 书 网：www.golden-book.com
封底无防伪标均为盗版 机工教育服务网：www.cmpedu.com

"十三五"国家重点出版物出版规划项目
卓越工程能力培养与工程教育专业认证系列规划教材
（电气工程及其自动化、自动化专业）
编审委员会

主任委员

  郑南宁 中国工程院 院士，西安交通大学 教授，中国工程教育专业认证协会电子信息与电气工程类专业认证分委员会 主任委员

副主任委员

  汪槱生 中国工程院 院士，浙江大学 教授
  胡敏强 东南大学 教授，教育部高等学校电气类专业教学指导委员会 主任委员
  周东华 清华大学 教授，教育部高等学校自动化类专业教学指导委员会 主任委员
  赵光宙 浙江大学 教授，中国机械工业教育协会自动化学科教学委员会 主任委员
  章　兢 湖南大学 教授，中国工程教育专业认证协会电子信息与电气工程类专业认证分委员会 副主任委员
  刘进军 西安交通大学 教授，教育部高等学校电气类专业教学指导委员会 副主任委员
  戈宝军 哈尔滨理工大学 教授，教育部高等学校电气类专业教学指导委员会 副主任委员
  吴晓蓓 南京理工大学 教授，教育部高等学校自动化类专业教学指导委员会 副主任委员
  刘　丁 西安理工大学 教授，教育部高等学校自动化类专业教学指导委员会 副主任委员
  廖瑞金 重庆大学 教授，教育部高等学校电气类专业教学指导委员会 副主任委员
  尹项根 华中科技大学 教授，教育部高等学校电气类专业教学指导委员会 副主任委员
  李少远 上海交通大学 教授，教育部高等学校自动化类专业教学指导委员会 副主任委员
  林　松 机械工业出版社 编审 副社长

委　　员（按姓氏笔画排序）

于海生	青岛大学 教授		吴成东	东北大学 教授
王　平	重庆邮电大学 教授		吴美平	国防科技大学 教授
王　超	天津大学 教授		谷　宇	北京科技大学 教授
王再英	西安科技大学 教授		汪贵平	长安大学 教授
王志华	中国电工技术学会 教授级高级工程师		宋建成	太原理工大学 教授
			张　涛	清华大学 教授
王明彦	哈尔滨工业大学 教授		张卫平	北方工业大学 教授
王保家	机械工业出版社 编审		张恒旭	山东大学 教授
王美玲	北京理工大学 教授		张晓华	大连理工大学 教授
韦　钢	上海电力学院 教授		黄云志	合肥工业大学 教授
艾　欣	华北电力大学 教授		蔡述庭	广东工业大学 教授
李　炜	兰州理工大学 教授		穆　钢	东北电力大学 教授
吴在军	东南大学 教授		鞠　平	河海大学 教授

序

工程教育在我国高等教育中占有重要地位，高素质工程科技人才是支撑产业转型升级、实施国家重大发展战略的重要保障。当前，世界范围内新一轮科技革命和产业变革加速进行，以新技术、新业态、新产业、新模式为特点的新经济蓬勃发展，迫切需要培养、造就一大批多样化、创新型卓越工程科技人才。目前，我国高等工程教育规模世界第一。我国工科本科在校生约占我国本科在校生总数的 1/3，近年来我国每年工科本科毕业生约占世界总数的 1/3 以上。如何保证和提高高等工程教育质量，如何适应国家战略需求和企业需要，一直受到教育界、工程界和社会各方面的关注。多年以来，我国一直致力于提高高等教育的质量，组织并实施了多项重大工程，包括卓越工程师教育培养计划（以下简称卓越计划）、工程教育专业认证和新工科建设等。

卓越计划的主要任务是探索建立高校与行业企业联合培养人才的新机制，创新工程教育人才培养模式，建设高水平工程教育教师队伍，扩大工程教育的对外开放。计划实施以来，各相关部门建立了协同育人机制。卓越计划要求试点专业要大力改革课程体系和教学形式，依据卓越计划培养标准，遵循工程的集成与创新特征，以强化工程实践能力、工程设计能力与工程创新能力为核心，重构课程体系和教学内容；加强跨专业、跨学科的复合型人才培养；着力推动基于问题的学习、基于项目的学习、基于案例的学习等多种研究性学习方法，加强学生创新能力训练，"真刀真枪"做毕业设计。卓越计划实施以来，培养了一批获得行业认可、具备很好的国际视野和创新能力、适应经济社会发展需要的各类型高质量人才，教育培养模式改革创新取得突破，教师队伍建设初见成效，为卓越计划的后续实施和最终目标的达成奠定了坚实基础。各高校以卓越计划为突破口，逐渐形成各具特色的人才培养模式。

2016 年 6 月 2 日，我国正式成为工程教育"华盛顿协议"第 18 个成员，这标志着我国工程教育真正融入世界工程教育，人才培养质量开始与其他成员达到了实质等效，同时，也为以后我国参加国际工程师认证奠定了基础，为我国工程师走向世界创造了条件。专业认证把以学生为中心、以产出为导向和持续改进作为三大基本理念，与传统的内容驱动、重视投入的教育形成了鲜明对比，是一种教育范式的革新。通过专业认证，把先进的教育理念引入了我国工程教育，有力地推动了我国工程教育专业教学改革，逐步引导我国高等工程教育实现从课程导向向产出导向转变、从以教师为中心向以学生为中心转变、从质量监控向持续改进转变。

在实施卓越计划和开展工程教育专业认证的过程中，许多高校的电气工程及其自动化、自动化专业结合自身的办学特色，引入先进的教育理念，在专业建设、人才培养模式、教学内容、教学方法、课程建设等方面积极开展教学改革，取得了较好的效果，建设了一大批优质课程。为了将这些优秀的教学改革经验和教学内容推广给广大高校，中国工程教育专业认证协会电子信息与电气工程类专业认证分委员会、教育部高等学校电气类专业教学指导委员会、教育部高等学校自动化类专业教学指导委员会、中国机械工业教育协会自动化学科教学委员会、中国机械工业教育协会电气工程及其自动化学科教学委员会联合组织规划了"卓越工程能力培养与工程教育专业认证系列规划教材（电气工程及其自动化、自动化专业）"。本套教材通过国家新闻出版广电总局的评审，入选了"十三五"国家重点图书。本套教材密切联系行业和市场需求，以学生工程能力培养为主线，以教育培养优秀工程师为目标，突出学生工程理念、工程思维和工程能力的培养。本套教材在广泛吸纳相关学校在"卓越工程师教育培养计划"实施和工程教育专业认证过程中的经验和成果的基础上，针对目前同类教材存在的内容滞后、与工程脱节等问题，紧密结合工程应用和行业企业需求，突出实际工程案例，强化学生工程能力的教育培养，积极进行教材内容、结构、体系和展现形式的改革。

经过全体教材编审委员会委员和编者的努力，本套教材陆续跟读者见面了。由于时间紧迫，各校相关专业教学改革推进的程度不同，本套教材还存在许多问题。希望各位老师对本套教材多提宝贵意见，以使教材内容不断完善提高。也希望通过本套教材在高校的推广使用，促进我国高等工程教育教学质量的提高，为实现高等教育的内涵式发展贡献一份力量。

卓越工程能力培养与工程教育专业认证系列规划教材
（电气工程及其自动化、自动化专业）
编审委员会

前　言

本书第3版自2015年问世以来，得到许多学校师生的喜爱。经过4年的教学实践，对学生了解智能电网、大数据应用、知识自动化、云计算、智慧城市和智慧企业等自动化前沿研究、应用热点起到了有益的作用。在广泛采纳兄弟院校师生建议的基础上，为了响应教育部高等学校自动化类专业教学指导委员会的教改精神，及科技部于2017年发布的《新一代人工智能发展规划》，对教材内容做了进一步修订，新增加了人工智能研究的前沿热点问题综述和人工智能在机器学习领域的应用两篇文章。

编写本书的指导思想是："内容新颖、全面、系统、精炼，既重视学科基础知识又反映学科发展前沿动态"，同时新增了专业介绍、学科相关科技文献网站、自动化专业的科技前沿、学术会议等辅助内容。"见多才能识广"，希望本书对培养有开拓精神、综合素质强的科技创新型人才有所帮助。

全书包括电气与电子工程基础、控制理论、计算机控制技术、过程控制系统、网络化与信息化控制、自动化技术的综合应用6部分共30个单元，覆盖9万余字的专业词汇量。

本书由河北工业大学王宏文教授担任主编，河北工业大学李练兵教授、刘作军教授担任副主编。书中PART 1的UNIT 1由天津工业大学綦建教授编写；PART 1的UNIT 2、UNIT 3和PART 2的UNIT 6由天津工业大学王萍教授编写；PART 2的UNIT 1~5由河北工业大学李练兵教授编写；PART 3的UNIT 1A、UNIT 2A由李练兵教授和河北科技大学陈志军教授编写；PART 3的UNIT 1B、UNIT 2B、UNIT 4A由陈志军教授和河北科技大学薛忠辉教授编写；PART 1的UNIT 5B、UNIT 6B由河北理工大学孙进生教授编写；PART 1的UNIT 4、UNIT 5A、UNIT 6A由河北工业大学江春冬副教授编写；PART 2的UNIT 7由河北工业大学孙昊副教授编写；PART 3的UNIT 3、UNIT 5由河北工业大学岳大为副教授编写；PART 3的UNIT 4B由河北工业大学孙曙光副教授编写；PART 4由河北工业大学耿昕硕士编写；PART 5的UNIT 1、UNIT 2由河北工业大学梁涛教授编写；PART 5的UNIT 3由河北工业大学雷兆明副教授编写；PART 5的UNIT 4由河北工业大学李洁博士编写；PART 6的UNIT 1、UNIT 2、UNIT 3由河北工业大学王宏文教授编写；PART 4的UNIT 4C由河北工业大学林燕研究馆员、刘作军教授编写；PART 5的UNIT 1C、UNIT 2C由林燕研究馆员编写；PART 4的UNIT 1C、UNIT 3C由河北工业大学暴永辉硕士编写；其余章节的C部分均由刘作军教授编写。王宏文教授对全书进行总编和修改更正，河北工业大学杨鹏教授、沈阳大学李彦平教授担任主审。

在此对参加本书第1版编写工作的天津理工大学陈在平教授，以及为本书第1版的出版提供大力帮助的孙鹤旭教授、杨鹏教授表示由衷的感谢！河北工业大学2004级硕士研究生陈悦，2005级硕士研究生刘通学、黄伟杰、王艳霞、刘丽，2013级硕士研究生吴红星、曹泽华、侯美杰、孟立新，2014级硕士研究生郭章亮、雷盼云、宁乐参加了本书部分章节的计算机绘图与文字校对工作，在此一并表示感谢。

  本书有完备的参考译文供任课教师使用，通信方式：18522018700@163.com，联系人：王宏文。也可登录机械工业出版社教育服务网（www.cmpedu.com）获取相关资源。

编 者

目 录
CONTENTS

序
前言

PART 1　Electrical and Electronic Engineering Basics ················· 1

UNIT 1　A　Electrical Networks ··· 1
　　　　B　Three-Phase Circuits ··· 4
　　　　C　专业英语(Specified English)概述 ································· 7
UNIT 2　A　The Operational Amplifier ·· 9
　　　　B　Transistors ··· 12
　　　　C　专业简介 ·· 15
UNIT 3　A　Logical Variables and Flip-Flop ································· 17
　　　　B　Binary Number System ··· 20
　　　　C　专业英语的翻译标准 ·· 24
UNIT 4　A　Power Semiconductor Devices ··································· 25
　　　　B　Power Electronic Converter ······································ 31
　　　　C　专业英语的词汇特点 ·· 34
UNIT 5　A　Types of DC Motors ·· 36
　　　　B　Closed-Loop Control of DC Drivers ·························· 40
　　　　C　理解与表达 ·· 43
UNIT 6　A　AC Machines ··· 45
　　　　B　Induction Motor Drive ·· 49
　　　　C　长句的翻译 ·· 53
UNIT 7　A　Electric Power System ·· 55
　　　　B　Power System Automation ······································· 59
　　　　C　被动句的翻译 ··· 64

PART 2　Control Theory ·· 66

UNIT 1　A　The World of Control ··· 66
　　　　B　The Transfer Function and the Laplace Transformation ··· 70
　　　　C　否定句的翻译 ··· 73
UNIT 2　A　Stability and the Time Response ······························· 75
　　　　B　Steady State ·· 80
　　　　C　名词的翻译 ·· 83
UNIT 3　A　The Root Locus ·· 85
　　　　B　The Frequency Response Methods：Nyquist Diagrams ···· 89

		C	动词的翻译 ···	94
UNIT 4		A	The Frequency Response Methods: Bode Plots ······················	96
		B	Nonlinear Control Systems ···	99
		C	形容词的翻译 ···	104
UNIT 5		A	Introduction to Modern Control Theory ·································	106
		B	State Equations ··	109
		C	词性的转换 ···	112
UNIT 6		A	Controllability, Observability, and Stability ··························	114
		B	Optimum Control Systems ···	117
		C	语法成分的转换 ···	120
UNIT 7		A	Conventional and Intelligent Control ····································	122
		B	Artificial Neural Networks ···	126
		C	增词译法 ··	130

PART 3 Computer Control Technology ·· 132

UNIT 1		A	Computer Structure and Function ··	132
		B	Fundamentals of Computers and Networks ····························	137
		C	减词译法 ··	140
UNIT 2		A	Interfaces to External Signals and Devices ····························	142
		B	The Applications of Computers ··	146
		C	常用数学符号和公式的读法 ··	150
UNIT 3		A	PLC Overview ···	152
		B	PACs for Industrial Control, the Future of Control ···················	156
		C	科技论文的结构与写作 ··	160
UNIT 4		A	Fundamentals of Single-Chip Microcomputers ························	162
		B	Understanding DSP and Its Uses ···	165
		C	论文的标题和摘要 ··	169
UNIT 5		A	A First Look at Embedded Systems ·····································	171
		B	Embedded Systems Design ··	174
		C	电子邮件 ···	177

PART 4 Process Control ·· 179

UNIT 1		A	A Process Control System ··	179
		B	Fundamentals of Process Control ··	181
		C	通知 ··	183
UNIT 2		A	Sensors and Transmitters ···	185
		B	Final Control Elements and Controllers ································	187
		C	简历 ··	189
UNIT 3		A	P Controllers and PI Controllers ···	191

		B	PID Controllers and Other Controllers	194
		C	面试	198
UNIT 4		A	Indicating Instruments	200
		B	Control Panels	202
		C	自动化专业信息检索	205

PART 5 Control Based on Network and Information ······ 206

UNIT 1	A	Automation Networking Application Areas	206
	B	Evolution of Control System Architecture	211
	C	国内自动化专业主要期刊	215
UNIT 2	A	Fundamental Issues in Networked Control Systems	217
	B	Stability of NCSs with Network-Induced Delay	220
	C	国外自动化专业主要期刊	224
UNIT 3	A	Fundamentals of the Database System	228
	B	Virtual Manufacturing—A Growing Trend in Automation	231
	C	自动化专业的科技前沿	233
UNIT 4	A	Overview of Artificial Intelligence	235
	B	Applications of Artificial Intelligence in Machine Learning	239
	C	自动化专业的学术会议	243

PART 6 Synthetic Applications of Automatic Technology ······ 245

UNIT 1	A	Scanning the Issue and Beyond: Toward ITS Knowledge Automation	245
	B	Automation or Interaction: What's Best for Big Data?	251
	C	说明书常用术语	257
UNIT 2	A	Smart Grid Standards for Home and Building Automation	259
	B	Cloud Computing for Industrial Automation Systems—A Comprehensive Overview	264
	C	合同与协议书常用术语和句型	269
UNIT 3	A	Smart City and the Applications	273
	B	Knowledge Management System Design Model for Smart Enterprises	280
	C	广告	287

参考文献 ······ 289

PART 1

Electrical and Electronic Engineering Basics

UNIT 1

A Electrical Networks

An *electrical circuit or network* is composed of elements such as resistors, inductors, and capacitors connected together in some manner. If the network contains no energy sources, such as batteries or electrical generators, it is known as a *passive network*. On the other hand, if one or more energy sources are present, the resultant combination is an *active network*. In studying the behavior of an electrical network, we are interested in determining the voltages and currents that exist within the circuit. Since a network is composed of passive circuit elements, we must first define the electrical characteristics of these elements.

In the case of a resistor, the voltage-current relationship is given by Ohm's law, which states that the voltage across the resistor is equal to the current through the resistor multiplied by the value of the resistance.[1] Mathematically, this is expressed as

$$u = iR \qquad (1\text{-}1\text{A}\text{-}1)$$

where u = voltage, V; i = current, A; R = resistance, Ω.

The voltage across a pure inductor is defined by Faraday's law, which states that the voltage across the inductor is proportional to the rate of change with time of the current through the inductor. Thus we have

$$u = L\frac{di}{dt} \qquad (1\text{-}1\text{A}\text{-}2)$$

where di/dt = rate of change of current, A/s; L = inductance, H.

The voltage developed across a capacitor is proportional to the electric charge q accumulating on the plates of the capacitor. Since the accumulation of charge may be expressed as the summation, or integral, of the charge increments dq, we have the equation

$$u = \frac{1}{C}\int dq \qquad (1\text{-}1\text{A}\text{-}3)$$

where the capacitance C is the proportionality constant relating voltage and charge. By definition, current equals the rate of change of charge with time and is expressed as $i = dq/dt$. Thus an increment of charge dq is equal to the current multiplied by the corresponding time increment, or $dq = i\, dt$. Eq. (1-1A-3) may then be written as

$$u = \frac{1}{C}\int i\mathrm{d}t \qquad (1\text{-}1\text{A-}4)$$

where C = capacitance, F.

A summary of Eqs. (1-1A-1), (1-1A-2) and (1-1A-4) for the three forms of passive circuit elements is given in Fig. 1-1A-1. Note that conventional current flow is used; hence the current in each element is shown in the direction of decreasing voltage.

Active electrical devices involve the conversion of energy to electrical form. For example, the electrical energy in a battery is derived from its stored chemical energy. The electrical energy of a generator is a result of the mechanical energy of the rotating armature.

Fig. 1-1A-1 Passive circuit elements
a) Resistor   b) Inductor   c) Capacitor

Active electrical elements occur in two basic forms: *voltage sources* and *current sources*. In their ideal form, voltage sources generate a constant voltage independent of the current drawn from the source. The aforementioned battery and generator are regarded as voltage sources since their voltage is essentially constant with load. On the other hand, current sources produce a current whose magnitude is independent of the load connected to the source. Although current sources are not as familiar in practice, the concept does find wide use in representing an amplifying device, such as the transistor, by means of an equivalent electrical circuit. Symbolic representations of voltage source and current source are shown in Fig. 1-1A-2.

A common method of analyzing an electrical network is mesh or loop analysis. The fundamental law that is applied in this method is Kirchhoff's first law, which states that the algebraic sum of the voltages around a closed loop is 0, or, in any closed loop, the sum of the voltage rises must equal the sum of the voltage drops. Mesh analysis consists of assuming that currents—termed loop currents—flow in each loop of a network, algebraically summing the voltage drops around each loop, and setting each sum equal to 0.

Consider the circuit shown in Fig. 1-1A-3a, which consists of an inductor and resistor connected in series to a voltage source e. Assuming a loop current i, the voltage drops summed around the loop are

$$-e + u_R + u_L = 0 \qquad (1\text{-}1\text{A-}5)$$

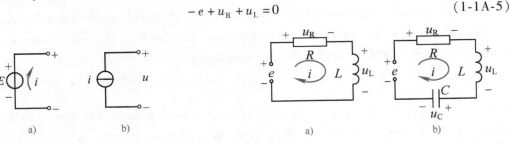

Fig. 1-1A-2 Voltage source and current source
a) Voltage source   b) Current source

Fig. 1-1A-3 Series circuits containing R, L and C

The input voltage is summed negatively since, in the direction of assumed current, it represents an increase in voltage. The drop across each passive element is positive since the current is in the direction of the developed voltage.

Using the equations for the voltage drops in a resistor and inductor, we have

$$L\frac{di}{dt} + Ri = e \qquad (1\text{-}1A\text{-}6)$$

Eq. (1-1A-6) is the differential equation for the current in the circuit.

It may be that the inductor voltage rather than the current is the variable of interest in the circuit.[2]

As noted in Fig. 1-1A-1, $i = \frac{1}{L}\int u_L dt$. Substituting this integral for i in Eq. (1-1A-6) gives

$$u_L + \frac{R}{L}\int u_L dt = e \qquad (1\text{-}1A\text{-}7)$$

After differentiation with respect to time, Eq. (1-1A-7) becomes

$$\frac{du_L}{dt} + \frac{R}{L}u_L = \frac{de}{dt} \qquad (1\text{-}1A\text{-}8)$$

which is the differential equation for the inductor voltage.

Fig. 1-1A-3b shows a series circuit containing a resistor, inductor, and capacitor. Following the mesh-analysis method outlined above, the circuit equation is

$$L\frac{di}{dt} + Ri + \frac{1}{C}\int i dt = e \qquad (1\text{-}1A\text{-}9)$$

Recalling that current $i = dq/dt$, a substitution of this variable may be made to eliminate the integral from the equation. The result is the second-order differential equation

$$L\frac{d^2q}{dt^2} + R\frac{dq}{dt} + \frac{q}{C} = e$$

WORDS AND TERMS

network n. 网络，电路
resistor n. 电阻器
inductor n. 电感器
capacitor n. 电容器
passive network 无源网络
active network 有源网络
characteristic adj. 特性（的）；n. 特性曲线
Ohm n. 欧姆
Faraday n. 法拉第
electric charge 电荷
integral n. 积分
increment n. 增量
armature n. 电枢，衔铁，加固

aforementioned adj. 上述的，前面提到的
represent v. 代表，表示，阐明
amplify v. 放大
symbolic adj. 符号的，记号的
mesh n. 网孔
Kirchhoff's first law 基尔霍夫第一定律
loop current 回路电流
voltage drop 电压降
in series 串联
differential adj. 微分的；n. 微分
variable n. 变量
outline n. 轮廓；v. 提出……的要点
eliminate v. 消除，对消

NOTES

[1] In the case of a resistor, the voltage-current relationship is given by Ohm's law, which states that the voltage across the resistor is equal to the current through the resistor multiplied by the value of the resistance.

就电阻来说,电压–电流的关系由欧姆定律决定。欧姆定律指出:电阻两端的电压等于电阻上流过的电流乘以电阻值。

  in the case of:就……来说,就……而论
  in case (of):假如;万一;在……的情况下
  in that case:那么,既然是那样
  in this case:既然是这样
  in any case:无论如何,总之
  in all case:就一切情况而论

[2] It may be that the inductor voltage rather than the current is the variable of interest in the circuit.

或许在电路中,人们感兴趣的变量是电感电压而不是电感电流。

  M rather than N:是 M 而不是 N
  of interest:有价值的;使人感兴趣的;有意义的

B Three-Phase Circuits

A three-phase circuit is merely a combination of three single-phase circuits. Because of this fact, current, voltage, and power relations of balanced three-phase circuits may be studied by the application of single-phae rules to the component parts of the three-phase circuit. Viewed in this light, it will be found that the analysis of three-phase circuits is little more difficult than that of single-phase circuits[1].

Reasons for Use of Three-Phase Circuits

In a single-phase circuit, the power is of a pulsating nature. At unity power factor, the power in a single-phase circuit is zero twice each cycle[2]. When the power factor is less than unity, the power is negative during parts of each cycle. Although the power supplied to each of the three phases of a three-phase circuit is pulsating, it may be proved that the total three-phase power supplied a balanced three-phase circuit is constant. Because of this, the characteristics of three-phase apparatus, in general, are superior to those of similar single-phase apparatus.

Three-phase machinery and control equipment are smaller, lighter in weight, and more efficient than single-phase equipment of the same rated capacity. In addition to the above-mentioned advantages offered by a three-phase system, the distribution of three-phase power requires only three fourths as much line copper as does the single-phase distribution of the same amount of power.

Generation of Three-Phase Voltages

A three-phase electric circuit is energized by three alternating emfs of the same frequency and differing in time phase by 120 electrical degrees. Three such sine-wave emfs are shown in Fig. 1-1B-1. These emfs are generated in three separate sets of armature coils in an AC generator. These three sets of coils are mounted 120 electrical degrees apart on the generator armature. The coil ends may all be brought out of the generator to form three separate single-phase circuits. However, the coils are ordinarily interconnected either internally or externally to form a three-wire or four-wire three-phase system.

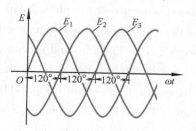

Fig. 1-1B-1 Three sine-wave emfs differing in phase by 120 electrical degrees are used for energizing a three-phase circuit

There are two ways of connecting the coils of three-phase generators, and in general, there are two ways of connecting devices of any sort to a three-phase circuit. These are the *wye-connection* and the *delta-connection*. Most generators are wye-connected, but loads may be either wye-connected or delta-Connected.

Voltage Relations in a Wye-Connected Generator

Fig. 1-1B-2a represents the three coils or phase windings of a generator. These windings are so spaced on the armature surface that the emfs generated in them are 120° apart in time phase. Each coil ends lettered S and F (start and finish). In Fig. 1-1B-2a, all the coil ends marked S are connected to a common point N, called the neutral, and the three coil ends marked F are brought out to the line terminals A, B, and C to form a three-wire three-phase supply. This type of connection is called the wye-connection. Often the neutral connection is brought out to the terminal board, as shown by the dotted line in Fig. 1-1B-2a, to form a four-wire three-phase system.

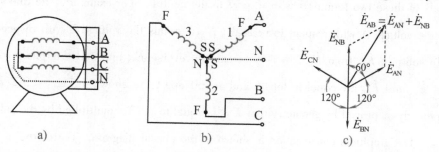

Fig. 1-1B-2 a) Connection of the phase windings in a wye-connection generator
b) Conventional diagram of a wye-connection
c) Phasor diagram showing the relation between phase and line voltages

The voltages generated in each phase of an AC generator are called the *phase voltages* (symbol E_p). If the neutral connection is brought out of the generator, the voltage from any one of the line terminals A, B, or C to the neutral connection N is a phase voltage. The voltage between any two of the three line terminals A, B, or C is called line-to-line voltage or, simply, a *line voltage* (symbol E_L).

The order in which the three voltages of a three-phase system succeed one another is called the phase sequence or the phase rotation of the voltages. This is determined by the direction of rotation of the generator but maybe reversed outside the generator by interchanging any two of the three line wires (not a line wire and a neutral wire).

It is helpful when drawing circuit diagrams of wye connection to arrange the three phases in the shape of a Y as shown in Fig. 1-1B-2b. Note that the circuit of Fig. 1-1B-2b is exactly the same as that of Fig. 1-1B-2a, with the S end of each coil connected to the neutral point and the F end brought out to the terminal in each case. After a circuit diagram has been drawn with all intersections lettered, a phasor diagram may be drawn as in Fig. 1-1B-2c. The phasor diagram shows the three phase voltages $\dot{E}_{AN}$, $\dot{E}_{BN}$, and $\dot{E}_{CN}$ which are 120° apart.

It should be noted in Fig. 1-1B-2 that each phasor is lettered with two subscripts. The two letters indicate the two points between which the voltage exists, and the order of the letters indicates the relative polarity of the voltage during its positive half-cycle. For example, the symbol $\dot{E}_{AN}$ indicates a voltage between the points A and N with the point A being positive with respect to point N during its positive half-cycle. In the phasor diagram shown, it has been assumed that the generator terminals were positive with respect to the neutral during the positive half-cycle. Since the voltage reverses every half-cycle, either polarity may be assumed if this polarity is assumed consistently for all three phases. It should be noted that if the polarity of point A with respect to N ($\dot{E}_{AN}$) is assumed for the positive half-cycle, then $\dot{E}_{NA}$ when used in the same phasor diagram should be drawn opposite to, or 180° out of phase with, $\dot{E}_{AN}$[3].

The voltage between any two line terminals of wye-connected generator is the difference between the potentials of these two terminals with respect to the neutral. For example, the line voltage $\dot{E}_{AB}$ is equal to the voltage A with respect to neutral ($\dot{E}_{AN}$) minus the voltage B with respect to neutral ($\dot{E}_{BN}$). To subtract $\dot{E}_{BN}$ from $\dot{E}_{AN}$, it is necessary to reverse $\dot{E}_{BN}$ and add this phase to $\dot{E}_{AN}$. The two phasors $\dot{E}_{AN}$ and $\dot{E}_{NB}$ are equal in length and are 60° apart, as shown in Fig. 1-1B-2c. It may be shown graphically or proved by geometry that $\dot{E}_{AB}$ is equal to 1.73, multiplied by the value of either $\dot{E}_{AN}$ or $\dot{E}_{NB}$. The graphical construction is shown in the phasor diagram. Therefore, in a balanced wye connection

$$E_L = 1.73 E_P$$

Current Relations in a Wye-Connected Generator

The current flowing out to the line wires from the generator terminals A, B, and C (Fig. 1-1B-2) must flow from the neutral point N, out through the generator coils. Thus, the current each line wire (I_L) must equal the current in the phase (I_P) to which it is connected. In a wye connection

$$I_L = I_P$$

WORDS AND TERMS

pulsate　v. 脉动，跳动，振动
apparatus　n. 一套仪器，装置
rated　adj. 额定的，设计的，适用的
coil　n. 绕组，线圈；v. 盘绕
distribution　n. 分配，分布，配电
generator　n. 发生器，发电机
emf（electromotive force）　电动势
interconnect　v. 互相连接
wye　n. Y形联结，星形联结，三通
delta　n. 希腊字母 Δ（δ），三角形（物）

geometry　n. 几何学，几何形状
winding　adj. 缠绕的；n. 线圈，绕组
polarity　n. 极性
neutral　adj. 中性的；n. 中性线
subscript　n. 下标，脚注，索引
succeed　v. 继……之后，接替
intersection　n. 相交，逻辑乘法
phase sequence　相序
reverse　v., n. 反转；adj. 变换极性的

NOTES

[1] Viewed in this light, it will be found that the analysis of three-phase circuits is little more difficult than that of single-phase circuits.

这样看来，三相电路的分析比单相电路的分析难不了多少。

viewed in this light：从这个意义上来看

that：指代 analysis

[2] At unity power factor, the power in a single-phase circuit is zero twice each cycle.

在功率因数为 1 时，单相电路里的功率值每个周波有两次为零。

twice each cycle：每个周波有两次（为零）。twice 和 each cycle 都做状语。

[3] It should be noted that if the polarity of point A with respect to N ($\dot{E}_{AN}$) is assumed for the positive half-cycle, then $\dot{E}_{NA}$ when used in the same phasor diagram should be drawn opposite to, or 180° out of phase with, $\dot{E}_{AN}$.

应该注意，如果是在电压的正半周定义 A 点相对于 N 的极性（$\dot{E}_{AN}$），那么 $\dot{E}_{NA}$ 在用于同一相量图中时就应该画得同 $\dot{E}_{AN}$ 相反，即相位差为 180°。

with respect to：相对于，关于

C　专业英语（Specified English）概述

大学生在经过基础英语的学习后，基本上已掌握了英语的常用语法，并具有 4000 以上的词汇量，具备了较扎实的英语基础。进入三年级后，随着专业课的进一步学习，学生的专

业知识技能也开始逐步加强。具备了以上两个条件后，学生应进行专业英语的训练，在保证 30 万词以上阅读量的基础上，对本专业英文资料的阅读达到基本的要求。换言之，掌握专业英语技能是大学基础英语学习的主要目的之一，是一种素质上的提高，直接关系到学生的求职和毕业后的工作能力。

专业英语的重要性体现在很多方面：大到日益广泛的国际科学技术交流，小到对产品说明书的翻译。而互联网为工程技术人员提供了更为巨大的专业信息量，作为主要网络语言的英语则对资料查询者提出了更高的要求。

尽管很多人在此之前已经进行了至少 8 年的基础英语学习，但专业英语的学习仍是很必要的。首先，专业英语在词义上具有不同于基础英语的特点和含义，如下例：

Connect the black pigtail with the dog-house.

错误译法：把黑色的猪尾巴系在狗窝上。

专业译法：将黑色的引出线接在高频高压电源屏蔽罩上。

通过上面的例子，我们不难认识到专业词汇的一些特点，即同一个词在日常生活中，在不同的专业中可能会有截然不同的含义。例如 bus 这个词，在日常生活中是"公共汽车"的意思，但在计算机中是指"总线"，在电力系统中是指"母线"。单靠日常用语进行望文生义的判断不仅会闹笑话，还有可能出事故。

其次，外文科技文章在结构上也具有很多自身的特点，如长句多、被动语态多、大量的名词化结构等，这都给对原文的理解和翻译带来了基础英语中所难以解决的困难。

再者，专业英语对听、说、读、写、译的侧重点不同，其最主要的要求在于"读"和"译"，也就是通过大量的阅读对外文资料进行正确的理解和翻译（interpretation & translation）；在读和译的基础上，对听、说、写进行必要的训练。此外，专业外文资料由于涉及许多科技内容而往往极为复杂而难以理解，加之这类文章的篇幅通常很长，所以只有经过一定的专业外语训练，才能完成从基础英语到专业英语的过渡，达到英语学以致用的最终目的。

专业翻译是指把科技文章由原作语言（source language）用译文语言（target language）忠实、准确、严谨、通顺、完整地再现出来的一种语言活动。它要求翻译者在具有一定专业基础知识和英语技能的前提下，借助于合适的英汉科技字典来完成整个翻译过程。专业翻译直接应用于科技和工程，因而对翻译的质量具有极高的要求。翻译上的失之毫厘，工程中就有可能差之千里，造成巨大的损失。例如，这样一个标志牌：

Control Center. Smoking Free.

它的意思是"控制中心，严禁吸烟"，free 在这里作"免除……的"讲；而如果理解为"随便的，自由的"，就会产生完全相反的意义。

UNIT 2

A   The Operational Amplifier

One problem with electronic devices corresponding to the generalized amplifiers is that the gains, A_U or A_I, depend upon internal properties of the two-port system (μ, β, R_i, R_o, etc.). [1] This makes design difficult since these parameters usually vary from device to device, as well as with temperature. The operational amplifier, or Op-Amp, is designed to minimize this dependence and to maximize the ease of design. An Op-Amp is an integrated circuit that has many component parts such as resistors and transistors built into the device. At this point we will make no attempt to describe these inner workings.

A totally general analysis of the Op-Amp is beyond the scope of some texts. We will instead study one example in detail, then present the two Op-Amp laws and show how they can be used for analysis in many practical circuit applications. These two principles allow one to design many circuits without a detailed understanding of the device physics. Hence, Op-Amps are quite useful for researchers in a variety of technical fields who need to build simple amplifiers but do not want to design at the transistor level. In the texts of electrical circuits and electronics they will also show how to build simple filter circuits using Op-Amps. The transistor amplifiers, which are the building blocks from which Op-Amp integrated circuits are constructed, will be discussed.

The symbol used for an ideal Op-Amp is shown in Fig. 1-2A-1. Only three connections are shown: the positive and negative inputs, and the output. Not shown are other connections necessary to run the Op-Amp such as its attachments to power supplies and to ground potential. The latter connections are necessary to use the Op-Amp in a practical circuit but are not necessary when considering the ideal Op-Amp applications we study in this unit. The voltages at the two inputs and the output will be represented by the symbols U^+, U^-, and U_o. Each is measured with respect to ground potential. Operational amplifiers are differential devices. By this we mean that the output voltage with respect to ground is given by the expression

$$U_o = A(U^+ - U^-) \tag{1-2A-1}$$

Fig. 1-2A-1   Operational amplifier

where A is the gain of the Op-Amp and U^+ and U^- the voltages at inputs. In other words, the output voltage is A times the difference in potential between the two inputs.

Integrated circuit technology allows construction of many amplifier circuits on a single composite "chip" of semiconductor material. One key to the success of an operational amplifier is the "cascading" of a number of transistor amplifiers to create a very large total gain. That is, the number A in Eq. (1-2A-1) can be on the order of 100,000 or more. (For example, cascading of five transistor amplifiers, each with a gain of 10, would yield this value for A.) A second important factor is that these circuits can be built in such a way that the current flow into each of the inputs is very small. A third important design feature is that the output resistance of the operational amplifier (R_o) is very small. This in turn means that the output of the device acts like an ideal voltage source.

We now can analyze the particular amplifier circuit given in Fig. 1-2A-2 using these characteristics. First, we note that the voltage at the positive input, U^+, is equal to the source voltage, $U^+ = U_s$. Various currents are defined in part b of the figure. Applying KVL around the outer loop in Fig. 1-2A-2b and remembering that the output voltage, U_o, is measured with respect to ground, we have

$$-I_1 R_1 - I_2 R_2 + U_o = 0 \quad (1\text{-}2\text{A-}2)$$

Since the Op-Amp is constructed in such a way that no current flows into either the positive or negative input, $I^- = 0$. Applying KCL at the negative input terminal then yields

$$I_1 = I_2$$

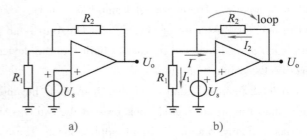

Fig. 1-2A-2 Operational amplifier circuits

Using Eq. (1-2A-2) and setting $I_1 = I_2 = I$,

$$U_o = (R_1 + R_2) I \quad (1\text{-}2\text{A-}3)$$

We may use Ohm's law to find the voltage at the negative input, U^-, noting the assumed current direction and the fact that ground potential is zero volts:

$$\frac{U^- - 0}{R_1} = I$$

So,

$$U^- = IR_1$$

and from Eq. (1-2A-3),

$$U^- = \left(\frac{R_1}{R_1 + R_2}\right) U_o$$

Since we now have expressions for U^+ and U^-, Eq. (1-2A-1) may be used to calculate the output voltage,

$$U_o = A(U^+ - U^-) = A\left(U_s - \frac{R_1 U_o}{R_1 + R_2}\right)$$

Gathering terms,

$$U_o\left(1 + \frac{AR_1}{R_1 + R_2}\right) = AU_s \quad (1\text{-}2\text{A-}4)$$

and finally,

$$A_U = \frac{U_o}{U_s} = \frac{A(R_1 + R_2)}{R_1 + R_2 + AR_1} \quad (1\text{-}2\text{A-}5\text{a})$$

This is the gain factor for the circuit. If A is a very large number, large enough that $AR_1 \gg (R_1 + R_2)$, the denominator of this fraction is dominated by the AR_1 term. The factor A, which is in both the numerator and denominator, then cancels out and the gain is given by the expression

$$A_U = \frac{R_1 + R_2}{R_1} \quad (1\text{-}2A\text{-}5b)$$

This shows that if A is very large, then the gain of the circuit is independent of the exact value of A and can be controlled by the choice of R_1 and R_2. This is one of the key features of Op-Amp design—the action of the circuit on signals depends only upon the external elements which can be easily varied by the designer and which do not depend upon the detailed character of the Op-Amp itself.[2] Note that if $A = 100,000$ and $(R_1 + R_2)/R_1 = 10$, the price we have paid for this advantage is that we have used a device with a voltage gain of $100,000$ to produce an amplifier with a gain of 10. In some sense, by using an Op-Amp we trade off "power" for "control".

A similar mathematical analysis can be made on any Op-Amp circuit, but this is cumbersome and there are some very useful shortcuts that involve application of the *two laws of Op-Amps* which we now present.

1) The first law states that in normal Op-Amp circuits we may assume that the voltage difference between the input terminals is zero, that is

$$U^+ = U^-$$

2) The second law states that in normal Op-Amp circuits both of the input currents may be assumed to be zero:

$$I^+ = I^- = 0$$

The first law is due to the large value of the intrinsic gain A. For example, if the output of an Op-Amp is 1V and $A = 100,000$, then $(U^+ - U^-) = 10^{-5}$ V. This is such a small number that it can often be ignored, and we set $U^+ = U^-$. The second law comes from the construction of the circuitry inside the Op-Amp which is such that almost no current flows into either of the two inputs.

WORDS AND TERMS

amplifier *n.* 放大器
integrated circuit 集成电路
building block 积木
potential *n.* （电）势
cascade *n.*, *v.* 串联; *adj.* 串联的

on the order of 属于同类的, 约为
trade off 换取
cumbersome *adj.* 麻烦的
intrinsic *adj.* 内在的
circuitry *n.* 电路

NOTES

[1] One problem with electronic devices corresponding to the generalized amplifiers is that the gains, A_U or A_I, depend upon internal properties of the two-port system.
对应于像广义放大器这样的电子器件存在的一个问题就是它们的增益 A_U 或者 A_I 取决于双端口系统的内部特性。

[2] This is one of the key features of Op-Amp design—the action of the circuit on signals depends only upon the external elements which can be easily varied by the designer and which do

not depend upon the detailed character of the Op-Amp itself.

这是运算放大器设计的重要特征之一——在信号作用下，电路的动作仅取决于能够容易被设计者改变的外部元件，而不取决于运算放大器本身的微观特性。

B   Transistors

Put very simply a semiconductor material is one which can be "doped" to produce a predominance of electrons or mobile negative charges (N-type); or "holes" or positive charges (P-type). [1] A single crystal of germanium or silicon treated with both N-type dope and P-type dope forms a semiconductor *diode*, with the working characteristics described. Transistors are formed in a similar way but like two diodes back-to-back with a common middle layer doped in the opposite way to the two end layers, thus the middle layer is much thinner than the two end layers or *zones*.

Two configurations are obviously possible, PNP or NPN (Fig. 1-2B-1). These descriptions are used to describe the two basic types of transistors. Because a transistor contains elements with two different polarities (i.e., "P" and "N" zones), it is referred to as a bipolar device, or *bipolar transistor*.

A transistor thus has three elements with three leads connecting to these elements. To operate in a working circuit it is connected with two external voltage or polarities. One external voltage is working effectively as a diode. A transistor will, in fact, work as a diode by using just this connection and forgetting about the top half. An example is the substitution of a transistor for a diode as the detector in a simple radio. It will work just as well as a diode as it *is* working as a diode in this case.

The diode circuit can be given forward or reverse bias. Connected with forward bias, as shown in Fig. 1-2B-2, drawn for a PNP transistor, current will flow from the bottom P to the middle N. If a second voltage is applied to the top and bottom sections of the transistor, with the *same* polarity applied to the bottom, the electrons already flowing through the middle N section will promote a flow of current through the transistor bottom-to-top.

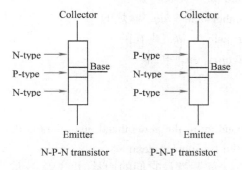

Fig. 1-2B-1 Two basic types of transistor

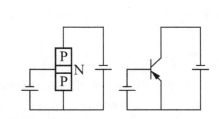

Fig. 1-2B-2 PNP transistor circuit

By controlling the degree of doping in the different layers of the transistor during manufacture, this ability to conduct current through the second circuit through a resistor can be very marked.

Effectively, when the bottom half is forward biased, the bottom section acts as a generous source of free electrons (and because it emits electrons it is called the *emitter*). These are collected readily by the top half, which is consequently called the *collector*, but the actual amount of current which flows through this particular circuit is controlled by the bias applied at the center layer, which is called the *base*.

Effectively, therefore, there are two separate "working" circuits when a transistor is working with correctly connected polarities (Fig. 1-2B-3). One is the loop formed by the bias voltage supply encompassing the emitter and base. This is called the *base* circuit or *input* circuit. The second is the circuit formed by the collector voltage supply and all three elements of the transistor. This is called the *collector* circuit or *output* circuit. (Note: this description applies only when the emitter connection is common to both circuits—known as *common emitter* configuration.) This is the most widely used way of connecting transistors, but there are, of course, two other alternative configurations—*common base* and *common collector*. But, the same principles apply in the working of the transistor in each case.

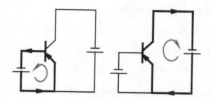

Fig. 1-2B-3　Base circuit and collector circuit

The particular advantage offered by this circuit is that a relatively small *base* current can control and instigate a very much larger *collector* current (or, more correctly, a small input *power* is capable of producing a much larger *output* power). In other words, the transistor works as an *amplifier*.

With this mode of working the base-emitter circuit is the input side; and the emitter through base to collector circuit the output side. Although these have a common path through base and emitter, the two circuits are effectively separated by the fact that as far as polarity of the base circuit is concerned, the base and upper half of the transistor are connected as a *reverse biased* diode. Hence there is no current flow from the base circuit into the collector circuit.

For the circuit to work, of course, polarities of both the base and collector circuits have to be correct (forward bias applied to the base circuit, and the collector supply connected so that the polarity of the common element (the emitter) is the same from both voltage sources). This also means that the polarity of the voltages must be correct for the type of transistor. In the case of a PNP transistor as described, the emitter voltage must be *positive*. It follows that both the base and collector are negatively connected with respect to the emitter. The symbol for a PNP transistor has an arrow on the emitter indicating the direction of *current flow*, always *towards* the base. ("P" for positive, with a PNP transistor).

In the case of an NPN transistor, exactly the same working principles apply but the *polarities* of both supplies are reversed (Fig. 1-2B-4). That is to say, the emitter is

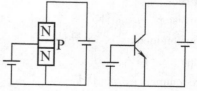

Fig. 1-2B-4　NPN transistor circuit

always made negative relative to base and collector ("N" for negative in the case of an NPN transistor). This is also inferred by the reverse direction of the arrow on the emitter in the symbol for an NPN transistor, i.e., current flow *away* from the base.

While transistors are made in thousands of different types, the number of *shapes* in which they are produced is more limited and more or less standardized in a simple code—TO (Transistor Outline) followed by a number.

TO1 is the original transistor shape—a cylindrical "can" with the three leads emerging in triangular pattern from the bottom. Looking at the base, the upper lead in the "triangle" is the *base*, the one to the right (marked by a color spot) the *collector* and the one to the left the *emitter*.[2] The collector lead may also be more widely spaced from the base lead than the emitter lead.

In other TO shapes the three leads may emerge in similar triangular pattern (but not necessarily with the same positions for base, collector and emitter), or in-line. Just to confuse the issue there are also sub-types of the same TO number shape with different lead designations. The TO92, for example, has three leads emerging in line parallel to a flat side on an otherwise circular "can" reading 1, 2, 3 from top to bottom with the flat side to the right looking at the base.

With TO92 sub-type a (TO92a):　　1 = emitter
　　　　　　　　　　　　　　　　2 = collector
　　　　　　　　　　　　　　　　3 = base
With TO92 sub-type b (TO92b):　　1 = emitter
　　　　　　　　　　　　　　　　2 = base
　　　　　　　　　　　　　　　　3 = collector

To complicate things further, some transistors may have only two emerging leads (the third being connected to the case internally); and some transistor outline shapes are found with more than three leads emerging from the base. These, in fact, are integrated circuits (ICs), packaged in the same outline shape as a transistor. More complex ICs are packaged in quite different form, e.g., flat packages.

Power transistors are easily identified by shape. They are metal cased with an elongated bottom with two mounting holes. There will only be two leads (the emitter and base) and these will normally be marked. The collector is connected internally to the can, and so connection to the collector is via one of the mounting bolts or bottom of the can.

WORDS AND TERMS

transistor　*n.*　晶体管
semiconductor　*n.*　半导体
dope　*v.*　掺入
predominance　*n.*　优势
crystal　*n.*　晶体
germanium　*n.*　锗

silicon　*n.*　硅
bipolar　*adj.*　双向的
lead　*n.*　引线
substitution　*n.*　代替
detector　*n.*　探测器
bias　*n.*　偏压

polarity   *n.* 极性
encompass   *v.* 包含
more or less   或多或少
cylindrical   *adj.* 圆柱形的

can   *n.* 密封外壳
triangular   *adj.* 三角的
elongate   *v.* 延长，拉长

NOTES

[1] Put very simply a semiconductor material is one which can be "doped" to produce a predominance of electrons or mobile negative charges (N-type); or "holes" or positive charges (P-type).

简单地说，半导体是这样一种物质，它能够通过"掺杂"来产生多余的电子，又称自由电子（N型）；或者产生"空穴"，又称正电荷（P型）。

[2] TO1 is the original transistor shape—a cylindrical "can" with the three leads emerging in triangular pattern from the bottom. Looking at the base, the upper lead in the "triangle" is the *base*, the one to the right (marked by a color spot) the *collector* and the one to the left the *emitter*.

TO1是最早的一种晶体管形状——即一个底部带有三个管脚的圆柱体"外罩"，这三个管脚在底部形成三角状。观察底部时"三角形"上面的管脚是基极，其右面的管脚（由一个彩色点标出）为集电极，其左面的管脚为发射极。

C 专 业 简 介

自动化专业是一门综合性、交叉性学科，主要研究控制科学的理论与技术及其工程应用。近年来，随着计算机技术、网络技术、智能算法、信息化技术等基础学科的发展，自动化技术也进入了新的发展阶段。在本科教育阶段，专业的名称为自动化；而在硕士和博士教育阶段，专业的一级学科名称为控制科学与工程（学科编号0811），其下包括5个二级学科，分别是：控制理论与控制工程（081101）、检测技术与自动化装置（081102）、系统工程（081103）、模式识别与智能系统（081104）和导航、制导与控制（081105）。而根据国家自然科学基金委员会自动化学科分类体系，则包括控制理论及其应用、系统科学与系统工程、检测技术及自动化装置、导航与制导理论及其应用、模式识别理论与方法及其应用、人工智能理论与算法及系统、先进制造理论及相关技术、认知及其信息处理、生物信息学、机器人学等几个分支。国际自动控制联合会（IFAC）的体系则将自动化科学与技术分为9个学术领域，分别是：制造与检测、设计方法、系统工程与管理、生命支持系统、系统与信号、工业应用领域、计算机控制、运输系统与运载工具、自动化的整体和教育效果。

自动化专业本科阶段的必修课程（compulsory/required subjects）和选修课程（elective subjects）除高等数学、英语、政治等大学基础课程外，还包括以下主要专业基础课和专业课：

电路理论（Theory of Circuit）
模拟电子技术（Analog Electronics Technology）
数字电子技术（Digital Electronics Technology）

电力电子技术（Power Electronics Technology）
电磁场（Electromagnetic Field）
电机与拖动（Electric Motor and Electric Drive）
电力拖动自动控制系统（Electric Drive Automatic Control System）
自动控制理论（Automatic Control Theory）
现代控制理论（Modern Control Theory）
智能控制（Intelligent Control）
微机原理（Principle of Microcomputer）
计算机接口技术（Computer Interface Technology）
计算机控制技术（Computer Control Technology）
过程控制系统（Process Control System）
过程检测及仪表（Process Measurement and Instrument）
单片机原理与应用（Principle and Application of Single-Chip Computer）
可编程序控制器系统（Programmable Logical Controller System）
现场总线技术（Fieldbus Technology）
嵌入式系统（Embedded System）
供电技术（Power Supplying Technology）
计算机仿真（Computer Simulation）
信号分析与处理（Signal Analyzing and Processing）
楼宇自动化（Building Automation）
机器人学（Robotics）

UNIT 3

A Logical Variables and Flip-Flop

Logical Variables

The two-valued variables which we have been discussing are often called *logical* variables, while the operations such as the OR operation and the AND operation are referred to as *logical* operations. We shall now briefly discuss the relevance of such terminology, and in so doing we shall bring out the special aptness of the designations "true" and "false" to identify the possible values of a variable. [1]

By way of example, suppose that you and two pilots are aloft in an airplane. You remain in the cabin, while the pilots, A and B, are in the cockpit. At some time, A joins you. This development causes you no concern. Suppose, however, that while you and A are in the cabin, you look up to find that B has also joined you. On the basis of your ability to reason *logically* you *deduce* that the plane is pilotless; and, presumably you sound an alarm so that one of the pilots will respond promptly to the urgency of the situation.

Alternatively, suppose that there had been attached to each pilot's seat an electronic device that provided an output voltage which is V_1 when the seat is occupied and V_2 when the seat is not occupied. Let us attach the designation "true" to the voltage level V_2 so that the level V_1 is "false". Let us further construct an electric circuit with two sets of input terminals and one set of output terminals. The circuit is to have the property that the output voltage will be V_2 if and only if both inputs, i.e., one input AND simultaneously the other, are at the level V_2. Otherwise the output is V_1. Finally let us connect the inputs to the devices on the chairs of pilots A and B and arrange that an alarm bell, connected to the output Z, respond when the output is V_2 ("true") and not otherwise. We have then constructed a circuit which performs the AND operation and is capable of making the *logical deduction* that the plane is unpiloted when, indeed, both pilots leave the cockpit.

To recapitulate, the situation is as follows: let the symbols A, B, and Z stand for the *propositions*

A = It is true (T) that pilot A has left his seat.
B = It is true (T) that pilot B has left his seat.
Z = It is true (T) that the plane is pilotless and in danger.

Of course, $\bar{A}$, $\bar{B}$, and $\bar{Z}$ then represent the contrary propositions, respectively. For example, $\bar{A}$ represents the proposition that it is false (F) that pilot A has left the cockpit, etc. The relationship among the propositions can now be written as

$$Z = AB \qquad (1\text{-}3\text{A-}1)$$

We have chosen to represent the *logical variables* A, B, and Z by electric voltages. But it must be noted that actually Eq. (1-3A-1) is a *relationship among propositions* and quite independent of the exact manner in which we choose to represent them or even whether we have any physical representation at all. Eq. (1-3A-1) says that proposition Z is true if propositions A and B are both true and that otherwise proposition Z is false.

This algebra of propositions of which Eq. (1-3A-1) is an example, is known as *Boolean algebra*. Just as other algebras deal with variables which have a numerical significance, Boolean algebra deals with propositions and is an effective tool for analyzing the relationships between propositions which allow only two mutually exclusive alternatives. [2]

SR Flip-Flop

The circuit of Fig. 1-3A-1, showing a pair of cross-coupled NOR gates, is called a *flip-flop*. It has a pair of input terminals S and R, standing for "set" and "reset", respectively. We shall use these symbols S and R not only to designate the terminals but also to specify the logical level at the terminals. Thus, $S = 1$ indicates that a voltage corresponding to logic level 1 is normally present at terminal S. Similarly, the output terminals and the corresponding output logic levels are Q and $\overline{Q}$. In this notation we have explicitly taken account of the fact that in normal operation, as we shall see, the logic levels at the outputs are complementary.

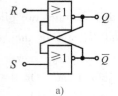

S	R	Q_{n+1}	$\overline{Q}_{n+1}$
0	0	Q_n	$\overline{Q}_n$
0	1	0	1
1	0	1	0

a)   b)

Fig. 1-3A-1 SR flip-flop

The fundamental, most important characteristic of a flip-flop is that it has a "memory". That is, given that the present logic levels at S and R, are 0 and 0, it is possible, from an examination of the output, to determine what the logic levels were at S and R immediately before they attained these present levels.

Terminology

In connection with the discussion to follow it is convenient to introduce some useful terminology, and it will be helpful to be aware of an attitude generally prevalent among logic-system designers.

In NAND and NOR gates (as well as an AND and OR gates), when it serves our purpose to do so, we can arbitrarily select one input terminal and view it as an *enable-disable input*. Thus, consider a NOR or an OR gate. If the one selected input is at logic 1, the output of the gate is independent of all other inputs. This one selected input takes control of the gate and the gate is disabled with respect to any other input. (The term "inhibit" is used synonymously with "disable"). Alternatively, if the selected input is at logic 0, it does not take control, and the gate is enabled to respond to the other inputs. In a NAND or AND gate, a selected input takes over control and disables the gate when this input goes to logic 0. For with one input at logic 0, the gate output cannot respond to other inputs. The difference between the NOR and OR gate, on the one hand, and the NAND or AND gate, on the other, is to be noted. In the first case, the control input achieves its control when it goes to logic 1; in the second case, it achieves its control when it goes to logic 0.

There is a generally prevailing attitude in digital systems to view logic 0 as a basic, undisturbed, unperturbed, quiescent state and to view the logic 1 state as the excited, active, effective state, i. e., the state arrived at "after something has happened".[3] Thus, when an effect has been produced, the inclination is to define the resultant state as one corresponding to which some logical variable has gone to logic 1. The logical variable is at logic 0 when "nothing has happened". Similarly, if an effect is to be produced by a change in a logical variable, it is preferred that the logical variable so involved be defined in such manner that the effect is achieved when the logic variable goes to logic 1. We shall see examples of this attitude in our discussion of flip-flops.

WORDS AND TERMS

flip-flop   *n.* 触发器
relevance   *n.* 关联
terminology *n.* 术语
aptness   *n.* 恰当
pilot   *n.* 飞行员
aloft   *adv.* 高高地
cockpit   *n.* 坐舱

deduce   *v.* 演绎
simultaneously *adv.* 同时地
Boolean algebra   布尔代数
gate   *n.* 门，门电路
prevalent   *adj.* 流行的
inhibit   *v.* 抑制

NOTES

[1] We shall now briefly discuss the relevance of such terminology, and in so doing we shall bring out the special aptness of the designations "true" and "false" to identify the possible values of a variable.

现在我们将简要地讨论一下这些术语之间的关联，并在此过程中阐明用标示"真"和"假"来识别一个变量的可能值的特殊用途。

[2] Just as other algebras deal with variables which have a numerical significance, Boolean algebra deals with propositions and is an effective tool for analyzing the relationships between propositions which allow only two mutually exclusive alternatives.

和其他处理有数字意义的变量的代数一样，布尔代数处理的是命题，而且布尔代数对于分析仅有两个互反变量的命题之间的关系是一种有效的工具。

[3] There is a generally prevailing attitude in digital systems to view logic 0 as a basic, undisturbed, unperturbed, quiescent state and to view the logic 1 state as the excited, active, effective state, i. e., the state arrived at "after something has happened".

在数字系统中，普遍的观点是把逻辑0看成一个基本的、无干扰的、稳定的、静止的状态，把逻辑1看成一个激励的、活跃的、有效的状态，也就是说，这种状态是发生在"某种操作动作之后。"

B Binary Number System

General Induction

In an algebra proposed by George Boole about 1850, the variables are permitted to have only two values *true* or *false*, usually written as 1 and 0, and the algebraic operations on the variables are limited to those defined as AND, OR, NOT.

Shannon, in 1938, recognized the parallels between this form of algebra and the functioning of electrical switching systems, in that switches are two-state, on-and-off devices. The reasoning processes called for by Boolean algebra are implemented through switches, acting as electronic *logic circuits*.

A great variety of integrated-circuit forms are available for logic operations on pulsed signals. These pulsed signals employ the *binary number system*, using cutoff and conduction of electronic devices as the two states for the number system.

The Binary Number System and Other Codes

To count directly in decimal numbers with transistors would require that they recognize the 10 states 0, 1, ..., 9, and this action would necessitate an accuracy not inherent in electronic devices. Such devices operate well in a two-state or binary system, using conduction and cutoff as the operating states, and as a result the binary number system is generally employed in internal operations in digital computers.[1]

In the decimal system the base or radix is 10, and each position to the left or right of the decimal point represents an increased or decreased weight as a power of 10. In the binary system the radix is 2, and the positions to the left or right of the binary point carry weights increasing or decreasing in powers of 2. Numbers are coded into chains of two-level pulses, with the levels usually designated as 1 and 0, as shown in Fig. 1-3B-1.

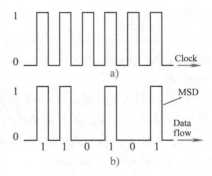

Fig. 1-3B-1 Chains of two-level pulses

The pulse chain of Fig. 1-3B-1b can be translated as:

Binary: $1 \times 2^5 + 0 \times 2^4 + 1 \times 2^3 + 0 \times 2^2 + 1 \times 2^1 + 1 \times 2^0 = 101011$
Decimal: $32 + 0 + 8 + 0 + 2 + 1 = 43$

In the inverse process of conversion of decimal 43 to binary form, we perform successive divisions by 2. The remainder of 1 or 0 after each division becomes a digit of the binary number. For conversion of decimal 43:

	Remainder	
43/2 = 21	1	Least significant digit
21/2 = 10	1	
10/2 = 5	0	
5/2 = 2	1	
2/2 = 1	0	
1/2 = 0	1	Most significant digit

The binary equivalent for decimal 43 is 101011.

While binary numbers require only two signal levels, this simplicity is achieved at the cost of additional digits. To represent n decimal digits in a system of radix r requires m digits, where

$$m = \frac{n}{\lg r}$$

The right side is an integer, or the next larger integer is chosen. For a number having 10 decimal digits, we have $m = 33.2$ and so must use 34 binary digits. Binary digits are referred to as *bits*.

A binary fraction written as 0.1101 means

$$0.1101 = 1 \times 2^{-1} + 1 \times 2^{-2} + 0 \times 2^{-3} + 1 \times 2^{-4}$$
$$= 1/2 + 1/4 + 0 + 1/16$$

In decimal numbers the binary number 0.1101 is

$$0.500 + 0.250 + 0.062 = 0.812$$

The conversion of a decimal number less than unity is performed by successive multiplications by 2. For each step that results in a 1 to the left of the decimal point, record a binary 1, and carry on with the fractional portion of the decimal number. With the result having a 0 to the left of the decimal point, record a binary 0 and carry on. To convert decimal 0.9375 to binary form, we operate as follows:

	Binary	
0.9375 × 2 = 1.8750	1	Most significant digit
0.8750 × 2 = 1.7500	1	
0.7500 × 2 = 1.5000	1	
0.5000 × 2 = 1.0000	1	
0.0000 × 2 = 0.0000	0	Least significant digit

The binary equivalent of decimal 0.9375 is written as 0.11110. The largest digit is the first binary bit obtained, and it is placed to the right of the binary point.

A table of binary equivalents from decimal 0 to decimal 15 is:

Decimal	Binary	Decimal	Binary
0	0	8	1000
1	1	9	1001
2	10	10	1010
3	11	11	1011
4	100	12	1100
5	101	13	1101
6	110	14	1110
7	111	15	1111

Given the basic idea of a chain of positive and negative, or positive and zero, or zero and negative pulses as representing binary 1s and 0s, there are many possible codes in which the pulses might be transmitted.[2] One of the most common for computer input is the *binary-coded-decimal* (BCD) *code*, requiring four pulses or bits per decimal digit. For this code each decimal digit is translated into its binary equivalent as given in the preceding table. That is, decimal 827 will appear in BCD form as

$$1000 \quad 0010 \quad 0111$$

The computer can readily translate such input to pure binary form by arithmetic operations. Decoders are also available to convert BCD to decimal form.

The BCD code can be extended to decimal 15 without requiring additional bits for transmittal; it then becomes a *hexadecimal code*, usually employing the letters a, b,..., f for the decimal numbers 10 through 15.

Another code that is employed in some computer operations is the *octal* or radix-8 system. The permitted symbols are 0, 1, 2,..., 7, and the decimal number 24 is written as octal 30 ($3 \times 8^1 + 0 \times 8^0$). Binary coding of octal numbers requires only the three least-significant bits of the BCD table, and binary coding of octal 30 is 011000.

Since decimal 24 is written 11000 in pure binary form, and 011000 in octal coded form, an easy means of conversion from binary numbers to octal numbers is indicated. By setting off the binary number in groups of three bits, each group appears as an equivalent octal-coded number. For instance, decimal 1206 in binary is 10010110110. In groups of three bits we have:

$$\text{Binary:} \quad 010 \quad 010 \quad 110 \quad 110$$
$$\text{Octal:} \quad 2 \quad 2 \quad 6 \quad 6$$

and the octal number is 2266.

The Gray code is often employed in translation of rotary or linear position to binary numbers by use of brushes on conductive segments, or by optical readers or code wheels. Because of alignment errors, two bits cannot change simultaneously and uncertainty is introduced. A Gray code is designed to eliminate this problem by requiring only one bit to change at each binary number step. One form of code is

Decimal	Gray code
0	0000
1	0001
2	0011
3	0010
4	0110
5	0111
6	0101
7	0100
8	1100
9	1101

Other codes are designed to reduce transmission errors, in which a 1 becomes changed to a 0 or vice versa. In general, a code that will detect single errors can be obtained by addition of a checking bit to the original code form. The resultant code will have a number of 1s either even or odd, and these codes are known as *even-parity* or *odd-parity codes*. For instance, 0000 will become 10000 in odd parity; an error in any digit will make the result have an even number of 1s, and the receiving equipment then calls for a correction.

Multiple errors can be found by more complex code forms.

WORDS AND TERMS

binary   *adj.* 二进制的
parallel   *n.* 类似
decimal   *adj.* 十进制的
radix   *n.* 权
chain   *n.* 串
remainder   *n.* 余数

digit   *n.* 位数
fractional   *adj.* 小数的
hexadecimal   *adj.* 十六进制的
octal   *adj.* 八进制的
alignment   *n.* 组合

NOTES

[1] Such devices operate well in a two-state or binary system, using conduction and cutoff as the operating states, and as a result the binary number system is generally employed in internal operations in digital computers.

将导通和关断作为工作状态，这样的装置可以在两态即二进制系统中运行，因此数字计算机中的内部操作一般采用二进制系统。

[2] Given the basic idea of a chain of positive and negative, or positive and zero, or zero and negative pulses as representing binary 1s and 0s, there are many possible codes in which the pulses might be transmitted.

给出一串正脉冲和负脉冲，或正脉冲和零，或者零和负脉冲来表示二进制的 1 及 0 时，就会有许多这些脉冲可以传递的码。

C 专业英语的翻译标准

翻译标准是评价译文质量的尺度，综合起来有以下三点：

第一，翻译首先要做到忠实、准确，要"信"。也就是说译文应避免漏译或错译，忠实、正确地转达原文的内容，既不歪曲，也不任意增减。同时，在表达上保持原作的风格和文体。例如：

The importance of computer in the use of automatic control can not be overestimated.

初译：计算机在自动控制应用上的重要性不能被估计过高。

更正：对计算机在自动控制应用上的重要性怎么估计也不会过高。

第二，要做到通顺、流畅，要"达"。这是指译文应通顺易懂，符合汉语的规范，要和原作同样的流利自如。同时，在翻译中要避免"死译"和"硬译"，以便于他人理解。例如：

This possibility was supported to a limited extend in the tests.

初译：这一可能性在试验中在有限的程度上被支持了。

更正：这一可能性在试验中于一定程度上得到了证实。

The differences between single-board microcomputer and single-chip microcomputer do not stop there.

初译：单板机和单片机的差别不停留在那里。

更正：单板机和单片机的差别不仅于此。

忠实和通顺是辩证统一的关系。要防止对"忠实"的片面理解。一味追求形式上的相似，会造成逐字翻译，产生翻译上的形式主义，例如：

Amplification means the transformation of little currents into big ones, without distortion of the shape of current fluctuation.

初译：放大意味着由小电流到大电流的转变，而电流起伏的形状没有歪曲。

更正：所谓放大，就是把小电流变为大电流，而又不使电流波形失真。

另一方面，也要防止片面理解"通顺"的要求，过分强调译文的流畅而不受原文的约束，添枝加叶，造成翻译上的自由主义。如：

This technique provides a solution with the longest range and the maximum data rate for user.

初译：这一技术提供给用户具有目前最为广阔的传输范围和最大的数据传输速度的最为有效的解决方案。

更正：这一技术为用户提供了具有最大的数据传输距离和速度的方案。

第三，译文在忠实和通顺的基础上还应注意文采，要"雅"。要讲修辞，使译文在逻辑上严谨而流畅，语言上优美而易懂。否则就会因"言之无文，行之不远"而失去读者。例如：在译文中适当地添加一些承上启下的转折词；灵活使用汉语中的成语或俗语，如 vice versa 译为"反之亦然"，It is well known that… 译为"众所周知……"等等。

此外，无论翻译或阅读，都希望有较高的翻译速度。速度越快，对信息的获取量就越大，效率也就越高。

UNIT 4

A   Power Semiconductor Devices

Power semiconductor devices constitute the heart of modern power electronic apparatus. They are used in power electronic converters in the form of a matrix of on-off switches. And the switching mode power conversion gives high efficiency.

Totay's power semiconductor devices are almost exclusively based on silicon material and can be classified as follows:

Diode

Thyristor or silicon-controlled rectifier (SCR)

Triac

Gate turn-off thyristor (GTO)

Bipolar junction transistor (BJT or BPT)

Power MOSFET

Static induction transistor (SIT)

Insulated gate bipolar transistor (IGBT)

MOS-controlled thyristor (MCT)

Integrated gate-commutated thyristor (IGCT)

Diodes

Power diodes provide uncontrolled rectification of power and are used in applications such as electroplating, anodizing, battery charging, welding, power supplies (DC and AC), and variable-frequency drives.[1] They are also used in feedback and the freewheeling functions of converters and snubbers. A typical power diode has P-I-N structure, that is, it is a P-N junction with a near intrinsic semiconductor layer (I-layer) in the middle to sustain reverse voltage.

Fig. 1-4A-1 shows the diode symbol and its volt-ampere characteristics. In the forward-biased condition, the diode can be represented by a junction offset drop and a series-equivalent resistance that gives a positive slope in the V-I characteristics. The typical forward conduction drop is 1.0V. This drop will cause conduction loss, and the device must be cooled by the appropriate heat sink to limit the junction temperature. In the reverse-biased condition, a small leakage current flows due to minority carriers, which

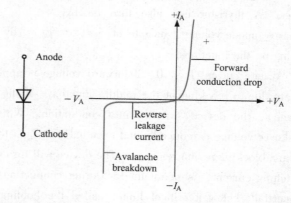

Fig. 1-4A-1   Diode symbol and its volt-ampere characteristics

gradually increase with voltage. If the reverse voltage exceeds a threshold value, called the breakdown voltage, the device goes through avalanche breakdown, which is when reverse current becomes large and the diode is destroyed by heating due to large power dissipation in the junction.

Power diodes can be classified as follows:

Standard or slow-recovery diode

Fast-recovery diode

Schottky diode

Thyristors

Thyristors, or silicon-controlled rectifiers (SCRs) have been the traditional workhorses for bulk power conversion and control in industry. The modern era of solid-state power electronics started due to the introduction of this device in the late 1950s. The term "thyristor" came from its gas tube equivalent, thyratron. Often, it is a family name that includes SCR, triac, GTO, MCT, and IGCT. Thyristors can be classified as standard, or slow phase-control-type and fast-switching, voltage-fed inverter-type. The inverter-type has recently become obsolete.

Fig. 1-4A-2 shows the thyristor symbol and its volt-ampere characteristics. Basically, it is a three-junction P-N-P-N device, where P-N-P and N-P-N component transistors are connected in regenerative feedback mode. The device blocks voltage in both the forward and reverse direction (symmetric blocking). When the anode is positive, the device can be triggered into conduction by a short positive gate current pulse; but once the device is conducting, the gate loses its control to turn off the device. A thyristor can also turn on by excessive anode voltage, its rate of rise (dv/dt), by a rise in junction temperature, or by light shining on the junctions.

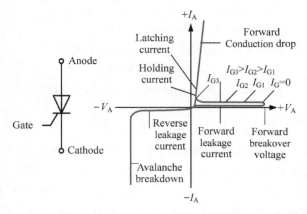

Fig. 1-4A-2 Thyristor symbol and its volt-ampere characteristics

At gate current $I_G = 0$, if forward voltage is applied on the device, there will be a leakage current due to blocking of the middle junction. If the voltage exceeds a critical limit (breakover voltage), the device switches into conduction. With increasing magnitude of I_G, the forward breakover voltage is reduced, and eventually at I_{G3}, the device behaves like a diode with the entire forward blocking region removed. The device will turn on successfully if a minimum current, called a latching current, is maintained. During conduction, if the gate current is zero and the anode current falls below a critical limit, called the holding current, the device reverts to the forward blocking state. With reverse voltage, the end P-N junctions of the device become reverse-biased. Modern thyristors are available with very large voltage (several kV) and current (several kA) ratings.

Triacs

A triac has a complex multiple-junction structure, but functionally, it is an integration of a pair of phase-controlled thyristors connected in inverse-parallel on the same chip. Fig. 1-4A-3 shows the triac symbol. The three-terminal device can be trigged into conduction in both positive and negative half-cycles of supply voltage by applying gate trigger pulses. In I + mode, the terminal T_2 is positive and the device is switched on by positive gate current pulse. In III- mode, the terminal T_1 is positive and it is switched on by negative gate current pulse. A triac is more economical than a pair of thyristors in anti-parallel and its control is simpler, but its integrated construction has some disadvantages. The gate current sensitivity of a triac is poorer and the turn-off time is longer due to the minority carrier storage effect. For the same reason, the reapplied dv/dt rating is lower, thus making it difficult to use with inductive load. A well-designed RC snubber is essential for a triac circuit. Triacs are used in light dimming, heating control, alliance-type motor drives, and solid-state relays with typically 50/60 Hz supply frequency.

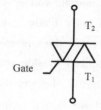

Fig. 1-4A-3   Triac symbol

GTOs

A gate turn-off thyristor (GTO), as the name indicates, is basically a thyristor-type device that can be turned on by a small positive gate current pulse, but in addition, has the capability of being turned off by a negative gate current pulse.[2] The turn-off capability of a GTO is due to the diversion of P-N-P collector current by the gate, thus breaking the P-N-P/N-P-N regenerative feedback effect. GTOs are available with asymmetric and symmetric voltage-blocking capabilities, which are used in voltage-fed and current-fed converters, respectively. The turn-off current gain of a GTO, defined as the ratio of anode current prior to turn-off to the negative gate current required for turn-off, is very low, typically 4 or 5. This means that a 6,000A GTO requires as high as 1,500A gate current pulse. However, the duration of the pulsed gate current and the corresponding energy associated with it is small and can easily be supplied by low-voltage power MOSFETs. GTOs are used in motor drives, static VAR compensators (SVCs), and AC/DC power supplies with high power ratings. When large-power GTOs became available, they ousted the force-commutated, voltage-fed thyristor inverters. Fig. 1-4A-4 shows the GTO symbol.

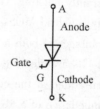

Fig. 1-4A-4   GTO symbol

Power MOSFETs

Unlike the devices discussed so far, a power MOSFET (metal-oxide semiconductor field-effect transistor) is a unipolar, majority carrier, "zero junction", voltage-controlled device. Fig. 1-4A-5 shows an N-type MOSFET symbol. If the gate voltage is positive and beyond a threshold value, an N-type conducting channel will be induced that permit current flow by majority carrier (electrons) between the drain and the source. Although the gate impedance is extremely high at steady state, the effective gate-source capacitance will demand a pulse current during turn-on and turn-off. The device has asymmetric voltage-blocking capability, and has an integral body diode, as shown, which can carry full current in the reverse direction. The diode is characterized by slow recovery and is often bypassed by an external fast-recovery diode in high-frequency applications.

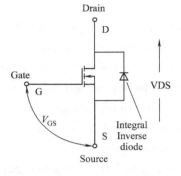

Fig. 1-4A-5 An N-type MOSFET symbol

While the conduction loss of a MOSFET is large for higher voltage devices, its turn-on and turn-off switching times are extremely small, causing low switching loss. The device does not have the minority carrier storage delay problem associated with a bipolar device. Although a MOSFET can be controlled statically by a voltage source, it is normal practice to drive it by a current source dynamically followed by a voltage source to minimize switching delays. MOSFETs are extremely popular in low-voltage, low-power, and high-frequency (hundreds of kHz) switching applications. Application examples include switching mode power supplies (SMPS), brushless DC motors (BLDMs), stepper motor drives, and solid-state DC relays.

IGBTs

The introduction of insulated gate bipolar transistors (IGBTs) in the mid-1980s was an important milestone in the history of power semiconductor devices. They are extremely popular devices in power electronics up to medium power (a few kW to a few MW) range and are applied extensively in DC/AC drives and power supply systems. They ousted BJTs in the upper range, and are currently ousting GTOs in the lower power range. An IGBT is basically a hybrid MOS-gated turn-on/off bipolar transistor that combines the advantages of both a MOSFET and BJT. Fig. 1-4A-6 shows the IGBT symbol. Its architecture is essentially similar to that of a MOSFET, except an additional P^+ layer has been added at the collector over the N^+ drain layer of the MOSFET. The device has the high-input impedance of a MOSFET, but BJT-like conduction characteristics. If the gate is positive with respect to the emitter, an N-channel is induced in the P region. This forward-biases the base-emitter junction of the P-N-P transistor, turning it on and causing conductivity modulation of the N^- region, which gives a significant reduction of conduction drop over that of a MOSFET. At the on-condition, the driver MOSFET in the equivalent circuit of the IGBT carries most of the total terminal current. The thyristor-like latching action caused by the parasitic N-P-N transistor is prevented by sufficiently reducing the resistivity of the P^+ layer and diverting most of the current through the MOSFET. The device is turned off by reducing the gate voltage to zero or negative, which shuts off the conducting channel in the P region. The device has higher current density than that of a BJT or MOSFET. Its

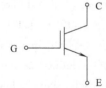

Fig. 1-4A-6 IGBT symbol

input capacitance (C_{iss}) is significantly less than that of a MOSFET. Also, the ratio of gate-collector capacitance to gate-emitter capacitance is lower, giving an improved Miller feedback effect.

MCTs

An MOS-controlled thyristor (MCT), as the name indicates, is a thyristor-like, trigger-into-conduction hybrid device that can be turned on or off by a short voltage pulse on the MOS gate. The device has a microcell construction, where thousands of microdevices are connected in parallel on the same chip. The cell structure is somewhat complex. Fig. 1-4A-7 shows the MCT symbol. It is turned on by a negative voltage pulse at the gate with respect to the anode and is turned off by a positive voltage pulse. The MCT has a thyristor-like P-N-P-N structure, where the P-N-P and N-P-N transistor components are connected in regenerative feedback. However, unlike a thyristor, it has unipolar (or asymmetric) voltage-blocking capability. If the gate of an MCT is negative with respect to the anode, a P-channel is induced in the P-FET, which causes forward-biasing of the N-P-N transistor. This also forward-biases the P-N-P transistor and the device goes into saturation by positive feedback effect. At conduction, the drop is around one volt (like a thyristor). If the gate is positive with respect to the anode, the N-FET will saturate and short-circuit the emitter-base junction of the P-N-P transistor. This will break the positive feedback loop for thyristor operation and the device will turn off. The turn-off occurs purely by recombination effect and therefore the tail time of the MCT is somewhat large. The device has a limited SOA, and therefore a snubber circuit is mandatory in an MCT converter. Recently, the device has been promoted for "soft-switched" converter applications, where the SOA is not utilized. In spite of complex geometry, the current density of an MCT is high compared to a power MOSFET, BJT and IGBT, and therefore it needs a smaller die area.

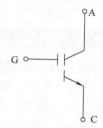

Fig. 1-4A-7　MCT symbol

The MCT was commercially introduced in 1992, and currently, medium-power devices are available commercially. The future acceptance of the device remains uncertain at this point.

IGCTs

The integrated gate-commutated thyristor (IGCT) is the newest member of the power semiconductor family at this time, and was introduced by ABB in 1997. Fig. 1-4A-8 shows the IGCT symbol. Basically, it is a high-voltage, high-power, hard-driven, asymmetric-blocking GTO with unity turn-off current gain. This means that a 4,500 V IGCT with a controllable anode current of 3,000 A requires turn-off negative gate current of 3,000 A. Such a gate current pulse of very short duration and very large di/dt has small energy content and can be supplied by multiple MOSFETs in parallel with ultra-low leakage inductance in the drive circuit.[3] The gate drive circuit is built-in on the device module. The device is fabricated with a monolithically integrated anti-parallel diode. The conduction drop, turn-

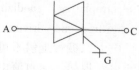

Fig. 1-4A-8　IGCT symbol

on di/dt, gate driver loss, minority carrier storage time, and turn-off dv/dt of the device are claimed to be superior to GTO. Faster switching of the device permits snubberless operation and higher-than-GTO switching frequency. Multiple IGCTs can be connected in series or in parallel for higher power applications. The device has been applied in power system intertie installations (100 MVA) and medium-power (up to 5 MW) industrial drives.

WORDS AND TERMS

converter　n. 转换器，换流器，变流器
matrix　n. 模型，矩阵
diode　n. 二极管，半导体二极管
thyristor　n. 晶闸管
triac　n. 三端双向晶闸管
GTO　门极可关断晶闸管
BJT　双极结型晶体管
power MOSFET　电力MOS场效应晶体管
SIT　静态感应晶体管
IGBT　绝缘栅双极型晶体管
MCT　MOS控制晶闸管
IGCT　集成门极换向晶闸管
rectification　n. 整流
feedback　n. 反馈
freewheeling　n. 自由停车
snubber　n. 缓冲器，减振器
intrinsic　adj. 固有的，体内的，本征的
forward biased　正向偏置
conduction　n. 导电，传导
reverse biased　反向偏置

leakage current　漏电流
threshold　n. 门限，阈限，极限
breakdown　n. 击穿
recovery　n. 恢复
Schottky diode　肖基特二极管
workhorse　n. 重载，重负荷
thyratron　n. 闸流管
breakover　n. 导通
latching current　闭锁电流
holding current　保持电流
phase controlled　相控的
asymmetric　adj. 不对称的
symmetric　adj. 对称的
force commutated　强制换向的
SMPS　开关电源
BLDM　无刷直流电动机
stepper motor　步进电动机
hybrid　n. 混合
emitter　n. 发射极
saturation　n. 饱和

NOTES

[1] Power diodes provide uncontrolled rectification of power and are used in applications such as electroplating, anodizing, battery charging, welding, power supplies (DC and AC), and variable-frequency drives.

电力二极管提供不可控的整流电源，这些电源有很广的应用，如：电镀、电极氧化、电池充电、焊接、交直流电源和变频驱动。

[2] A gate turn-off thyristor (GTO), as the name indicates, is basically a thyristor-type device that can be turned on by a small positive gate current pulse, but in addition, has the capability of being turned off by a negative gate current pulse.

门极可关断晶闸管，顾名思义，是一种晶闸管类型的器件。同其他晶闸管一样，它可以

由一个小的正门极电流脉冲触发,但除此之外,它还能被负门极电流脉冲关断。

[3] Such a gate current pulse of very short duration and very large di/dt has small energy content and can be supplied by multiple MOSFETs in parallel with ultra-low leakage inductance in the drive circuit.

这样一个持续时间非常短、di/dt 非常大、能量又较小的门极电流脉冲可以由多个并联的 MOSFET 来提供,并且驱动电路中的漏感要特别低。

B Power Electronic Converters

Power electronic converters can convert power from AC-to-DC (rectifier), DC-to-DC (chopper), DC-to-AC (inverter), AC-to-AC at the same (AC controller) or different frequencies (cycloconverter). They are four types of power electronic converters. Converters are widely used in applications such as heating and lighting controls, AC and DC power supplies, electrochemical processes, DC and AC motor drives, static VAR generation, active harmonic filtering, etc.

Rectifiers

The rectifiers can convert AC power to DC power. They can be made up of diodes, thyristors, GTOs, IGBTs, IGCTs, ect. Diodes and phase-controlled rectifiers constitute the largest segment of power electronics that interface to the electric utility system today. The efficiency of the rectifiers is very high, typically in the vicinity of 98%, because device conduction loss is low and switching loss is practically negligible.[1] However, the disadvantage is that they generate harmonics in the utility system creating a power quality problem for other consumers. Besides, thyristor converters constitute a low lagging power factor load on the utility system.

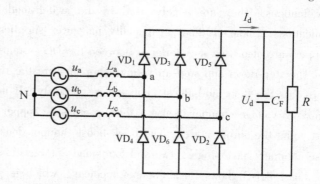

Fig. 1-4B-1 Three-phase diode bridge rectifier with RC load

Diode rectifiers are the simplest and possibly the most important power electronics circuits. They are rectifiers because power can flow only from the AC side to the DC side. The most important circuit configurations include the single-phase diode bridge and three-phase diode bridge. And, the commonly used loads include resistance, resistance-inductance, and capacitance-resistance. Fig. 1-4B-1 shows three-phase diode bridge rectifier with RC load. Other device rectifiers circuits are also as diode ones.

Inverters

Inverter is a device which receives DC voltage at one side and converts it to AC voltage on the other side. The AC voltage and frequency may be variable or constant depending on the application.

Inverter can be classified voltage-fed inverter and current-fed inverter. A voltage-fed inverter should have a stiff voltage source at the input, that is, its Thevenin impedance should ideally be zero. A large capacitor can be connected at the input if the source is not stiff. The DC voltage may be fixed or variable, and may be obtained from a utility line or rotation AC machine through a rectifier and filter. A current-fed or current-source inverter (CFI or CSI), as the name indicates, likes to see a stiff DC current source at the input, which is in contrast to a stiff voltage source, which is desirable in a voltage-fed inverter. A variable voltage source can be converted to a variable current source by connecting a large inductance in series and controlling the voltage within a feedback current control loop.[2] They are both used extensively. The semiconductor devices in them may be IGBTs, power MOSFETs and IGCTs etc.

Fig. 1-4B-2 gives one common circuit of three-phase bridge voltage-fed inverter.

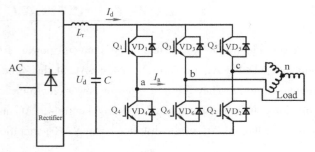

Fig. 1-4B-2 Three-phase bridge voltage-fed inverter with L load

Choppers

The choppers convert a DC power source to another DC source with different terminal specifications. They are widely used in the switch-mode power supplies and DC motor drive applications. Some of them, especially in power supplies, have an isolation transformer. The choppers are also used as interfaces between the DC systems of different voltage levels.

The step-down and step-up voltage choppers are the two basic chopper topologies. These are referred as the Buck and Boost choppers respectively. It must, however, be kept in mind that a step-down voltage chopper is also a step-up current chopper and vice versa because the input power must equal the output power. The Buck-boost chopper does both stepping up and down action. All these choppers have single, two and four quadrant variations in topologies.

Fig. 1-4B-3 shows Buck chopper topology, which is a voltage step-down and current step-up chopper. The two-position switch is synthesized from a switch and a diode. The switch is turned on for a time τ periodically at a rate $1/T_s$. The voltage waveform of Buck chopper is shown in Fig. 1-4B-4. So the average output voltage is $\bar{U}_o = U_s \dfrac{\tau}{T_s} = DU_s$, and the average current is $\bar{I}_o = \dfrac{I_s}{D}$.

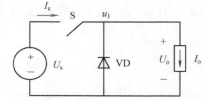

Fig. 1-4B-3 Buck chopper

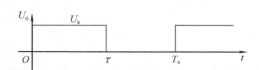

Fig. 1-4B-4 The voltage waveform of Buck chopper

D is called the duty ratio, which has a range of 0-1. I_s is the average current from the DC source.

Cycloconverters

A cycloconverter is a frequency changer that converts AC power at one input frequency to output power at a different frequency with a one-stage conversion process. The phase-controlled thyristor converters can be easily extended for cycloconverter operation. Self-controlled AC switches, usually based on IGBT, can also be used in high-frequency link cycloconverters. In large power industrial applications, thyristor phase-controlled cycloconverters are widely used.

Fig. 1-4B-5 shows the block diagram of a cycloconverter. In an industrial cycloconverter driving an AC motor, the input 50/60 Hz power is converted to variable-frequency, variable-voltage AC at the output to control the motor speed. The output frequency may vary from zero (rectifier operation) to an upper limit, which is always lower than the input frequency (step-down cycloconverter), and the power flow can be in either direction for four-quadrant motor speed control. In a VSCF system, the input power is usually generated by a synchronous machine that is coupled to a variable-speed turbine. The generated voltage can be regulated if the synchronous machine is wound field, but the output frequency is always proportional to the turbine speed. The function of the cycloconverter is to regulate the output frequency to be constant (typically 60 or 400 Hz). Fig. 1-4B-5 shows alternate schemes of frequency conversion. Fig. 1-4B-5a is a commonly used scheme where the input AC is rectified to DC and then inverted to variable-frequency AC through an inverter. In Fig. 1-4B-5b, the input AC is converted to high-frequency AC through a step-up cycloconverter, and then converted to variable-frequency AC by a step-down cycloconverter.

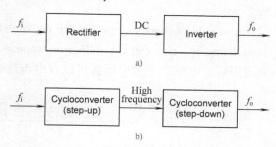

Fig. 1-4B-5 Alternate schemes of frequency conversion
a) DC link b) High-frequency link

WORDS AND TERMS

rectifier *n.* 整流器
chopper *n.* 斩波器
inverter *n.* 逆变器
cycloconverter *n.* 周波变换器
electrochemical *adj.* 电化学的
VAR 静态无功功率
harmonics *n.* 谐波
lagging *n.* 滞后，迟滞
power factor 功率因数

configuration *n.* 轮廓，格局
voltage-fed inverter 电压源型逆变器
current-fed inverter 电流源型逆变器
stiff voltage source 恒压源
stiff current source 恒流源
Thevenin impedance 戴维南电路等效阻抗
filter *n.* 滤波器
isolation transformer 隔离变压器
buck chopper 降压式变压器

boost chopper　升压式变压器
quadrant　*n.* 象限

duty ratio　占空比，功率比

NOTES

[1] The efficiency of the rectifiers is very high, typically in the vicinity of 98%, because device conduction loss is low and switching loss is practically negligible.

由于器件开通时损耗低，且其开关损耗几乎可忽略不计，故该类整流器的效率很高，典型值约为98%。

[2] A variable voltage source can be converted to a variable current source by connecting a large inductance in series and controlling the voltage within a feedback current control loop.

通过串联大电感，可变电压源可以在电流反馈控制回路的控制下转换为可变电流源。

C　专业英语的词汇特点

一、词汇的专业性含义强

科技文章中涉及大量的专业术语，这些词汇有的仅是某专业的专用词，不具有其他含义，如：diode 二极管，capacitor 电容，tachometer 转速表，Internet 互联网；有些则除了具有本专业含义外，还在日常生活中和其他专业中具有不同的含义，如：memory 内存/记忆，bus 总线/公共汽车，monitor 监视器/班长，order 阶次/命令/订货。后一类词在翻译或阅读中往往会引起误解，需要通过积累并根据其语言环境和专业知识判断其含义。

二、缩略词多

缩略词或缩写词经常大量出现在科技文章中，掌握一些常见的缩略词对阅读外文资料是十分必要的。除一些常用的缩略词外，有时某一篇论文的作者还会将仅在该文中使用的术语转换为缩略词：

AC（alternate current）交流电　　　　　IC（integrated circuit）集成电路
r.p.m.（revolutions per minute）转数/分　FET（field effect transistor）场效应管
CPU（central processor unit）中央处理器　Fig.（figure）图
MODEM（modulator & demodulator）调制解调器　Eq.（equation）等式
w.w.w.（world wide web）万维网　　　　E-mail（electronic mail）电子邮件

三、复合词和派生词多

大量使用复合词和派生词也是科技文章的一个特点。随着技术发展的需要，经常会复合或派生出一些新的词汇，而有些词可能无法在字典中查到。对这类词汇要通过分析掌握其含义。

1. 复合词，例如：　feedback 反馈　　　　　open-loop 开环
　　　　　　　　　　forward-gain 前向增益　　bandwidth 带宽
　　　　　　　　　　single-phase half-wave circuit 单相半波电路

2. 派生词，一般组成为：前缀＋词根＋后缀＝新词

例：un ＋ pre ＋ ced ＋ ent ＋ ed ＝ unprecedented
　　　无　　先　　进行（表事物）的　　　无先例的
　　　conduct　　conductor　　semiconductor
　　　传导 v.　　~ or 导体 n.　　semi（半）~ 半导体 n.

多记词根和前后缀并注意积累，就会收到举一反三，甚至以一当十的效果，从而有效地扩充词汇量。例如：

electro（电，电的）-　electronic　电子的
　　　　　　　　　　electromechanical　机电的
　　　　　　　　　　electrical　电气的
　　　　　　　　　　electrician　电气技师
micro（微小的）-　microcomputer　微机
　　　　　　　　microprocessor　微处理器
　　　　　　　　microdetector　灵敏电流计

UNIT 5

A Types of DC Motors

The types of commercially available DC motors basically fall into four categories: ①permanent-magnet DC motors, ②series-wound DC motors, ③shunt-wound DC motors, and ④compound-wound DC motors. Each of these motors has different characteristics due to its basic circuit arrangement and physical properties.[1]

Permanent-Magnet DC Motors

The permanent-magnet DC motor, as shown in Fig. 1-5A-1, is constructed in the same manner as its DC generator counterpart. The permanent-magnet DC motor is used for low-torque applications. When this type of motor is used, the DC power supply is connected directly to the armature conductors through the brush/commutator assembly. The magnetic field is produced by permanent magnets mounted on the stator. The rotor of permanent magnet motors is a wound armature.

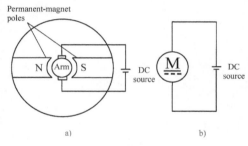

Fig. 1-5A-1 Permanent-magnet DC motor
a) Pictorial diagram b) Schematic diagram

This type of motor ordinarily uses either alnico or ceramic permanent magnets rather than field coils. The alnico magnets are used with high-horsepower applications. Ceramic magnets are ordinarily used for low-horsepower slow-speed motors. Ceramic magnets are highly resistant to demagnetization, yet they are relatively low in magnetic-flux level. The magnets are usually mounted in the motor frame and then magnetized prior to the insertion of the armature.

The permanent-magnet motor has several advantages over conventional types of DC motors. One advantage is reduced operational cost. The speed characteristics of the permanent-magnet motor are similar to those of the shunt-wound DC motor. The direction of rotation of a permanent-magnet motor can be reversed by reversing the two power lines.

Series-Wound DC Motors

The manner in which the armature and field circuits of DC motors are connected determines its basic characteristics. Each of the types of DC motors is similar in construction to the type of DC generator that corresponds to it. The only difference, in most cases, is that the generator acts as a voltage source while the motor functions as a mechanical power conversion device.[2]

The series-wound motor, as shown in Fig. 1-5A-2, has the armature and field circuits connected in a series arrangement. There is only one path for current to flow from the DC voltage source. Therefore, the field is wound of relatively few turns of large diameter wire, giving the field a low resistance. Changes in load applied to the motor shaft cause changes in the current through the field. If the mechanical load increases, the current also increases. The increased current creates a stronger magnetic field. The speed of a series motor varies from very fast at no load to very slow at heavy loads. Since large currents may flow through the low-resistance field, the series motor produces a high-torque output. Series motors are used where heavy loads must be moved and speed regulation is not important. A typical application is for automobile starter motors.

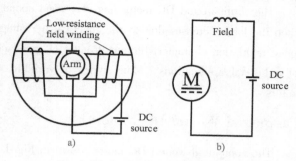

Fig. 1-5A-2 Series-wound DC motor
a) Pictorial diagram b) Schematic diagram

Shunt-Wound DC Motors

Shunt-wound DC motors are more commonly used than any other type of DC motor. As shown in Fig. 1-5A-3, the shunt-wound DC motor has field coils connected in parallel with its armature. This type of DC motor has field coils which are wound for many turns of small-diameter wire and have a relatively high resistance. Since the field is a high-resistant parallel path of the circuit of the shunt motor, a small amount of current flows through the field. A strong electromagnetic field is produced due to the many turns of wire that form the field windings.

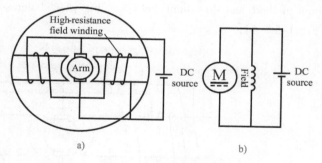

Fig. 1-5A-3 Shunt-wound DC motor
a) Pictorial diagram b) Schematic diagram

A large majority (about 95%) of the current drawn by the shunt motor flows in the armature circuit. Since the current has little effect on the strength of the field, motor speed is not affected appreciably by variations in load current. The relationship of the currents that flow through a DC shunt motor is as follows:

$$I_L = I_a + I_f$$

Where I_L—total current drawn from the power source;

I_a—armature current;

I_f—field current.

The field current may be varied by placing a variable resistance in series with the field windings. Since the current in the field circuit is low, a low-wattage rheostat may be used to vary the speed of the motor due to the variation in field resistance. As field resistance increases, field current will decrease. A decrease in field current reduces the strength of the electromagnetic field. When the field flux is decreased, the armature will rotate faster, due to reduced magnetic-field interaction.[3] Thus the speed of a DC shunt motor may be easily varied by using a field rheostat.

The shunt-wound DC motor has very good speed regulation. The speed does decrease slightly when the load increases due to the increase in voltage drop across the armature. Due to its good speed regulation characteristic and its ease of speed control, the DC shunt motor is commonly used for industrial applications. Many types of variable-speed machine tools are driven by DC shunt motors.

Compound-Wound DC Motors

The compound-wound DC motor shown in Fig. 1-5A-4, has two sets of field windings, one in series with the armature and one in parallel. This motor combines the desirable characteristics of the series-wound and shunt-wound motors. There are two methods of connecting compound motors: cumulative and differential. A cumulative compound DC motor has series and shunt fields that aid each other. Differential compound DC motors have series and shunt fields that oppose each other. There are also two ways in which the series windings are placed in the circuit. One method is called a short shunt (Fig. 1-5A-4), in which the shunt field is placed across the armature. The long-shunt method has the shunt field winding placed across both the armature and the series field (Fig. 1-5A-4).

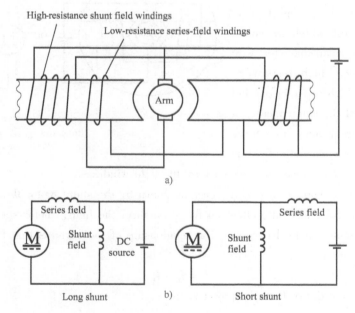

Fig. 1-5A-4　Compound-wound DC motor
a) Pictorial diagram　b) Schematic diagram

Compound motors have high torque similar to a series-wound motor, together with good speed regulation similar to a shunt motor. Therefore, when good torque and good speed regulation are needed, the compound-wound DC motor can be used. A major disadvantage of a compound-wound motor is its expense.

DC Motor Speed-Torque Characteristics

In many applications, DC motors are used to drive mechanical loads. Some applications require that the speed remain constant as the mechanical load applied to the motor changes. On the other hand, some applications require that the speed be controlled over a wide range. An engineer who wishes to use a DC motor for a particular application must therefore know the relation between torque and speed of the machine.

Initially our remarks are confined to the shunt motor, but a similar line of reasoning applies for the others. For our purposes the two pertinent equations are those for torque and current. Thus

$$T = K_T \Phi I_a$$

and

$$I_a = \frac{U_t - E_a}{R_a} = \frac{U_t - K_E \Phi n}{R_a}$$

So

$$n = \frac{U_t}{K_E \Phi} - \frac{R_a}{K_E \Phi} I_a = \frac{U_t}{K_E \Phi} - \frac{R_a}{K_E K_T \Phi^2} T$$

Fig. 1-5A-5 shows the general shapes of the speed-torque characteristics as they apply for the shunt, cumulatively compounded, and series motors. For the sake of comparison the curves are drawn through a common point of rated torque and speed.

The only variables involved are the speed n and the armature current I_a. At rated output torque the armature current is at its rated valve and so, too, is the speed. As the load torque is removed, the

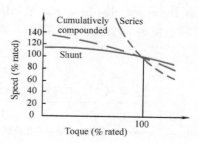

Fig. 1-5A-5 Typical speed-torque curves of DC motors

armature current becomes correspondingly smaller, making the numerator term of n larger. This results in higher speeds. The extent to which the speed increases depends upon how large the armature circuit resistance drop is in comparison to the terminal voltage. It is usually around 5 percent to 10 percent.

WORDS AND TERMS

commercially   *adv.* 商业地
permanent-magnet DC motor   永磁直流电动机
series-wound DC motor   串励直流电动机
shunt-wound DC motor   并励直流电动机
compound-wound DC motor   复励直流电动机
counterpart   *n.* 对应物，配对物
brush   *n.* 电刷
commutator   *n.* 换向器，整流器

assembly   *n.* 装置，构件
stator   *n.* 定子
rotor   *n.* 转子
alnico   *n.* 铝镍钴合金，铝镍钴永磁合金
ceramic   *adj.* 陶瓷的
horsepower   *n.* 马力，功率
frame   *n.* 机壳，机座
demagnetization   *n.* 去磁，退磁

flux　*n*.　磁通
mechanical power　机械功率
shaft　*n*.　转轴
field winding　*n*.　励磁绕组
turn　*n*.　匝数
rheostat　*n*.　变阻器
cumulative　*adj*.　累积的
differential　*adj*.　差的，差别的

NOTES

[1] Each of these motors has different characteristics due to its basic circuit arrangement and physical properties.
每种类型的电动机由于其基本电路的不同而具有不同的特征和物理特性。

[2] The only difference, in most cases, is that the generator acts as a voltage source while the motor functions as a mechanical power conversion device.
很多情况下，二者的唯一区别在于发电机常作为电压源，而电动机常作为机械能转换装置。

[3] A decrease in field current reduces the strength of the electromagnetic field. When the field flux is decreased, the armature will rotate faster, due to reduced magnetic-field interaction.
励磁电流的减小会使磁场减弱。当磁通减少时，转子会由于与减弱的磁场相互作用而加速旋转。

B　Closed-Loop Control of DC Drivers

A basic scheme of the closed-loop speed control system employing *current limit control*, also known as *parallel current control*, is shown in Fig. 1-5B-1. ω_m^* sets the speed reference. A signal proportional to the motor speed is obtained from the speed sensor. The speed sensor output is filtered to remove the AC ripple and compared with the speed reference. The speed error is processed through a speed controller. The output of the speed controller u_c adjusts the rectifier firing angle α to make the actual

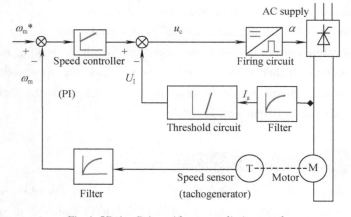

Fig. 1-5B-1　Drive with current limit control

speed close to the reference speed. The speed controller is usually a PI (proportional and integral) controller and serves three purposes—stabilizes the drive and adjusts the damping ratio at the desired value, makes the steady-state speed-error close to zero by integral action, and filters out noise again due to the integral action.

The drive employs current limit control, the purpose of which is to prevent the current from exceeding safe values. As long as $I_A < I_x$, where I_x is the maximum permissible value of I_A, the current control loop does not affect the drive operation. If I_A exceeds I_x, even by a small amount, a large output signal is produced by the threshold circuit, the current control overrides the speed control, and the speed error is corrected essentially at a constant current equal to the maximum permissible value. When the speed reaches close to the desired value, I_A falls below I_x, the current control goes out of action and speed control takes over. Thus in this scheme, at any given time the operation of the drive is mainly controlled either by the speed control loop or the current control loop, and hence it is also called parallel current control.

Another scheme of closed-loop speed control is shown in Fig. 1-5B-2. It employs an *inner current control loop* within an outer speed loop. The output of the speed controller e_c is applied to a current limiter which sets the current reference I_a^* for the inner current control loop. The output

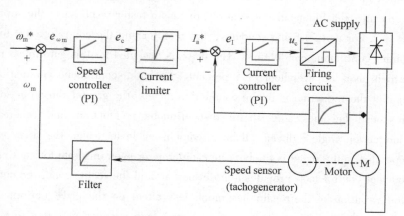

Fig. 1-5B-2 Drive with inner current control loop

of the current controller u_c adjusts the converter firing angle such that the actual speed is brought to a value set by the speed command ω_m^*. Any positive speed error, caused by either an increase in the speed command or an increase in the load torque, produces a higher current reference I_a^*.[1] The motor accelerates due to an increase in I_a, to correct the speed error and finally settles at a new I_a^* which makes the motor torque equal to the load torque and the speed error close to zero. For any large positive speed error, the current limiter saturates and the current reference I_a^* is limited to a value I_{am}, and the drive current is not allowed to exceed the maximum permissible value. The speed error is corrected at the maximum permissible armature current until the speed error becomes small and the current limiter comes out of saturation.[2] Now the speed error is corrected with I_a less than the permissible value.

A negative speed error will set the current reference I_a^* at a negative value. Since the motor current cannot reverse, a negative I_a^* is of no use. It will however "charge" the PI controller. When the speed error becomes positive the "charged" PI controller will take a longer time to respond, causing unnecessary delay in the control action. The current limiter is therefore arranged to set a zero-current reference for negative speed errors.

Since the speed control loop and the current control loop are in cascade, the inner current control is also known as *cascade control*. It is also called *current guided control*. It is more commonly used than the current-limit control because of the following advantages:

1. It provides faster response to any supply voltage disturbance. This can be explained by

considering the response of two drives to a decrease in the supply voltage. A decrease in the supply voltage reduces the motor current and torque. In the current-limit control, the speed falls because the motor torque is less than the load torque that has not changed. The resulting speed error is brought to the original value by setting the rectifier firing angle at a lower value. In the case of inner current control, the decrease in motor current, due to the decrease in the supply voltage, produces a current error which changes the rectifier firing angle to bring the armature current back to the original value. The transient response is now governed by the electrical time constant of the motor. Since the electrical time constant of a drive is much smaller compared to the mechanical time constant, the inner current control provides a faster response to the supply voltage disturbances.

2. For certain firing schemes, the rectifier and the control circuit together have a constant gain under continuous conduction. The drive is designed for this gain to set the damping ratio at 0.707, which gives an overshoot of 5 percent. Under discontinuous conduction, the gain reduces. The higher the reduction is in the conduction angel, the greater the reduction is in the gain. The drive response becomes sluggish in discontinuous conduction and progressively deteriorates as the conduction angle reduces. If an attempt is made to design the drive for discontinuous conduction operation, the drive is likely to be oscillatory or even unstable for continuous conduction. The inner current control loop provides a close loop around the rectifier and the control circuit, and therefore, the variation of their gain has much less affect on the drive performance. Hence, the transient response of the drive with the inner current loop is superior to that with the current-limit control.

3. In the current-limit control, the current must first exceed the permissible value before the current-limit action can be initiated. Since the firing angle can be changed only at discrete intervals, substantial current overshoot can occur before the current limiting becomes effective.

Small motors are more tolerant to high transient current. Therefore, to obtain a fast transient response, much higher transient currents are allowed by selecting a larger size rectifier. The current regulation is then needed only for abnormal values of current. In such cases because of the simplicity, current-limit control is employed.

Both the schemes have different responses for the increase and decrease in the speed command. A decrease in speed command at the most can make the motor torque zero; it cannot be reversed as braking is not possible. The drive decelerates mainly due to the load torque. When load torque is low, the response to a decrease in the speed command will be slow. These drives are therefore suitable for applications with large load torque, such as paper and printing machines, pumps, and blowers.

WORDS AND TERMS

scheme *n.* 方法，形式，示意图
proportional to 与……成正比的
sensor *n.* 传感器
filter *v.* 滤波

ripple *n.* 波纹，波动
firing angle 触发角
damping *n.* 阻尼；*adj.* 阻尼的
steady-state 稳态

prevent... from doing 使……不……
threshold *n.* 阈
override *v., n.* 超过，压倒
converter *n.* 逆变器，整流器
accelerate *v.* 加速
saturation *n.* 饱和
disturbance *n.* 扰动，干扰

transient *adj.* 暂态的，瞬态的，过渡的
overshoot *n.* 超调量
deteriorate *v.* 恶化，变坏
sluggish *adj.* 惰性的，缓慢的
oscillatory *adj.* 振荡的
tolerant *adj.* 容许的，容忍的
at the most 至多

NOTES

[1] Any positive speed error, caused by either an increase in the speed command or an increase in the load torque, produces a higher current reference I_a^*.

由速度给定或负载转矩的增加所引起的任何正的速度偏差，都会产生更大的参考电流值 I_a^*。

either... or 是连接词，作"或……或""是……是""不是……就是"讲。

[2] The speed error is corrected at the maximum permissible armature current until the speed error becomes small and the current limiter comes out of saturation.

在最大允许电枢电流下纠正速度偏差，直到速度偏差减小且限流装置退出饱和状态。

当主句为肯定句时，until 可译为"到……为止"；当主句为否定句时，until 的含义是"直到……才……"。

C　理解与表达

翻译首先要充分理解原文。这个过程要求结合上下文，推敲所在语言环境中英文单词的词义（一个单词在字典中可查到多种含义），并且辨明语法，搞清逻辑关系、主谓宾定补状的结构，同时根据所学专业知识来完成。理解之后，才可以进行表达，也就是用汉语将所理解的内容恰如其分地叙述出来。理解是表达的前提和基础，不理解、理解不透彻或理解错误的东西，根本无法正确地表达；表达是理解的结果和目的，错误的表达会使理解的东西无法实现。二者相互联系、相辅相成，理解的同时就在寻找表达内容和选择表达方式；而表达时又会进一步加深理解。这两个过程往往要反复多次，才能完成译文。

表达时可以根据需要采用直译或者意译，但切忌机械照搬的死译和望文生义的乱译。举例如下：

1. Computer viruses are programs designed to replicate and spread, sometimes without indicating that they exist.

分析：Computer viruses 是主语，系动词 are 做谓语，programs 做表语，designed to replicate and spread 做 programs 的定语，sometimes without indicating that they exist 是独立主格结构。分析清语法结构后，可以很容易地理解和表达。原文可直译为：计算机病毒是可以复制和传播的程序，有时它们的存在并不为人所知。

2. The year, 2012, saw great development in the field of Mobile Internet.

分析：本句较好理解，但如按原文直译表达则不符汉语习惯，可意译为：移动互联网领域在 2012 年有了很大的发展。

3. Control systems frequently employ components of many types.

分析：employ 一般作"雇佣"讲，可理解为"使用"，在表述时可译作"要用到……"。

4. Automatic control systems are physical systems which have dynamic behavior.

分析：behavior 一般作"行为""举止"讲，可理解成"习性""性质"，表述时译作"特性"；dynamic 一般作"生机勃勃的""活动的"讲，结合上下文可理解成"动作的"，表述时根据所学专业概念把它同 behavior 合译为"动态特性"。

5. A transistor has three electrodes, the E (emitter), the B (base) and the C (collector).

分析：base 根据上下文和所学专业知识可以很容易地理解和表达为"基极"；同时根据对全句内容的理解，可以很容易地判断生词 transistor 的含义是"晶体管"或"三极管"。

6. Digital computers are finding extensive application in industrial process control systems.

分析：find 译作"发现"显然不通。理解全句意思后可以意译为：数字计算机在工业过程控制系统中获得了广泛的应用。

7. Continuous control readily lends itself to an understanding of feedback control theory using relatively uncomplicated mathematics.

分析：本句的语法关系较复杂，较难理解和翻译。首先 using relatively uncomplicated mathematics 可理解为原因状语"由于使用的数学相对简单"。而 readily lends itself to 可理解为"有助于"。全句可译为：由于使用的数学相对简单，故连续控制有助于对反馈控制理论的理解。或意译为：由于使用的数学相对简单，故连续控制的反馈控制理论更容易理解。

UNIT 6

A   AC Machines

Introduction

The electrical machine that converts electrical energy into mechanical energy, and vice versa, is the workhorse in a drive system. A machine is a complex structure electrically, mechanically, and thermally. Although machines were introduced more than one hundred years ago, the research and development in this area appears to be never-ending. However, the evolution of machines has been slow compared to that of power semiconductor devices and power electronic converters. Traditionally, AC machines with a constant frequency sinusoidal power supply have been used in constant-speed applications, whereas DC machines were preferred for variable-speed drives. But in the last two or three decades, we have seen extensive research and development efforts for variable-frequency, variable-speed AC machine drive technology, and they will progressively replace DC drives. In most cases, new applications use AC drives.

AC machines can generally be classified as follows:

Induction machines: cage or wound rotor (doubly-fed), rotating or linear;

Synchronous machines: rotating or linear, reluctance, wound field or permanent magnet, radial or axial gap (disk), surface magnet or interior (buried) magnet, sinusoidal or trapezoidal;

Variable reluctance machines: switched reluctance, stepper.

Induction Machines

Among all types of AC machines, the induction machine, particularly the cage type, is most commonly used in industry. These machines are very economical, rugged, and reliable, and are available in the ranges of fractional horse power (FHP) to multi-megawatt capacity. Low-power FHP machines are available in single-phase, but poly-phase (three-phase) machines are used, most often in variable-speed drives. Fig. 1-6A-1 shows an idealized three-phase, two-pole induction motor where each phase winding in the stator and rotor is represented by a concentrated coil. The three-phase windings are distributed sinusoidally and embedded in slots. In a wound-rotor machine, the rotor winding is similar to that of the stator, but in a cage machine, the rotor has a squirrel cage-like

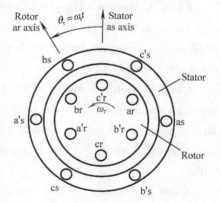

Fig. 1-6A-1   Idealized three-phase, two pole induction motor

structure with shorted end rings. Basically, the machine can be looked upon as a three-phase transformer with a rotating and short-circuited secondary. Both stator and rotor cores are made with laminated ferromagnetic steel sheets. The air gap in the machine is practically uniform (non-salient pole).

One of the most fundamental principles of induction machines is the creation of a rotating and sinusoidally distributed magnetic field in the air gap. Neglecting the effect of slots and space harmonics due to non-ideal winding distribution, it can be shown that a sinusoidal three-phase balanced power supply in the three-phase stator winding creates a synchronously rotating magnetic field.[1] The rotational speed can be given as equation (1-6A-1). N_e is called synchronous speed in rpm and $f_e = \omega_e/2\pi$ is the stator frequency in Hz. P is the pole numbers of a machine.

$$N_e = \frac{120 f_e}{P} \tag{1-6A-1}$$

The rotor winding will be subjected to a sweeping magnetic field, and have inducing current in the short-circuited rotor.[2] The interaction of air gap flux and rotor mmf produces torque, make the rotor rotate. But the speed of the rotor is less than synchronous speed. So it called induction machine or asynchronous machine. To meet the various starting and running requirements of a variety of industrial applications, several standard designs of squirrel-cage motors are available from manufacturers' stock. The torque-speed characteristics of the most common designs, readily available and standardized in accordance with the criteria established by the National Electrical Manufacturers' Association (NEMA), are shown in Fig. 1-6A-2. The most significant design variable in these motors is the effective resistance of the rotor cage circuits.

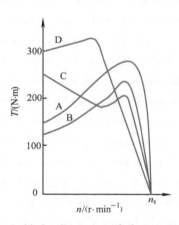

Fig. 1-6A-2 Torque-speed characteristics for different class of induction motors

Class A Motors These machines are suitable for applications where the load torque is low at start (such as fan or pump loads) so that full speed is achieved rapidly, thereby eliminating the problem of overheating during starting. In large machines, low-voltage starting is required to limit the starting current.

Class B Motors Motors of this class are good general-purpose motors and have a wide variety of industrial applications. They are particularly suitable for constant-speed drives, where the demand for starting torque is not severe. Examples are drives for fans, pumps, blowers, and motor-generator sets.

Class C Motors Class C motors are suitable for driving compressors, conveyors, and so forth.

Class D Motors These motors are suitable for driving intermittent loads requiring rapid acceleration and high-impact loads such as punch presses or shears. In the case of impact loads, a flywheel is fitted to the system. As the motor speed falls appreciably with load impact, the flywheel delivers some of its kinetic energy during the impact.

Synchronous Machines

A synchronous machine, as the name indicates, must rotate at synchronous speed, that is, the speed is uniquely related to the supply frequency, as indicated in Equation (1-6A-1). It is a serious competitor to the induction machine invariable-speed drive applications.

Fig. 1-6A-3 shows an idealized three-phase, two-pole wound field synchronous machine. The stator winding of the machine is identical to that of the induction machine, but the rotor has a winding that carries DC current and produces flux in the air gap that helps the stator-induced rotating magnetic field to drag the rotor along with it. The DC field current is supplied to the rotor from a static rectifier through slip rings and brushes, or by brushless excitation. Since the rotor always moves at synchronous speed, the synchronously rotating d^e-q^e axes are fixed with the rotor, where the d^e axis corresponds to the north pole. There is no stator-induced induction in the rotor, and therefore, the rotor mmf is supplied exclusively by the field

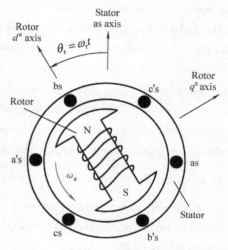

Fig. 1-6A-3 Idealized three-phase, two pole synchronous machine

winding. This permits the machine to run at an arbitrary power factor at the stator terminal, that is, leading, lagging, or unity. On the other hand, in an induction machine, the stator supplies the rotor excitation that makes the machine power factor always lagging.

The mechanism of torque production is somewhat similar to that of an induction machine. The machine shown is characterized as a salient pole because of the nonuniform air gap around the rotor, which contributes to asymmetrical magnetic reluctance in the d and q axes. This is in contrast to a machine with a cylindrical rotor structure having a uniform air gap (such as an induction motor), defined as a nonsalient pole machine.[3] For example, low-speed synchronous generators in hydro-electric power stations use salient pole machines, whereas high-speed generators in steam-power stations use nonsalient pole machines. In addition to field winding, the rotor usually contains an amortisseur, or damper winding, which is like short-circuited squirrel cage bar in an induction motor. The machine is more expensive but efficiency is somewhat higher. Wound field machines are normally used for high-power (multi-megawatt) drives.

Variable Reluctance Machine

A variable or double reluctance machine (VRM), as the name indicates, has double saliency, meaning it has saliency in the stator as well as in the rotor. As mentioned before, the VRM has two classifications: switched reluctance machine (SRM) and stepper motor. The stepper motor is basically a digital motor, i.e., it moves by a fixed step or angle with a digital pulse.

Small stepper motors are widely used for computer peripheral-type applications. However, since the machine is not suitable for variable-speed applications, there will not be any further discussion of it.

There has been interest in switched reluctance motor drives in the literature, and recently, great effort has been made to commercialize them in competition with induction motors. Fig. 1-6A-4 shows the cross-section of a four-phase machine with four stator-pole pairs and three rotor-pole pairs (8/6 motor). The machine rotor does not have any winding or PM. The stator poles have concentrated winding (instead of sinusoidal winding), and each stator-pole pair winding, as shown in the figure, is excited by a converter phase. For example, the stator-pole pair A-A′ is energized when the rotor pole-pair a-a′ approaches it to produce the torque by magnetic pull, but is de-energized when pole alignment occurs.[4] All four machine phases are excited sequentially and synchronously with the help of a rotor position encoder to get unidirectional torque. The magnitude of torque can be given as: $T_e = \frac{1}{2}mi^2$. Where m = inductance slope and i = instantaneous current. The current i can be maintained constantly by adjusting the inductance slope. At high speeds, the rotor-induced CEMF is high.

The favorable attributes of this electronic motor are simplicity and robustness of construction; potentially, it is somewhat cheaper than other classes of machines. However, the torque generation is pulsating in nature and there are serious acoustic noise problems.

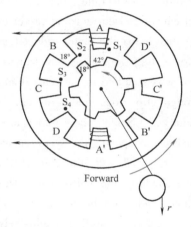

Fig. 1-6A-4   Construction of switched reluctance machine

WORDS AND TERMS

sinusoidal   *adj.* 正弦的
constant-speed   *adj.* 恒速的
variable-speed   *adj.* 变速的
induction machine   感应电机
synchronous machine   同步电机
VRM   变磁阻电机
switched reluctance machine   开关磁阻电机
rugged   *adj.* 结实的，耐用的
fractional   *adj.* 分数的
concentrated coil   集中绕组
distributed   *adj.* 分散的，分布的
slot   *n.* 槽
wound-rotor   *n.* 绕线转子

cage   *n.* 笼子，笼形
core   *n.* 铁心
laminated   *adj.* 分层的，叠片的
ferromagnetic   *adj.* 铁磁性的，铁磁体的
air gap   气隙
salient   *adj.* 凸起的，突出的
synchronous speed   同步转速
leading   *adj.* 超前的
hydro-electric   *adj.* 水力发电的
nonsalient   *adj.* 非凸起的，隐藏的
amortisseur   *n.* 阻尼器
damper winding   阻尼绕组
encoder   *n.* 编码器

NOTES

[1] Neglecting the effect of slots and space harmonics due to non-ideal winding distribution, it can be shown that a sinusoidal three-phase balanced power supply in the three-phase stator winding creates a synchronously rotating magnetic field.

如果忽略槽和由于非理想分布的绕组产生的空间谐波的影响，可以证明，在三相定子绕组中能以三相对称电源建立一个同步旋转的旋转磁场。

[2] The rotor winding will be subjected to a sweeping magnetic field, and have inducing current in the short-circuited rotor.

转子绕组切割磁场，就会在短路的转子中产生感应电流。

[3] This is in contrast to a machine with a cylindrical rotor structure having a uniform air gap (such as an induction motor), defined as a nonsalient pole machine.

与其（凸极式同步电机）对应的另一种电机是有均匀气隙的圆柱体形转子结构的电机（与异步电机相似），定义为隐极式同步电机。

[4] For example, the stator-pole pair A-A' is energized when the rotor pole-pair a-a' approaches it to produce the torque by magnetic pull, but is de-energized when pole alignment occurs.

例如，当转子极对 a-a'接近定子极对 A-A'时，定子极对 A-A'被通电，通过磁拉力产生转矩，当两个极对重合时，定子极对 A-A'断电。

B Induction Motor Drive

The speed of an induction motor is determined by the synchronous speed and slip of the motor. The synchronous speed is related to the supply frequency and the slip can be controlled by the regulation of voltage or current supplied to the motor. There are several mechanisms for controlling the speed of induction motor. These are: ① variable-voltage constant-frequency or stator voltage control, ② variable-voltage variable-frequency control, ③ variable-current variable-frequency control, and ④ regulation of slip power. One of these methods, the *variable-voltage variable-frequency control*, is described as follows.

Square-Wave Inverter Drive

The voltage-fed inverters (also called Voltage Source Inverters, VSI) are generally classified into two types: *square-wave inverter* and *pulse-width modulation inverter*. This type of inverter was introduced from the beginning of nineteen-sixties when elegant force-commutation techniques were introduced. Fig. 1-6B-1 shows the conventional power circuit of a square-wave inverter drive. A three-phase bridge rectifier converts AC to variable-voltage DC, which is impressed at the input of a force-commutated bridge inverter. The inverter generates a variable-voltage variable-frequency power supply to control the speed of the motor. The inverter is called voltage-fed because a large filter capacitor provides a stiff voltage supply to the inverter and the inverter output voltage waves are not affected by the nature of the load. Normally, each thyristor of an inverter leg conducts for 180° to generate a square-wave voltage at the machine phase with respect to the fictitious center point of the DC supply. The line to line voltage can be shown to be a six-stepped wave as shown in the figure. Since the induction motor constitutes a lagging power factor load, the inverter thyristors require forced commutation. The feedback diodes help circulation of load reactive power with the filter capacitor and maintain the output voltage waves clamped to the level of DC link voltage. The diodes also participate in the commutation and braking process.

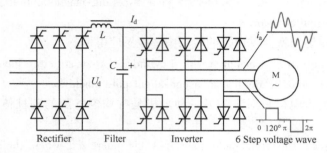

Fig. 1-6B-1　Variable-voltage variable-frequency square-wave induction motor drive
(Inverter forced commutation is not shown)

The theory of variable-voltage variable-frequency speed control method can be explained with the help of Fig. 1-6B-2 and Fig. 1-6B-3. The motor used in this type of drive has low slip characteristic, which results in improvement of efficiency. The speed of the motor can be varied by simply varying its synchronous speed, i.e., by varying the inverter frequency. However, as the frequency is increased, the machine airgap flux falls causing low developed torque capability.[1] The airgap flux can be maintained constant as in a DC shunt motor if the voltage is varied with frequency so that the ratio remains constant. Fig. 1-6B-2 shows the desired voltage-frequency relationship of the motor. Below the base (1.0 PU) frequency, the airgap flux is maintained constant by the constant volts/Hz ratio, which results in constant torque capability. At very low frequency, the stator resistance dominates over the leakage inductance, and therefore additional voltage is impressed to compensate this effect. At base frequency, the full motor voltage is established as permitted by fully advancing the rectifier firing angle. Beyond this point, as frequency increases, the torque declines because of loss of airgap flux, and the machine operates in constant horsepower as shown in the figure. This is analogous to the field-weakening mode of speed control of a DC motor. The motor torque-speed curves for constant torque and constant power regions are shown in Fig. 1-6B-3 where each torque-speed curve corresponds to a particular voltage

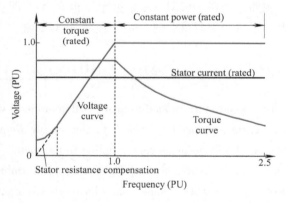

Fig. 1-6B-2　Voltage-frequency relation of induction motor

and frequency combination at the machine terminal. Two steady state operating points A and B which correspond to constant and variable load, respectively, are shown in the figure. The machine can accelerate from zero speed at maximum available torque and then approach the steady state points either at constant flux slip control mode or at constant slip flux control mode. Regulation of both flux and slip for steady state operation adds improvement of machine efficiency.

The voltage-fed square-wave drives are normally used in low to medium horsepower industrial applications where the speed ratio is usually limited to 10 : 1. Recently, this type of drive has largely been superseded by PWM drives which will be described in the next subsection. The voltage-fed inverters are easily adaptable to multi-motor drives where speed of a number of induction motors can be closely tracked.

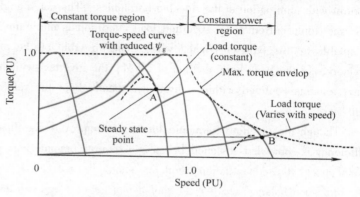

Fig. 1-6B-3 Torque-speed curves of induction motor with variable-voltage variable-frequency power supply

PWM Inverter Drive

In the variable-voltage variable-frequency inverter drive described in the previous subsection, the DC link voltage can be kept uncontrolled by a diode rectifier and the fundamental frequency output voltage can be controlled electronically within the inverter by using a pulse-width modulation technique. In this method, the thyristors are switched on and off many times within a half-cycle to generate a variable-voltage output which is normally low in harmonic content. Among several PWM techniques, *sinusoidal PWM* is common and it is explained in Fig. 1-6B-4. An isosceles triangle carrier wave is compared with the sine wave signal and the crossover points determine the points of commutation. Except at low frequency range, the carrier is synchronized with the signal, and an even integral (multiple of three) ratio is maintained to improve the harmonic content. The fundamental output voltage can be varied by variation of the modulation index. It can

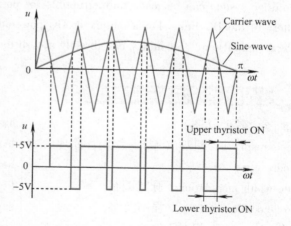

Fig. 1-6B-4 Principle of pulse width modulation

be shown that if the modulation index is less than unity, only carrier frequency harmonics with fundamental frequency related side bands appear at the output. Such a waveform causes considerably less harmonic heating and torque pulsation compared to that of a square-wave. The voltage can be increased beyond the modulation index of unity until maximum voltage is obtained in square wave mode. Therefore, PWM voltage control is applicable in the constant torque region (see Fig. 1-6B-2), whereas in the constant power region the operation is identical to that of square-wave drive.

The technique of selected harmonic elimination PWM has received wide attention recently. In this method, notches are created at predetermined angles of the square-wave which permits voltage control with elimination of the selected harmonics. The notch angles can also be programmed so that the rms ripple current for a specified load condition is minimum. The microcomputer is especially adaptable to this type of PWM, where look-up table of the angles can be stored in the ROM memory. In the bang bang method of PWM, the inverter switching is controlled so that the current wave remains confined within a hysteresis band about the reference wave resulting in low ripple current.

Though the machine harmonic losses are improved significantly in PWM drive, the inverter efficiency is somewhat lessened because of many commutations per halfcycle. In a well-designed PWM drive, the commutation frequency should be increased as permitted by the devices so as to obtain a good balance between the increase of inverter loss and decrease of machine loss.[2] The use of a simple and economical diode rectifier in the front end improves line distortion and power factor, reduces the filter size, and improves the reliability of system operation. Since the DC link voltage is relatively constant, the commutation of thyristors is satisfactory in the whole range of fundamental voltage. In addition, low harmonics and minimal torque pulsation in the low frequency region permit wide range speed control practically from standstill with the full torque capability of the machine. Since the DC link voltage is not controlled, a number of inverters with independent control can be operated with a single rectifier supply resulting in considerable saving of rectifier cost. The drive system can be made uninterruptible for possible AC line power failure by switching in a battery in the DC link. For a battery or DC powered driver system, such as, electric vehicle or subway propulsion, the supply can directly absorb the regenerative braking power.

WORDS AND TERMS

slip   n. 转差（率）
mechanism   n. 机理（制），作用，原理
square-wave   n. 方波
pule-width modulation   脉宽调制
force-commutation   n. 强制换向
with respect to   相对于
fictitious   adj. 假想的
line to line voltage   线电压
lagging   n. 滞后
clamp   v. 箝位，定位

airgap = air gap   气隙
leakage   n. 漏
horsepower   n. 功率
field-weakening   n. 弱磁
supersede   v. 取代
isosceles   adj. 等腰的
carrier   n. 载波，载体
even   adj. 偶数的
notch   n. 换相点，换级点
rms = root-mean-square   有效值，方均根

look-up table　查询表
confine　*v.* 限制（在……范围内）

distortion　*n.* 失真

NOTES

[1] However, as the frequency is increased, the machine airgap flux falls causing low developed torque capability.

然而随着频率增加，电机的气隙磁通下降，导致产生的转矩降低。

句中 causing 引出的短语做结果状语。

[2] In a well-designed PWM drive, the commutation frequency should be increased as permitted by the devices so as to obtain a good balance between the increase of inverter loss and decrease of machine loss.

在设计完善的 PWM 传动系统中，应在器件允许的条件下增加换相频率，以便在逆变器损耗的增加和电机损耗的降低之间找到一个合适的平衡点。

so as to 接动词不定式，可译作"为的是……""以便……"；so...as to 接动词不定式，意思是"如此……以致"。

C　长句的翻译

科技英语中由于大量使用定语从句、状语从句和各种短语，使句子常具有较长的复杂结构，加之英汉两种语言在词法、句法、逻辑和思维等方面的差异，因而造成很多长句难懂、难译。对待长句，要分清层次、突出重点，搞清各分句的内容和结构。译文要逻辑严密、前后呼应，借助语法关系和逻辑承接语，使译文前后衔接、相互呼应，把原文复杂的概念准确、通顺地表达出来。长句的翻译方法通常有顺序法、变序法、分句法、并句法等典型的几种，这些方法可根据不同情况单独使用或结合使用。

1. 原句层次分明，和汉语语序相近时可采用顺序法。例如：

No such limitation is placed on an AC motor; here the only requirement is relative motion, and since a stationary armature and a rotating field system have numerous advantages, this arrangement is standard practice for all synchronous motor rated above a few kilovolt-amperes.

交流电机不受这种限制，唯一的要求是相对运动，而且由于固定电枢及旋转磁场系统具有很多优点，所以这种安排是所有容量在几千伏安以上的同步电机的标准做法。

2. 英语习惯前果后因和定、状语后置，译成汉语时可采用变序法，即先译后部，再依次向前，改变原句的语序，按汉语习惯的语序译出，从而便于表达和读者理解。例如：

The resistance of any length of a conducting wire is easily measured by finding the potential difference in volts between its ends when a known current is following.

已知导线中流过的电流，只要测出导线两端电位差的伏特值，就能很容易得出任何长度导线的电阻值。

3. 为了使译文结构清楚，合乎表达习惯，有时可用拆译法，即将原句分成几个独立的小句，顺序基本不变，保持前后连贯。例如：

This kind of two-electrodes tube consists of a tungsten filament, which gives off electrons when it is heated, and a plate toward which the electrons migrate when the field is in the right direction.

这种二极管由一根钨丝和一个极板组成。钨丝受热时放出电子,当电场方向为正时,电子就移向极板。(把两个 which 引导的定语从句从原句中分出,拆成两个独立分句,更符合汉语的表达习惯,意思明确,通俗易懂。)

4. 并句法是将原句中的某些部分合并翻译,以使译文简洁通顺。例如:

It is common practice that electric wires are made from copper.

电线通常是铜制的。(把主句和从句合二为一,避免了冗长复杂的表述方式。)

以上几种方法在不同情况下可以灵活掌握,结合使用。例如:

The computer performs a supervisory function in the liquid-level control system by analyzing the process conditions against desired performance criteria and determining the changes in process variables to achieve optimum operation.

在液位控制系统中,计算机执行一种监控功能。它根据给定的特性指标来分析各种过程条件,并决定各过程变量的变化以获得最佳运行。(拆译和分译相结合。)

UNIT 7

A Electric Power System

Introduction

Electric Power Systems, components that transform other types of energy into electrical energy and transmit this energy to a consumer. The production and transmission of electricity is relatively efficient and inexpensive, although unlike other forms of energy, electricity is not easily stored and thus must generally be used as it is being produced.

Components of an Electric Power System

A modern electric power system consists of six main components: ① the power station, ② a set of transformers to raise the generated power to the high voltages used on the transmission lines, ③ the transmission lines, ④ the substations at which the power is stepped down to the voltage on the distribution lines, ⑤ the distribution lines, and ⑥ the transformers that lower the distribution voltage to the level used by the consumer's equipment.

Power Station The power station of a power system consists of a prime mover, such as a turbine driven by water, steam, or combustion gases that operate a system of electric motors and generators. Most of the world's electric power is generated in steam plants driven by coal, oil, nuclear energy, or gas. A smaller percentage of the world's electric power is generated by hydroelectric (waterpower), diesel, and internal-combustion plants.

Transformers Modern electric power systems use transformers to convert electricity into different voltages. With transformers, each stage of the system can be operated at an appropriate voltage. In a typical system, the generators at the power station deliver a voltage of from 1,000 to 26,000 volts (V). Transformers step this voltage up to values ranging from 138,000 to 765,000 V for the long-distance primary transmission line because higher voltages can be transmitted more efficiently over long distances. At the substation the voltage may be transformed down to levels of 69,000 to 138,000 V for further transfer on the distribution system. Another set of transformers step the voltage down again to a distribution level such as 2,400 or 4,160 V or 15, 27, or 33 kilovolts (kV). Finally the voltage is transformed once again at the distribution transformer near the point of use to 240 or 120 V.

Transmission Lines The lines of high-voltage transmission systems are usually composed of wires of copper, aluminum, or copper-clad or aluminum-clad steel, which are suspended from tall latticework towers of steel by strings of porcelain insulators. By the use of clad steel wires and high towers, the distance between towers can be increased, and the cost of the transmission line thus reduced. In modern installations with essentially straight paths, high-voltage lines may be built with as few as six towers to the kilometer. In some areas high-voltage lines are suspended from tall wooden poles spaced more closely together.

For lower voltage distribution lines, wooden poles are generally used rather than steel towers. In cities and other areas where open lines create a safety hazard or are considered unattractive, insulated underground cables are used for distribution.[1] Some of these cables have a hollow core through which oil circulates under low pressure. The oil provides temporary protection from water damage to the enclosed wires should the cable develop a leak. Pipe-type cables in which three cables are enclosed in a pipe filled with oil under high pressure (14 kg per sq cm/200psi) are frequently used. These cables are used for transmission of current at voltages as high as 345,000 V (or 345 kV).

Supplementary Equipment Any electric-distribution system involves a large amount of supplementary equipment to protect the generators, transformers, and the transmission lines themselves. The system often includes devices designed to regulate the voltage or other characteristics of power delivered to consumers.

To protect all elements of a power system from short circuits and overloads, and for normal switching operations, circuit breakers are employed. These breakers are large switches that are activated automatically in the event of a short circuit or other condition that produces a sudden rise of current. Because a current forms across the terminals of the circuit breaker at the moment when the current is interrupted, some large breakers (such as those used to protect a generator or a section of primary transmission line) are immersed in a liquid that is a poor conductor of electricity, such as oil, to quench the current. In large air-type circuit breakers, as well as in oil breakers, magnetic fields are used to break up the current. Small air-circuit breakers are used for protection in shops, factories, and in modern home installations. In residential electric wiring, fuses were once commonly employed for the same purpose. A fuse consists of a piece of alloy with a low melting point, inserted in the circuit, which melts, breaking the circuit if the current rises above a certain value. Most residences now use air-circuit breakers.

Power Failures

In most parts of the world, local or national electric utilities have joined in grid systems. The linking grids allow electricity generated in one area to be shared with others. Each utility that agrees to share gains an increased reserve capacity, use of larger, more efficient generators, and the ability to respond to local power failures by obtaining energy from a linking grid.[2]

These interconnected grids are large, complex systems that contain elements operated by different groups. These systems offer the opportunity for economic savings and improve overall reliability but can create a risk of widespread failure. For example, the worst blackout in the history of the United States and Canada occurred August 14, 2003, when 61,800 megawatts of electrical power was lost in an area covering 50 million people. (One megawatt of electricity is roughly the amount needed to power 750 residential homes.) The blackout prompted calls to replace aging equipment and raised questions about the reliability of the national power grid.

Despite the potential for rare widespread problems, the interconnected grid system provides necessary backup and alternate paths for power flow, resulting in much higher overall reliability than is possible with isolated systems. National or regional grids can also cope with unexpected outages such as those caused by storms, earthquakes, landslides, and forest fires, or due to human error or deliberate acts of sabotage.

Power Quality

In recent years electricity has been used to power more sophisticated and technically complex manufacturing processes, computers and computer networks, and a variety of other high-technology consumer goods. These products and processes are sensitive not only to the continuity of power supply but also to the constancy of electrical frequency and voltage. Consequently, utilities are taking new measures to provide the necessary reliability and quality of electrical power, such as by providing additional electrical equipment to assure that the voltage and other characteristics of electrical power are constant.

Voltage Regulation Long transmission lines have considerable inductance and capacitance. When a current flows through the line, inductance and capacitance have the effect of varying the voltage on the line as the current varies. Thus the supply voltage varies with the load. Several kinds of devices are used to overcome this undesirable variation in an operation called regulation of the voltage. The devices include induction regulators and three-phase synchronous motors (called synchronous condensers), both of which vary the effective amount of inductance and capacitance in the transmission circuit.

Inductance and capacitance react with a tendency to nullify one another. [3] When a load circuit has more inductive than capacitive reactance, as almost invariably occurs in large power systems, the amount of power delivered for a given voltage and current is less than when the two are equal. The ratio of these two amounts of power is called the power factor. Because transmission-line losses are proportional to current, capacitance is added to the circuit when possible, thus bringing the power factor as nearly as possible to 1. For this reason, large capacitors are frequently inserted as a part of power-transmission systems.

World Electric Power Production Over the period from 1950 to 2003, the most recent year for which data are available, annual world electric power production and consumption rose from slightly less than 1 trillion kilowatt-hours (kW·h) to 15.9 trillion kW·h. A change also took place in the type of power generation. In 1950 about two-thirds of the world's electricity came from steam-generating sources and about one-third from hydroelectric sources. In 2003 thermal sources produced 65 percent of the power, but hydropower had declined to 17 percent, and nuclear power accounted for 16 percent of the total. The growth in nuclear power slowed in some countries, notably the United States, in response to concerns about safety. Nuclear plants generated 20 percent of U.S. electricity in 2003; in France, the world leader, the figure was 78 percent.

Conservation

Much of the world's electricity is produced from the use of nonrenewable resources, such as natural gas, coal, oil, and uranium. Coal, oil, and natural gas contain carbon, and burning these fossil fuels contributes to global emissions of carbon dioxide and other pollutants. Scientists believe that carbon dioxide is the principal gas responsible for global warming, a steady rise in Earth's surface temperature.

Consumers of electricity can save money and help protect the environment by eliminating unnecessary use of electricity, such as turning off lights when leaving a room. Other conservation methods include buying and using energy-efficient appliances and light bulbs, and using appliances, such as washing machines and dryers, at off-peak production hours when rates are lower. Consumers may also consider environmental measures such as purchasing "green power" when it is offered by a local utility. "Green power" is usually more expensive but relies on renewable and environmentally friendly energy sources, such as wind turbines and geothermal power plants.

WORDS AND TERMS

transformer *n.* 变压器
substation *n.* 变电站
prime mover 原动机
turbine *n.* 涡轮
aluminum *n.* 铝
copper-clad *n.* 镀铜
suspend *v.* 悬挂
latticework *n.* 格子
porcelain *adj.* 瓷制的
supplementary *adj.* 辅助的
circuit breaker 断路器
immerse *v.* 沉浸，浸入
quench *v.* 熄灭
wiring *n.* 配线
alloy *n.* 合金

grid *n.* 格子，网格
reserve capacity 储备功率
blackout *n.* （大区域的）停电
outage *n.* 暂时停电
landslide *n.* 泥石流
sabotage *n.* 破坏
sophisticated *adj.* 复杂精密的
continuity *n.* 连续性
constancy *n.* 恒定
synchronous condenser 同步调相机
nullify *v.* 无效
pollutant *n.* 污染物质
off-peak *adj.* 非高峰的
geothermal *adj.* 地热的

NOTES

[1] In cities and other areas where open lines create a safety hazard or are considered unattractive, insulated underground cables are used for distribution.
在城市和其他地区，明线存在安全隐患或者被认为影响美观，所以使用绝缘地下电缆进行配电。

[2] Each utility that agrees to share gains an increased reserve capacity, use of larger, more efficient generators, and the ability to respond to local power failures by obtaining energy from a linking grid.

同意共享的每个电力企业可以获得不断增加的储备功率，使用更大、效率更高的发电机，从电网中获取电能以应对局部电力故障。

[3] Inductance and capacitance react with a tendency to nullify one another.
电感和电容的作用能相互抵消。

B Power System Automation

Overview

Power providers constantly deal with demands to increase productivity and reduce costs. This translates into the need for administrators, engineers, operators, planners, field crews, and others to collect and act on decision-making information. Power system vendors are following a trend to make devices smarter so they can create and communicate this information. The term "power system" describes the collection of devices that make up the physical systems that generate, transmit, and distribute power. The term "instrumentation and control (I&C) system" refers to the collection of devices that monitor, control, and protect the power system.

Power system automation refers to using I&C devices to perform automatic decision making and control of the power system.

Data Acquisition Data acquisition refers to acquiring, or collecting, data. This data is collected in the form of measured analog current or voltage values or the open or closed status of contact points. Acquired data can be used locally within the device collecting it, sent to another device in a substation, or sent from the substation to one or several databases for use by operators, engineers, planners, and administrators.

Power System Supervision Computer processes and personnel supervise, or monitor, the conditions and status of the power system using this acquired data. Operators and engineers monitor the information remotely on computer displays and graphical wall displays or locally, at the device, on front-panel displays and laptop computers.

Power System Control Control refers to sending command messages to a device to operate the I&C and power system devices. Traditional supervisory control and data acquisition (SCADA) systems rely on operators to supervise the system and initiate commands from an operator console on the master computer. Field personnel can also control devices using front-panel push buttons or a laptop computer.

Power System Automation System automation is the act of automatically controlling the power system via automated processes within computers and intelligent I&C devices. The processes rely on data acquisition, power system supervision, and power system control all working together in a coordinated automatic fashion. The commands are generated automatically and then transmitted in the same fashion as operator initiated commands.

I&C System IEDs I&C devices built using microprocessors are commonly referred to as intelligent electronic devices (IEDs). Microprocessors are single chip computers that allow the devices into which they are built to process data, accept commands, and communicate information like a computer. Automatic processes can be run in the IEDs, and communications are handled through a serial port like the communications ports on a computer. IEDs are found in the substation and on the pole-top.

Equipments for Power System Automation

Power system automation includes a variety of equipments. The principal items are listed and briefly described below.

Instrument Transformers Instrument transformers are used to sense power system current and voltage values. They are physically connected to power system apparatus and convert the actual power system signals, which include high voltage and current magnitudes, down to lower signal levels.

Transducers Transducers convert the analog output of an instrument transformer from one magnitude to another or from one value type to another, such as from an ac current to dc voltage.

Remote Terminal Unit As the name implies, a remote terminal device, RTU, is an IED that can be installed in a remote location, and acts as a termination point for field contacts. A dedicated pair of copper conductors are used to sense every contact and transducer value. These conductors originate at the power system device, are installed in trenches or overhead cable trays, and are then terminated on panels within the RTU. The RTU can transfer collected data to other devices and receive data and control commands from other devices through a serial port. User programmable RTUs are referred to as "smart RTUs."

Communications Port Switch A communications switch is a device that switches between several serial ports when it is told to do so. The remote user initiates communications with the port switch via a connection to the substation, typically a leased line or dial-up telephone connection. Once connected, the user can route their communications through the port switch to one of the connected substation IEDs. The port switch merely "passes through" the IED communications.

Meters A meter is an IED that is used to create accurate measurements of power system current, voltage, and power values. Metering values such as demand and peak are saved within the meter to create historical information about the activity of the power system. [1]

Digital Fault Recorder A digital fault recorder (DFR) is an IED that records information about power system disturbances. It is capable of storing data in a digital format when triggered by conditions detected on the power system. Harmonics, frequency, and voltage are examples of data captured by DFRs.

Load Tap Changer (LTC) Load tap changers are devices used to change the tap position on transformers. These devices work automatically or can be controlled via another local IED or from a remote operator or process.

Recloser Controller Recloser controllers remotely control the operation of automated reclosers and switches. These devices monitor and store power system conditions and determine when to perform control actions. They also accept commands from a remote operator or process.

Time Synchronization Source A time synchronization source is an IED that creates a time-of-day value which is then broadcast to the IEDs in order to set all their clocks to the same time.

Protocol Gateway IEDs communicate over serial connections by speaking a particular language or protocol. A protocol gateway converts communications from one protocol to another. This task is often performed by software on a personal computer.

Human Machine Interface (HMI) The front panel display and push buttons or a personal computer act as interfaces to system data and controls for personnel in the substation.

Programmable Logic Controller As the name implies, a programmable logic controller (PLC), is an IED that can be programmed to perform logical control. As with the RTU, a dedicated pair of copper conductors for each contact and transducer value are terminated on panels within the PLC. Personnel familiar with the PLC development environment can program PLCs to create information from sensor data and perform automation. The PLC can transfer collected data to other devices and receive data and control commands from other devices through a serial port.

Protective Relay A protective relay is an IED designed to sense power system disturbances and automatically perform control actions on the I&C system and the power system to protect personnel and equipment. The relay has local termination so that the copper conductors for each contact do not have to be routed to a central termination panel associated with RTUs and PLCs. Transducers are not necessary since the relay accepts signals directly from the instrument transformers. Protective relays create metering information, collect system status information, and store historical records of power system operation.

Communications Processor A communications processor is a substation controller that incorporates the functions of many other I&C devices into one IED. It has many communications ports to support multiple simultaneous communications links. The communications processor performs data acquisition and control of the other substation IEDs and also concentrates the data it acquires for transmission to one or many masters inside and outside the substation. The communications processor incorporates features of many of the other IEDs including an RTU, a communications port switch, a protocol gateway, a time synchronization source, and a limited PLC functionality. The communications processor has locally terminated I/O and can perform dial-out to alert personnel or processes when a status changes.

Power System Automation

Power System Integration Power system integration is the act of communicating data to, from, or among IEDs in the I&C system and remote users. Substation integration refers to combining data from the IED's local to a substation so that there is a single point of contact in the substation for all of the I&C data. [2] Poletop devices often communicate to the substation via wireless or fiber connections. Remote and local substation and feeder control is passed through the substation controller acting as a single point of contact. Some systems bypass the substation controller by using direct connections to the poletop devices, such as RTUs, protective relays, and controllers.

Power System Automation Power system automation is the act of automatically controlling the power system via I&C devices. Substation automation refers to using IED data, control and automation capabilities within the substation, and control commands from remote users to control power system devices. Since true substation automation relies on substation integration, the terms are often used interchangeably.

Power system automation includes processes associated with generation and delivery of power. A subset of these processes deal with delivery of power at transmission and distribution levels, which is power delivery automation. Together, monitoring and control of power delivery systems in the substation and on the poletop reduce the occurrence of outages and shorten the duration of outages that do occur. The IEDs, communications protocols, and communications methods described in previous sections, work together as a system to perform power system automation.

Power Delivery Automation Though each utility is unique, most consider power delivery automation of transmission and distribution substations and feeders to include:
- Supervisory Control and Data Acquisition (SCADA) —operator supervision and control;
- Distribution Automation—fault location, auto-isolation, auto-sectionalizing, and auto-restoration;
- Substation Automation—breaker failure, reclosing, battery monitoring, dead substation transfer, and substation load transfer;
- Energy Management System (EMS) —load flow, VAR and voltage monitoring and control, generation control, transformer and feeder load balancing;
- Fault analysis and device maintenance.

Systems without automated control still have the advantages of remote monitoring and operator control of power system devices, which includes:
- Remote monitoring and control of circuit breakers and automated switches;
- Remote monitoring of non-automated switches and fuses;
- Remote monitoring and control of capacitor banks;
- Remote monitoring and voltage control;
- Remote power quality monitoring and control.

Power System Automation Features

IEDs described in the overview are used to perform power system integration and automation. Most designs require that one IED act as the substation controller and perform data acquisition and control of the other IEDs. The substation controller is often called upon to support system automation tasks as well. The communications industry uses the term client/server for a device that acts as a master, or client, retrieving data from some devices and then acts as a slave, or server, sending this data to other devices.[3] The client/server collects and forwards data dynamically. A data concentrator creates a substation database by collecting and concentrating dynamic data from several devices. In this fashion, essential subsets of data from each IED are forwarded to a master through one data transfer. The data concentrator database is used to pass data between IEDs that are not directly connected.

A substation archive client/server collects and archives data from several devices. The archive data is retrieved when it is convenient for the user to do so.

The age of the IEDs now in substations varies widely. Many of these IEDs are still useful but lack the most recent protocols. A communications processor that can communicate with each IED via a unique baud rate and protocol extends the time that each IED is useful. Using a communications processor for substation integration also easily accommodates future IEDs. It is rare for all existing IEDs to be discarded during a substation integration upgrade project.

Power System Automation Benefits to Utility

The benefits of monitoring, remote control, and automation of power delivery include improved employee and public safety, and deferment of the cost of purchasing new equipment. Also, reduced operation and maintenance costs are realized through improved use of existing facilities and optimized performance of the power system through reduced losses associated with outages and improved voltage profile. Collection of information can result in better planning and system design, and increased customer satisfaction will result from improved responsiveness, service reliability, and power quality.

WORDS AND TERMS

data acquisition 数据采集
personnel n. 人员，职员
console n. 控制台
pole-top n. 杆顶
instrument transformer 仪表（用）互感器
transducer n. 传感器，变换器
originate v. 发生
trench n. 电缆沟
tray n. 盘子
lease v. 出租
harmonic n. 谐波
load tap changer 负载抽头开关转换器
recloser n. 自动重合闸装置（开关）
time-of-day n. 日历时钟

relay n. 继电器
dial-out v. 拨叫
bypass n. 旁路；v. 设旁路
interchangeably adv. 可交换地
auto-isolation n. 自动隔离
auto-sectionalizing n. 自动分段
auto-restoration n. 自动恢复供电
dead substation transfer 故障变电站转移
substation load transfer 变电站负荷转移
load flow 潮流
archive v. 存档
deferment n. 延期，暂缓
responsiveness n. 响应

NOTES

[1] Metering values such as demand and peak are saved within the meter to create historical information about the activity of the power system.

测量值，如需求量和峰值，可以保存在仪表中，用于创建电力系统运行的历史信息。

[2] Substation integration refers to combining data from the IED's local to a substation so that

there is a single point of contact in the substation for all of the I&C data.

变电站集成指的是将局部和整个变电站的 IED 数据进行组合，于是对于变电站内所有 I&C 数据，只有一个单联系点。

［3］ The communications industry uses the term client/server for a device that acts as a master, or client, retrieving data from some devices and then acts as a slave, or server, sending this data to other devices.

通信行业对设备使用术语客户/服务器，该设备作为主设备或客户从其他设备得到数据，然后作为从设备或服务器向其他设备发送数据。

C 被动句的翻译

科技文章侧重描述和推理，强调客观、准确，所以谓语大量地采用被动语态，以避免过多使用第一、二人称而引起主观臆断的印象。英语和汉语都有被动语态，但这两种语言在运用和表达上却不尽相同。因此，翻译时，必须对被动句做适当的灵活处理。

1. 英语中某些着重被动动作的被动句，为突出其被动意义，可直接译为汉语的被动句。翻译时最常见的方式是在谓语动词前加"被"。但过多地使用同一个词会使译文缺乏文采，因而应根据汉语习惯采用一些其他的方式来表达被动态，如可用："由""给""受""加以""把""使"等。例如：

The machine tools are controlled by PLC.

机床由可编程序控制器控制。

2. 着重描述事物过程、性质和状态的英语被动句，实际上与系表结构很相近，往往可译成汉语的判断句，即将谓语动词放在"是……的"之间的结构。例如：

This kind of device is much needed in the speed-regulating system.

这种装置在调速系统中是很需要的。

3. 英语的被动语态有时可改译成汉语的主动语态。当原句中主语为无生命的名词，而又无由介词 by 引导的行为主体时，可将原句的主语仍译为主语，按汉语习惯表述成主动句。因为汉语中的很多情况下，表达被动的含义通常不需要加"被"字，而采用主动语态的形式。如果不顾汉语习惯，强要加上"被"字表被动，有时则会使译文看上去不像汉语。例如：

The quartz crystal does not vibrate at certain frequency until the voltage is applied.

直到电压加上去以后，石英晶体才会以某一频率振荡。

对于有些被动句，翻译时可将原主语译成宾语，而把原行为主体或相当于行为主体的介词宾语译成主语。例如：

Since numerical control was adopted at machine tools, the productivity has been raised greatly.

自从机床采用数控以来，生产率大大提高了。

有些原句中没有行为主体，翻译时可增添适当的主语使译文通顺流畅，如"人们""有人""大家""我们"等。例如：

A few decades ago it was thought unbelievable that the computer could have so high speed as well so small volume.

几十年前人们还认为计算机能具有如此高的运行速度和如此小的体积是一件难以置信的事。

Fuzzy control is found an effective way to control the systems without precise mathematic models.
人们发现,模糊控制是一种控制不具备精确数学模型系统的有效方法。

4. 当不需要或无法讲出动作的发出者时,如表明观点、要求、态度的被动句或描述某地发生、存在、消失了某事物的被动句,可将原句译为汉语的无主句,把原句的主语译作宾语。例如:

What kind of device is needed to make the control system simple?
需要什么装置使控制系统简化?

PART 2

Control Theory

UNIT 1

A   The World of Control

Introduction

The word *control* is usually taken to mean regulate, direct, or command. Control systems abound in our environment. In the most abstract sense it is possible to consider every physical object as a control system.

Control systems designed by humans are used to extend their physical capabilities, to compensate for their physical limitations, to relieve them of routine or tedious tasks, or to save money. In a modern aircraft, for example, the power boost controls amplify the force applied by the pilot to move the control surfaces against large aerodynamic forces. The reaction time of a human pilot is too slow to enable him or her to fly an aircraft with a lightly damped Dutch roll mode without the addition of a yaw damper system.[1] An autopilot (flight control system) relieves the pilot of the task of continuously operating the controls to maintain the desired heading, altitude, and attitude. Freed of this routine task, the pilot can perform other tasks, such as navigation and/or communications, thus reducing the number of crew required and consequently the operating cost of the aircraft.

In many cases, the design of control system is based on some theory rather than intuition or trail-and-error. Control theory is used for dealing with the dynamic response of a system to commands, regulations, or disturbances. The application of control theory has essentially two phases: dynamic analysis and control system design. The analysis phase is concerned with determination of the response of a plant (the controlled object) to commands, disturbances, and changes in the plant parameters. If the dynamic response is satisfactory, there need be no second phase. If the response is unsatisfactory and modification of the plant is unacceptable, a design phase is necessary to select the control elements (the controller) needed to improve the dynamic performance to acceptable levels.

Control theory itself has two categories: classical and modern. Classical control theory, which had its start during World War II, can be characterized by the transfer function concept with analysis and design principally in the Laplace and frequency domains. Modern control theory has arisen with the advent of high-speed digital computers and can be characterized by the state variable concept with emphasis on matrix algebra and with analysis and design principally in the time domain. As might be expected, each approach has its advantages and disadvantages as well as its proponents and detractors.

As compared to modern approach, the classical approach has the tutorial advantage of placing less emphasis on mathematical techniques and more emphasis on physical understanding. Furthermore, in many design situations the classical approach is not only simpler but may be completely adequate. In those more complex cases where it is not adequate, the classical approach solution may aid in applying the modern approach and may provide a check on the more complete and exact design. For these reasons the subsequent articles will introduce the classical approach in detail.

Classification and Terminology of Control Systems

Control systems are classified in terms that describe either the system itself or its variables:

Open-loop control system and *closed-loop control system* (Fig. 2-1A-1). An open-loop system is one in which the control action is independent of the output. A closed-loop system, however, the input of the plant is somehow dependent on the actual output. Since the output is fed back in a functional form determined by the nature of the feedback elements and then subtracted from the input,[2] a closed-loop system is often referred to as a negative feedback system or simply as a feedback system.

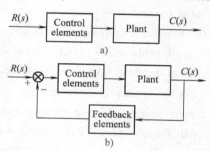

Fig. 2-1A-1 Open-loop control system and closed-loop control system

Continuous and *discrete systems*. The system that all its variables are continuous functions of time is called continuous-variable or analog system; the describing equations are differential equations. A discrete-variable or digital system has one or more variables known only at particular instants of time, as shown in Fig. 2-1A-2b; the equations are difference equations. If the time intervals are controlled, the system is termed a sampled-data system. Discrete variables occur naturally, as from a scanning radar that obtains position data once per scan or a data channel that transmits many pieces of information in turn. A discrete variable will obviously approach a continuous variable as the sampling interval is decreased. Discontinuous variables, such as shown in Fig. 2-1A-2c, occur in "on-off" or "bang-bang" control systems and are treated separately in a subsequent paper.

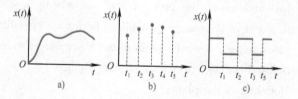

Fig. 2-1A-2 Continuous system and discrete system

Linear and *nonlinear systems*. A system is linear if all its elements are linear, and nonlinear if any element is nonlinear.

Time-invariant and *time-variant systems*. A time-invariant (or stationary) system is one whose parameters do not vary with time. The output of a stationary system is independent of the time at which an input is applied, and the coefficients of the describing differential equations are constants. A time-variant (or nonstationary) system is a system with one or more parameters that vary with time. The time at which an input is applied must be known, and the coefficients of the differential equations are time-dependent.

Lumped parameter and *distributed parameter systems*. Lumped parameter systems are those for which physical characteristics are assumed to be concentrated in one or more "lumps" and thus independent of any spatial distribution. In effect, bodies are assumed rigid and treated as point massed; springs are massless and electrical leads resistanceless, or suitable corrections are made to the system mass or resistance; temperatures are uniform; etc. In distributed parameter systems, the continuous spatial distribution of a physical characteristic is taken into account. Bodies are elastic, springs have a distributed mass, electrical leads have a distribute resistance, and temperatures vary across a body. Lumped parameter systems are described by ordinary differential equation; while distributed parameter systems are described by partial differential equations.

Deterministic and *stochastic systems*. A system or variable is deterministic if its future behavior is both predictable and repeatable within reasonable limits. If not, the system or variable is called stochastic or random. Analyses of stochastic systems and of deterministic system with stochastic inputs are based on probability theory.

Single-variable and *multivariable systems*. A single-variable system is defined as one with only one output for one reference or command input and is often referred to as a single input single output (SISO) system. A multivariable (MIMO) system has any number of inputs and outputs.

Control System Engineering Design Problem

Control system engineering consists of analysis and design of control configurations. Analysis is the investigation of the properties of an existing system. The design problem is one choice and arrangement of system components to perform a specific task.

Designing a control system is not a precise or well-defined process; rather, it is a sequence of interrelated events. A typical sequence might be

1) Modeling of the plant;
2) Linearization of the plant model;
3) Dynamic analysis of the plant;
4) Nonlinear simulation of the plant;
5) Establishment of the control philosophy & strategy;
6) Selection of the performance criteria and indices;
7) Design of the controller;
8) Dynamic analysis of the complete system;
9) Nonlinear simulation of the complete system;
10) Selection of the hardware to be used;
11) Construction and test of the development system;
12) Design of the production model;
13) Test of the production model.

This sequence is not rigid, all-inclusive, or necessarily sequential. It is given here to establish a rationale for the techniques developed and discussed in the subsequent units.

WORDS AND TERMS

regulate *v.* 调整
abound *v.* 大量存在
power boost 功率助推装置
aerodynamic *adj.* 空气动力学的
damp *v.* 阻尼，减幅，衰减
yaw *n.* 偏航
altitude *n.* 海拔
attitude *n.* 姿态
intuition *n.* 直觉
trail-and-error *n.* 试凑法
dynamic response 动态响应
disturbance *n.* 扰动
parameter *n.* 参数
modification *n.* 修正，修改
transfer function 传递函数
domain *n.* 域，领域
advent *n.* 出现
state variable 状态变量
matrix algebra 矩阵代数
approach *n.* 途径，方法；研究
proponent *n.* 提倡者
detractor *n.* 批评者
tutorial *adj.* 指导性的
subsequent *adj.* 后序的
open-loop *n.* 开环
closed-loop *n.* 闭环
discrete *adj.* 离散的
differential equation 微分方程
difference equation 差分方程
interval *n.* 间隔

sampled-data *n.* 采样数据
nonlinear *adj.* 非线性的
time-invariant *adj.* 时不变的
coefficient *n.* 系数
stationary *adj.* 静态的
lumped parameter 集中参数
distributed parameter 分散参数
spatial *adj.* 空间的
spring *n.* 弹簧
lead *n.* 导线
resistance *n.* 阻抗
uniform *adj.* 一致的
elastic *adj.* 有弹性的
ordinary differential equation 常微分方程
partial differential equation 偏微分方程
deterministic *adj.* 确定的
stochastic *adj.* 随机的
predictable *adj.* 可断定的
probability theory 概率论
multivariable *n.* 多变量
configuration *n.* 构造，结构
property *n.* 性质
model *n.* 模型；*v.* 建模
linearization *n.* 线性化
strategy *n.* 方法
performance criteria 性能指标
hardware *n.* 硬件
development system 开发系统
rationale *n.* 理论，原理的阐述

NOTES

[1] The reaction time of a human pilot is too slow to enable him or her to fly an aircraft with a lightly damped Dutch roll mode without a yaw damper system.

飞行员的反应速度太慢，如果不附加阻尼偏航系统，飞行员就无法通过轻微阻尼的侧倾转向方式来驾驶飞机。

[2] Since the output is fed back in a functional form determined by the nature of the feedback

elements and then subtracted from the input...

因为输出会以由反馈部件特性决定的函数形式反馈回来，然后从输入中减去……

B　The Transfer Function and the Laplace Transformation

The Transfer Function Concept

If the input-output relationship of the linear system of Fig. 2-1B-1 is known, the characteristics of the system itself are also known. The input-output relationship in the Laplace domain is called the transfer function (*TF* or *G* Gain). By definition, the transfer function of a component or system is the ratio of the transformed output to the transformed input:

$$G(s) = \frac{\text{output}(s)}{\text{input}(s)} = \frac{C(s)}{R(s)} \tag{2-1B-1}$$

This definition of the transfer function requires the system to be linear and stationary, with continuous variables and with zero initial conditions. The transfer function is most useful when the system is also lumped parameter and when transport lags are absent or neglected. Under these conditions the transfer function itself can be expressed as a ratio of two polynomials in the complex Laplace variable s, or

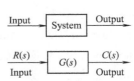

Fig. 2-1B-1　Transfer function

$$G(s) = \frac{N(s)}{D(s)} = \frac{b_m s^m + b_{m-1} s^{m-1} + \cdots + b_0}{a_n s^n + a_{n-1} s^{n-1} + \cdots + a_0} \tag{2-1B-2}$$

For physical systems, $N(s)$ will be of lower order than $D(s)$ since nature integrates rather than differentiates. It will be shown later that a frequency transfer function (*FTF*) for use in the frequency domain can be obtained by replacing the Laplace variable s in the transfer function by $j\omega t$.

In Eq. (2-1B-2) the denominator $D(s)$ of the transfer function is called the characteristic function since it contains all the physical characteristics of the system. The characteristic equation is formed by setting $D(s)$ equal to zero. The roots of the characteristic equation determine the stability of the system and the general nature of the transient response to any input. The numerator polynomial $N(s)$ is a function of how the input enters the system. Consequently, $N(s)$ does not affect the absolute stability or the number or nature of the transient modes. It does, however, along with the specific input, determine the magnitude and sign of each transient mode and thus establishes the shape of the transient response as well as the steady-state value of the output.

For a closed-loop system, the transfer function is:

$$W(s) = \frac{C(s)}{R(s)} = \frac{G(s)H(s)}{1 + G(s)H(s)} = \frac{N(s)}{D(s)} \tag{2-1B-3}$$

where $W(s)$ is the closed-loop transfer function; $G(s)H(s)$ is called the open-loop transfer function; $1 + G(s)H(s)$ is the characteristic function.

The transfer function can be obtained in several ways. One method is purely mathematical and consists of taking the Laplace transform of the differential equations describing the components or system and then solving for the transfer function; nonzero initial conditions, when they occur, are treated as additional inputs. A second method is experimental. A known input (sinusoids and steps are commonly used) is apply to the system, the output is measured, and the transfer function is constructed from operating data and curves. The transfer function for a subsystem or complete system is often obtained by proper combination of the known transfer functions of the individual elements. This combination or reduction process is termed block diagram algebra.

The Laplace Transformation

The Laplace transformation comes from the area of operational mathematics and is extremely useful in the analysis and design of linear systems. Ordinary differential equations with constant coefficients transform into algebraic equations that can be used to implement the transfer function concept. Furthermore, the Laplace domain is a nice place in which to work, and transfer functions may be easily manipulated, modified, and analyzed. The designer quickly becomes adept in relating changes in the Laplace domain to behavior in the time domain without actually having to solve the system equations.[1] When time domain solutions are required, the Laplace transform method is straightforward. The solution is complete, including both the homogeneous (transient) and particular (steady-state) solutions, and initial conditions are automatically included. Finally, it is easy to move from the Laplace domain into the frequency domain.

The Laplace transform is an evolution from the unilateral Fourier integral and is defined as

$$L[f(t)] = \int_0^\infty f(t) e^{-st} dt = F(s) \qquad (2\text{-}1B\text{-}4)$$

where $F(s)$ is the Laplace transform of $f(t)$. Conversely, $f(t)$ is the inverse transform of $F(s)$ and can be represented by the relationship

$$L^{-1}[F(s)] = f(t) \qquad (2\text{-}1B\text{-}5)$$

The symbol s denotes the Laplace variable and is a complex variable ($\sigma + j\omega$); consequently, s is sometimes referred to as a complex frequency and the Laplace domain is called the complex frequency domain.

Since the definite integral of Eq. (2-1B-4) is improper, not all functions are Laplace-transformable; fortunately, the functions of interest to control system designer usually are. The conditions for existence, proofs of theorems, and other uses of the Laplace transformation can be found in standard works on operational mathematics.

The definition of Eq. (2-1B-4) can be used to find the Laplace transform of the functions we are most likely to encounter or use. For convenience, we used to construct a table of transform pairs, which simplify transformation into and out of the Laplace domain.

There are certain theorems and properties of the Laplace transformation that either essential or helpful.

1. Linearity and superposition:

$$L[cf(t)] = cL[f(t)] = cF(s)$$
$$L[c_1 f(t) + c_2 f(t) + \cdots] = c_1 F_1(s) + c_2 F_2(s) + \cdots$$

where c and c_i are constants.

2. Theorems of differentiation and integration: The Laplace transformation of derivatives with respect to time can be shown to be

$$L\left[\frac{d}{dt}f(t)\right] = sF(s) - f(0); \quad L\left[\frac{d^2}{dt^2}f(t)\right] = s^2 F(s) - sf(0) - \frac{d}{dt}f(0); \cdots$$

where $f(0)$, $df(0)$, etc., are the initial conditions. If the initial conditions are zero, as is generally the case for control system analysis and design, the last equation reduces to

$$L\left[\frac{d^n}{dt^n}f(t)\right] = s^n F(s)$$

The Laplace transform of an integral is

$$L\left[\int_0^t f(t)dt\right] = \frac{1}{s}F(s) + \frac{1}{s}\int f(t)\,dt \bigg|_{t=0}$$

It can also be reduced to $F(s)/s$ with zero initial conditions.

3. The initial value and final value theorems: The initial value theorem states that

$$\lim_{t \to 0+} f(t) = f(0+) = \lim_{s \to \infty} sF(s)$$

and can be useful at times in the inverse transformation. The final value theorem states that

$$\lim_{t \to \infty} f(t) = f_{ss} = \lim_{s \to 0} sF(s)$$

where f_{ss} is the steady-state value of $f(t)$.

4. Shifting theorems: The first shifting theorem states that

$$L[e^{-at}f(t)] = L[f(t)]_{s \to s+a} = F(s+a) \quad \text{or} \quad L^{-1}[F(s+a)] = e^{-at}f(t) \quad (2\text{-}1\text{B-}6)$$

Eq. (2-1B-6) indicates that translation through a units in the Laplace domain results in multiplication by e^{-at} in the time domain. The second shifting theorem states that

$$L[f(t)u(t-a)] = e^{-as}L[f(t+a)] = e^{-as}F(s+a)$$

This theorem is useful in transforming delayed inputs and signals such as transport lags and piecewise continuous inputs that are represented by analytic functions.

Modeling

Analytical techniques require mathematical models. The transfer function is a convenient model form for the analysis and design of stationary linear systems with a limited number of differential equations and by block diagram algebra. From the deferential or integro-differential equations describing the behavior of a particular plant, process, or component, using the Laplace transformation and its properties can develop the transfer functions.

We can illustrate it by a simple example:

The output voltage u_c of the circuit indicated in Fig. 2-1B-2 is excited by the input voltage u. According to the Kirchhoff's laws, the relationship between u_c and u can be written as

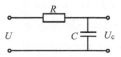

Fig. 2-1B-2　A electrical system

$$u = C\frac{du_c}{dt}R + u_c$$

Using the theorems, the transformed equation with zero initial conditions is

$$U(s) = RCsU_c(s) + U_c(s)$$

Solving for the ratio of the transformed output to the transformed input yields transfer function of the system

$$G(s) = \frac{U_c(s)}{U(s)} = \frac{1}{RCs+1}$$

WORDS AND TERMS

initial condition　初始条件
lag　*v.*, *n.* 延迟
polynomial　*n.* 多项式
order　*n.* 阶
integrate　*v.* 积分
differentiate　*v.* 微分
denominator　*n.* 分母
stability　*n.* 稳定性
transient response　暂态响应
numerator　*n.* 分子
magnitude　*n.* 幅值
sign　*n.* 符号
steady-state　*n.* 稳态
step　*n.* 阶跃（信号）
block diagram algebra　方块图计算（代数）
Laplace transformation　拉普拉斯变换
operational mathematics　工程数学

algebraic equation　代数方程
implement　*v.* 实现
manipulate　*v.* 处理
become adept in　熟练
homogeneous solution　通解
particular solution　特解
unilateral Fourier integral　单边傅里叶积分
inverse transform　反（逆）变换
improper integral　奇异（无理）积分
superposition　*n.* 叠加
initial value　初值
final value　终值
shifting theorem　平移定理
multiplication　*n.* 复合性
piecewise　*adj.* 分段的
integro-differential equation　微积分方程
yield　*v.* 推导出，得出

NOTES

[1] The designer quickly becomes adept in relating changes in the Laplace domain to behavior in the time domain without actually having to solve the system equations.

设计人员很快就会熟练地把拉普拉斯域的变化与时域状态联系起来，而不需真正地解系统方程（时域）。

C　否定句的翻译

否定形式在科技英语中应用很广，其使用非常灵活、微妙，而且在用词、语法和逻辑等方面都与汉语有很大的差别。在翻译否定句时，必须仔细揣摩，彻底理解其意义及其否定的重点，然后根据汉语的习惯来翻译。有些否定句在翻译时，稍有不慎就会造成是非混淆的错误。例如：

1. 有些句子在形式上是否定的，但语义实际上是肯定的意思。例如：

The importance of the project can hardly be exaggerated.

误：这项工程的重要性不能被夸大。

正：这项工程的重要性怎么夸大也不过分。

It is impossible to overestimate the value of the invention.

误：这项发明的价值不可能过高估计。

正：这项发明的价值怎样高估也不可能过分。

2. 英语否定句中否定的对象和意义需要经过分析辨清后才能正确表达。例如：

The computer is not valuable because it is expensive.

误：计算机因为其价格贵而没有价值。

正：计算机不是因为价格贵才有价值。

We do not consider conventional PID control is outdated.

误：我们不认为传统的 PID 控制过时了。

正：我们认为传统的 PID 控制并没有过时。

3. 英语的否定有全部否定和部分否定的差别。前者用 no，not，nothing，none 等否定词构成，表示对全句的否定，较容易理解和翻译；后者则由 all，every，both，always 等具有全体意义的词和否定词构成，只表示部分否定，翻译时要认真对待，避免误译。例如：

In a thermal power plant, all the chemical energy of the coal is not converted into electric power.

误：在热电站中，所有煤中的化学能都不能转化为电能。

正：在热电站中，煤中的化学能并不全都能转化为电能。

Everything is not straightened out.

误：每一个问题都没有弄清楚。

正：并非每一个问题都弄清楚了。

4. 有些形容词或形容词短语，如：little（少），few（少），free from（不受……影响），short of（缺少）等也具有否定意义，翻译时应突出其否定的语气。尤其注意区分 little 和 a little，few 和 a few 的差别：little 和 few 通常表示否定，而 a little 和 a few 则通常表示肯定。例如：

The analysis of three-phase circuits is little more difficult than that of single-phase circuit.

误：三相电路的分析比单相电路要难一点。

正：三相电路的分析比单相电路难不了多少。

此外，前几节中提到的"Smoking free"中的 free 也表示否定，全句应为"禁止吸烟"或"无烟区"的意思，切忌混淆。

UNIT 2

A Stability and the Time Response

Introduction

The stability of a continuous or discrete-time system is determined by its response to inputs or disturbance. Intuitively, a stable system is one that remains at rest (or in equilibrium) unless excited by an external source and returns to rest if all excitations are removed. The output will pass through a transient phase and settle down to a steady-state response that will be of the same form as, or bounded by, the input. If we apply the same input to an unstable system, the output will never settle down to a steady-state phase; it will increase in an unbounded manner, usually exponentially or with oscillations of increasing amplitude.

Stability can be precisely defined in terms of the impulse response $y_\delta(t)$ of a continuous system, or Kronecker delta response $y_\delta(k)$ of a discrete-time system, as follows:

A continuous (discrete-time) system is stable if its impulse response $y_\delta(t)$ (Kronecker delta response $y_\delta(k)$) approaches zero as time approaches infinity.

An acceptable system must at minimum satisfy the three basic criteria of stability, accuracy, and a satisfactory transient response. These three criteria are implied in the statement that an acceptable system must have a satisfactory time response to specified inputs and disturbances. So, although we work in the Laplace and frequency domains for convenience, we must be able to relate these two domains, at least qualitatively, to the time domain.

In fact, the Laplace domain can provides information about the transient response of both stable and unstable systems and about the steady-state response of stable systems. This article is concerned with relating the Laplace domain to the time response with emphasis on the transient response, and with establishing specific criteria in the Laplace domain for system stability. Accuracy will be treated in the next article, and the frequency response in the subsequent units.

The Characteristic Equation

The time response of a system to any input can be expressed as

$$c(t) = L^{-1}[c(s)] = c_{ss}(t) + c_{tr}(t)$$

where $c_{ss}(t)$ is the steady-state response and $c_{tr}(t)$ is the transient response. If the system is unstable, there will be no steady-state response, only a transient response.

Without transport lag, the transfer function of a system can be expressed as a ratio of polynomials in the complex Laplace variable s:

$$G(s) = \frac{C(s)}{R(s)} = \frac{N(s)}{D(s)} \qquad (2\text{-}2\text{A-}1)$$

The characteristic equation is formed by setting the denominator polynomial equal to zero

$$D(s) = a_n s^n + a_{n-1} s^{n-1} + \cdots + a_1 s^1 + a_0 = 0 \qquad (2\text{-}2\text{A-}2)$$

And can be written in factored form as

$$D(s) = \prod_{i=1}^{n} (s + r_i) = 0 \qquad (2\text{-}2\text{A-}3)$$

where $-r_i$ denotes the roots of the characteristic equation—the values of s that make $D(s)$ equal to zero. These roots may be real, complex, or equal to zero; if complex, they will always occur in conjugate pairs since the coefficients of the differential equations are real.

It can be shown that the transient response for n distinct roots in the Laplace domain is

$$C_{tr}(s) = \frac{C_1}{s + r_1} + \frac{C_2}{s + r_2} + \frac{C_3}{s + r_3} + \cdots + \frac{C_n}{s + r_n} \qquad (2\text{-}2\text{A-}4)$$

And in the time domain is

$$c_{tr}(t) = C_1 e^{-r_1 t} + C_2 e^{-r_2 t} + C_3 e^{-r_3 t} + \cdots + C_n e^{-r_n t} \qquad (2\text{-}2\text{A-}5)$$

Each term in the last equation is called transient mode. There is a transient mode for each root with a shape determined solely by the location of the root in the s plane.

So, a necessary and sufficient condition for the system to be stable is that the roots of the characteristic equation have negative real parts. This ensures that the impulse response will decay exponentially with time.

Routh Stability Criterion

The Routh criterion is a method for determining continuous system stability, for systems with an nth-order characteristic equation of the form:

$$a_n s^n + a_{n-1} s^{n-1} + \cdots + a_1 s^1 + a_0 = 0 \qquad (2\text{-}2\text{A-}6)$$

The criterion is applied using a Routh table defined as follows:

s^n	a_n	a_{n-2}	a_{n-4}	$\cdots$
s^{n-1}	a_{n-1}	a_{n-3}	a_{n-5}	$\cdots$
$\vdots$	b_1	b_2	b_3	$\cdots$
$\vdots$	c_1	c_2	c_3	$\cdots$
$\vdots$	$\cdots$	$\cdots$	$\cdots$	$\cdots$

where $a_n, a_{n-1}, \cdots, a_0$ are the coefficients of the characteristic equation and

$$b_1 = \frac{a_{n-1} a_{n-2} - a_n a_{n-3}}{a_{n-1}} \qquad b_2 = \frac{a_{n-1} a_{n-4} - a_n a_{n-5}}{a_{n-1}} \qquad \text{etc.}$$

$$c_1 = \frac{b_1 a_{n-3} - a_{n-1} b_2}{b_1} \qquad c_2 = \frac{b_1 a_{n-5} - a_{n-1} b_3}{b_1} \qquad \text{etc.}$$

The table is continued horizontally and vertically until only zeros are obtained.[1] Any row can be multiplied by a positive constant before the next row is computed without disturbing the properties of the table.

The Routh criterion: All the roots of the characteristic equation have negative real parts if and only if the elements of the first column of the Routh table have the same sign. Otherwise, the number of roots with positive real parts is equal to the number of changes of sign.

The Hurwitz criterion is another method for determining whether all the roots of the characteristic equation of a continuous system have negative real parts. It has the same principle with the Routh criterion in substantial although their forms or patterns are different, so they are commonly called Routh-Hurwitz criterion.

Simple Lag: First-Order System

With the transfer function in the form of Eq. (2-2A-1), the order of the system is defined as the order of the characteristic function $D(s)$, the highest power of s appearing in $D(s)$ establishes the order of the system.

For a simple first-order system, the transfer function $G(s) = \dfrac{1}{Ts+1}$, as shown in Fig. 2-2A-1, if the input is a unit step $R(s) = 1/s$, the output will be

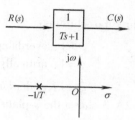

Fig. 2-2A-1 First-order system

$$C(s) = G(s)R(s) = \frac{1}{s(Ts+1)} = \frac{1/T}{s(s+1/T)} = \frac{K_1}{s} + \frac{K_2}{s+1/T}$$

$$K_1 = \left.\frac{1/T}{s+1/T}\right|_{s=0} = 1, \quad K_2 = \left.\frac{1/T}{s}\right|_{s=-\frac{1}{T}} = -1$$

hence the system response is $c(t) = 1 - e^{-t/T}$.

The first term is the forced solution, due to the input, and the second the transient solution, due to the system pole. Fig. 2-2A-2 shows this transient as well as $c(t)$. The transient is seen to be a decaying exponential, and the commonly used measure of the speed of decay is the time constant:

The *time constant* is the time in seconds for the decaying exponential transient to be reduced to $e^{-1} = 0.368$ of its initial value.

Since $e^{-t/T} = e^{-1}$ when $t = T$, it is seen that the time constant for a simple lag $1/(Ts+1)$ is T seconds. This is, in fact, the reason a simple lag transfer function is often written in this form. The coefficient of s then immediately indicates the speed of decay, and it takes $4T$ seconds for the transient to decay to 1.8% of its initial value.

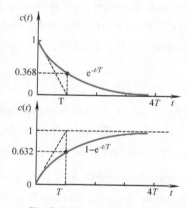

Fig. 2-2A-2 Transients of first-order systems

For a simple lag, two features are important:
1. Stability: For system stability, the system pole(s) must lie in the left half of the s-plane, so that the transient of the system decays instead of grows as t increases.
2. Speed of response: To speed up the response of the system (i.e., to reduce its time constant T), the pole $-1/T$ must be moved left.

Quadratic Lag: Second-Order System

This very common transfer function can always be reduced to the standard form

$$G(s) = \frac{\omega_n^2}{s^2 + 2\zeta\omega_n s + \omega_n^2} \tag{2-2A-7}$$

where ω_n is undamped natural frequency, ζ is damping ratio. The significance of these parameters will be discussed.

Depending on the damping ratio ζ, the roots (or the poles) of the system characteristic equation

$$s^2 + 2\zeta\omega_n s + \omega_n^2 = 0 \tag{2-2A-8}$$

have three possibilities:

$\zeta > 1$: overdamped: $s_{1,2} = -\zeta\omega_n \pm \omega_n\sqrt{\zeta^2 - 1}$

$\zeta = 1$: critically damped: $s_{1,2} = -\omega_n$

$\zeta < 1$: underdamped: $s_{1,2} = -\zeta\omega_n \pm j\omega_n\sqrt{1 - \zeta^2}$

Fig. 2-2A-3 shows the s-plane for plotting the pole positions.

For a unit step input $R(s) = 1/s$, the transform of the output is

$$C(s) = \frac{\omega_n^2}{s(s^2 + 2\zeta\omega_n s + \omega_n^2)}$$

For $\zeta > 1$, these poles are on the negative real axis, on both sides of $-\omega_n$; the transient is a sum of two decaying exponentials, each with its own time constant. The exponential corresponding to the pole closest to the origin has the largest time constant and takes the longest to decay. This pole is called the *dominating pole*. For $\zeta = 1$, both poles coincide at $-\omega_n$. For $\zeta < 1$, the poles move along a circle of radius ω_n centered at the origin. From the geometry in Fig. 2-2A-3, it is seen also that $\cos\phi = \zeta\omega_n/\omega_n = \zeta$. The output is

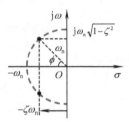

Fig. 2-2A-3 The pole positions on the s-plane

$$c(t) = 1 - \frac{1}{\sqrt{1 - \zeta^2}} e^{-\zeta\omega_n t} \sin\left(\omega_n\sqrt{1 - \zeta^2}\, t + \arctan\frac{\sqrt{1 - \zeta^2}}{\zeta}\right) \tag{2-2A-9}$$

Fig. 2-2A-4 shows a normalized plot of this response for different values of the damping ratio ζ. The transient term is an oscillation of damped natural frequency $\omega_n\sqrt{1 - \zeta^2}$, of which the amplitude decays according to $e^{-\zeta\omega_n}$.

Important performance criteria of the response are identified in Fig. 2-2A-5.

Settling time T_s is the time required for the response to come permanently within a 5% or 2% band around the steady-state value. $T_s = 3T$ (5%), or $T_s = 4T$ (2%). (The time constant

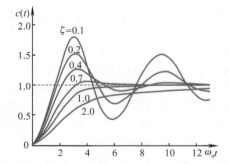

Fig. 2-2A-4 Normalized response plot of second-order systems for different damping ration ζ

T is the time in seconds for $e^{-\zeta\omega_n} = e^{-1}$, hence $T = 1/\zeta\omega_n$). The *maximum percentage overshoot* (P. O.) over the steady-state response is a critical measure of performance. Equating the derivative of $c(t)$ in Eq. (2-2A-9) to zero, to determine the extreme of the response, easily yields the equation

$$\tan\left(\omega_n\sqrt{1-\zeta^2}\,t + \arctan\frac{\sqrt{1-\zeta^2}}{\zeta}\right) = \frac{\sqrt{1-\zeta^2}}{\zeta} \qquad (2\text{-}2A\text{-}10)$$

This implies that at the peaks $\omega_n\sqrt{1-\zeta^2}\,t = i\pi$, $i = 1, 3, \cdots$ since then left and right sides are equal. Hence the time at the maximum peak ($i = 1$), the *peak time* T_p is

$$T_p = \frac{\pi}{\omega_n\sqrt{1-\zeta^2}} \qquad (2\text{-}2A\text{-}11)$$

If the tan of the angle in Eq. (2-2A-10) is $\sqrt{1-\zeta^2}/\zeta$, its sin is $\pm\sqrt{1-\zeta^2}$, and substituting Eq. (2-2A-11) into Eq. (2-2A-9) yields

$$\text{P.O.} = 100\exp\left(\frac{-\pi\zeta}{\sqrt{1-\zeta^2}}\right)$$

The *rise time* T_r, identified in Fig. 2-2A-5 as the time at which the response first reaches the steady-state level, is closely related to peak time T_p.

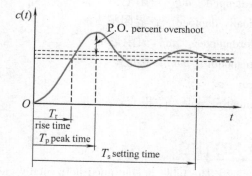

Fig. 2-2A-5 Performance criteria of response

It is noted that while T_s, T_p, and T_r depend on both ω_n and ζ, P.O. depends only on the damping ratio ζ (Fig. 2-2A-6). Permissible overshoot, and hence minimum acceptable ζ, depends on the application. For a machine tool slide, overshoot may cause the tools to gouge into the material being machined, so $\zeta \geqslant 1$ is required. But in most cases a limited overshoot is quite acceptable, and then $\zeta < 1$ is preferable, because it reduces peak time T_p and rise time T_r. For $\zeta = 0.7$ the overshoot is only 5%, and the response approaches steady state much sooner.

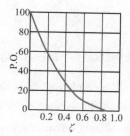

Fig. 2-2A-6 Relationship between P.O. and ζ

If the damping ratio could be held constant while ω_n is increased, the poles would move radically outward and both settling time and rise time would decrease. So, we can adjust the transient response by adjusting the poles of a closed-loop system.

WORDS AND TERMS

intuitively *adv.* 直观地
at rest 处于平衡状态
excitation *n.* 激励
phase *n.* 状态，相位
exponential *adj.* 指数的；*n.* 指数
oscillation *n.* 振荡
amplitude *n.* 振幅
impulse *v.* 冲激

criteria *n.* 判据
qualitatively *adv.* 定性地
complex *adj.* 复数的；*n.* 复数
characteristic equation 特征方程
factor *n.* 因子；*v.* 分解因式
decay *v.* 衰减
horizontally *adv.* 水平地
vertically *adv.* 垂直地

Routh criterion 劳斯判据
Hurwitz criterion 赫尔维茨判据
quadratic *adj.* 二次方的
significance *n.* 意义
overdamped *adj.* 过阻尼的
critically damped 临界阻尼
underdampted *adj.* 欠阻尼的
corresponding *adj.* 相应的
origin *n.* 原点
dominating pole 主极点

settling time 调节时间
overshoot *n.* 超调
derivation *n.* 导数
extreme *adj.* 极端的；*n.* 极端的事情/情况
peak time 峰值时间
substitute *n.* 代替
rise time 上升时间
gouge *v.* 挖
radically *adv.* 完全地

NOTES

[1] The table is continued horizontally and vertically until only zeros are obtained.
这张表向水平（向右）垂直（向下）方向延伸，直到得到的都是零为止。

B Steady State

The Steady-State Errors

A control system is designed to control the dynamic behavior (the time response) of a plant subjected to commands or disturbances. The designer should be fully aware, however, of the role of the steady equations and errors in the overall process, as well as their influence on the dynamic behavior of the plant.

An accuracy of a system is a measure of how well it follows commands. It is an important performance criterion; a guidance system that cannot place a spacecraft on a suitable trajectory is obviously useless no matter how well-behaved its transient response.

Accuracy is generally expressed in terms of acceptable *steady-state errors for specified inputs* (E_r) or *for disturbances* (E_d). The error $e(t)$ is defined as the difference between the desired output $r(t)$ and the actual output $c(t)$. Note that this error is not necessarily the actuating signal $\varepsilon(t)$ unless for unity feedback system. The error $e(t)$ will be the steady-state error e_{ss} of the system after all transients have died out. Using the final value theorem, the steady-state error in time domain can be written as

$$e_{ss} = \lim_{t \to \infty} e(t) = \lim_{s \to 0} sE_r(s) \tag{2-2B-1}$$

Steady-State Error for Specified Input

For a unity feedback system indicated in Fig. 2-2B-1, the closed-loop transfer function is

$$W(s) = \frac{C(s)}{R(s)} = \frac{G(s)}{1+G(s)} = \frac{G_c(s)G_p(s)}{1+G_c(s)G_p(s)} \tag{2-2B-2}$$

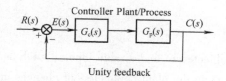

Fig. 2-2B-1 An unity feedback

where $G = G_c G_p$ is the open-loop TF.

The error E for specified input is

$$E_r(s) = R(s) - C(s) = \frac{1}{1+G(s)}R(s) = G_r(s)R(s) \qquad (2\text{-}2\text{B-}3)$$

where $G_r(s) = 1/[1+G(s)]$ is the error transfer function for specified input.

For the open-loop transfer function $G(s)$, the following general form is assumed:

$$G(s) = \frac{K a_k s^k + \cdots + a_1 s + 1}{s^n \, b_l s^l + \cdots + b_1 s + 1} \qquad (2\text{-}2\text{B-}4)$$

In this equation:

1) K as given, with the constant terms in numerator and denominator polynomials made unity, is formally the *gain* of the transfer function G. It should be distinguished from the *root locus gain*, defined in the next unit as that for which the hightest power coefficients are unity.

2) The *type number* of G is the value of the integer n. A factor s in the denominator represents an integration, so the type number is the number of integrators in G.

3) Gain $K = \lim_{s \to 0} s^n G(s)$, and a common practice associates the following names and notations with K, depending on n:

$n = 0$: K_p = *position error constant*
$n = 1$: K_v = *velocity error constant*
$n = 2$: K_a = *acceleration error constant*

Eq. (2-2B-4) shows that $\lim_{s \to 0} G(s) = \lim_{s \to 0}(K/s^n)$, combining Eq. (2-2B-3), so that Eq. (2-2B-1) can be written as follows:

$$e_{ss} = \lim_{s \to 0} \frac{sR(s)}{1 + (K/s^n)} \qquad (2\text{-}2\text{B-}5)$$

This readily yields Table 2-2B-1 for the steady-state errors corresponding to different type numbers and inputs.

Table 2-2B-1 Steady-state error

Type number	$n = 0$	$n = 1$	$n = 2$
Step $u(t)$; $R = 1/s$	$\dfrac{1}{1+K_p}$	0	0
Ramp t; $R = 1/s^2$	∞	$\dfrac{1}{K_v}$	0
Acceleration $t^2/2$; $R = 1/s^3$	∞	∞	$\dfrac{1}{K_a}$

Disturbance Errors

Actual systems are also subjected to undesirable inputs, such as noise in command inputs and disturbances arising from changes in the plant parameters or changes in the environment in which the plant is operating. Noise inputs that enter the system with the command inputs require filtering techniques to remove or suppress them without affecting the command input itself. We shall limit our discussion to disturbance inputs which enter the system at the plant rather than at the controller, as indicated in the generalized block diagram of Fig. 2-2B-2a. In Fig. 2-2B-2b the diagram is redrawn with the disturbance d as the principal input. Since the system is linear, the principle of superposition holds,[1] and we can assume r equal to zero. The disturbance input transfer function for an unity feedback system ($H(s) = 1$) can be written as

$$G_d(s) = \frac{C_d(s)}{D(s)} = \frac{G_p(s)}{1 + G_c(s)G_p(s)H(s)}\bigg|_{H=1} = \frac{G_p(s)}{1 + G_c(s)G_p(s)}$$

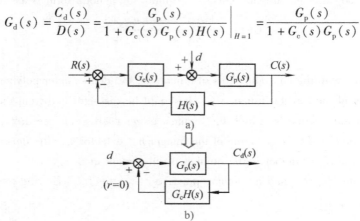

Fig. 2-2B-2 The equivalent transform of block diagram with disturbance d as the principal input

When this transfer function is compared with the usual input-output transfer function with $d = 0$, we see that characteristic equations are identical, as is to be expected, but that numerator functions are different. A disturbance input, therefore, will not affect the stability of the system but may change the shape of the transient response and introduce steady-errors that must be considered in determining the overall accuracy of the system.

Since any change in the output in response to a disturbance is undesirable, the disturbance error E_d is the actual output itself, represented by C_d.

$$E_d(s) = C_d(s) = \frac{G_p(s)D(s)}{1 + G_c(s)G_p(s)} \tag{2-2B-6}$$

The total steady-state error is the sum of the command error and the disturbance error:

$$e_{ss} = \lim_{s \to 0} \frac{sR + sG_p D}{1 + G_c G_p} \tag{2-2B-7}$$

It is often difficult to minimize both components of the error simultaneously. Obviously, it is necessary to have some knowledge as to the nature of probable disturbance inputs. Both error terms of Eq. (2-2B-7) can be set equal to zero by introducing an integrator into the controller. This additional integrator increases the type of the system (from 1 to 2, for example), thus eliminating the velocity error, and by being introduced *ahead* of the point of entry of the disturbance into the system, eliminates the steady-state error resulting from a step in the disturbance input.[2] This additional integrator must be accompanied by at least one zero if the system is to remain stable.

WORDS AND TERMS

guidance system　引导（导航）系统
trajectory　*n.* 轨迹
unity feedback system　单位反馈系统
general form　一般形式
root locus gain　根轨迹增益

filtering technique　滤波技术
suppress　*v.* 抑制
principal　*adj.* 主要的
minimize　*v.* （使）最小化

NOTES

[1] the principle of superposition holds,…
叠加原理成立……

[2] …, thus eliminating the velocity error, and by being introduced ahead of the point of entry of the disturbance into the system, eliminates the steady-state error resulting from a step in the disturbance.
……这样通过在系统扰动进入点之前引入（积分环节），可消除由扰动输入中的阶跃（成分）导致的稳态误差。

C　名词的翻译

　　几乎所有语言都存在一词多义的现象。在英汉字典中我们往往会查到一个英语单词的多种含义，因此在翻译科技文章时，必须结合语法知识和上下文的逻辑关系，尤其是结合所涉及的专业知识，才能对一个词的具体词义做出准确的判断。名词也不例外，在不同的专业和上下文中，往往具有不同的汉语翻译方式。例如：cell 在生物学中作"细胞"讲，在化工领域可作"电解槽"讲，在电学中是"电池"的意思；base 在日常作"基础"讲，机械中可作"底座"讲，化学中是"碱"的意思，三极管的 base 是"基极"，而三角形的 base 是"底边"。在一篇文章中，既会因为与其他学科的交叉，同时也会由于上下文不同而涉及以上问题。例如，order 这个词做名词使用时，就会在不同情况下有不同的意思：

operational order 操作命令，运算指令
order of a differential equation 微分方程的阶
order of matrix 矩阵的阶
technical order 技术说明，技术规程
be in/out of order 正常/发生故障
in order to 为了

order code 指令码
order of connection 接通（连接）次序
order of poles 极点的相重数
working order 工序，加工单
give an order for sth. 订货
order of magnitude 数量级

由上例可以看出，如要翻译正确，决不能仅凭借日常生活用语中相对狭窄的知识面和词汇量，而应结合本专业知识，在字典的帮助下，熟悉和积累常用英文专业词汇的汉语词义，切莫随意妄加推断。但词典不可能对每个词所有搭配的含义和翻译都一一列出，因而在必要时需对原文中的词汇做词义引申或意译的处理，以避免译文生硬晦涩、意思含糊，甚至误解。请看下面几个例句中较难翻译名词的翻译技巧：

1. The study of neural network is one of the last frontiers of artificial intelligence.

对神经网络的研究是人工智能的最新领域之一。last frontiers 从"最后的边疆"引申转译为"最新领域"。

2. All the wit and learning in this field are to be present at the symposium.

所有的这一领域的学者都将出席这个科学讨论会。the wit and learning 由原来的"智慧和学识"经过词义的具体化处理后译为"学者"。

3. The contributors in component technology have been in the semiconductor components.

元件技术中起主要作用的因素是半导体元件。contributors 由原来的"贡献者"经过词义的抽象化处理后译为"因素"。

4. IPC took over an immense range of tasks from worker's muscles and brains.

工控机取代了工人大量的体力和脑力劳动。muscles and brains 从"肌肉和大脑"引申意译为"体力和脑力劳动"。

5. The foresight and coverage shown by the inventor of this apparatus are impressing.

这种装置的发明者所表现的远见和渊博学识给人很深的印象。coverage 原意为"范围"，结合上下文可具体化为"知识范围"，译作汉语的常用语"渊博学识"则使译文更流畅。

这种通过对词汇做词义转译、具体化或抽象化处理等引申翻译的技巧不仅适用于名词，也同样适用于动词、形容词和副词及词组结构等。

UNIT 3

A The Root Locus

Introduction

The three basic performance criteria for a control system are stability, acceptable steady-state accuracy, and an acceptable transient response. With the system transfer function known, the Routh-Hurwitz criterion will tell us whether or not a system is stable. If it is stable, the steady-state accuracy can be determined for various types of inputs. To determine the nature of the transient response, we need to know the location in the s plane of the roots of the characteristic equation. Unfortunately, the characteristic equation is normally unfactored and of high order.

The root locus technique is a graphical method of determining the location of the roots of the characteristic equation as any single parameter, such as a gain or time constant, is varied from zero to infinity.[1] The root locus, therefore, provides information not only as to the absolute stability of a system but also as to its degree of stability, which is another way of describing the nature of the transient response. If the system is unstable or has an unacceptable transient response, the root locus indicates possible ways to improve the response and is a convenient method of depicting qualitatively the effects of any such changes.

The Angle and Magnitude Criteria

Without transport lag the transfer function of a system can be reduced to a ratio of polynomials such that

$$W(s) = \frac{C}{R} = \frac{N(s)}{D(s)} \qquad (2\text{-}3\text{A-}1)$$

The root locus technique is developed by expressing the characteristic function $D(s)$ as the sum of the integer unity and a new ratio of polynomials in s. The characteristic equation will be written as

$$D(s) = 1 + K\frac{Z(s)}{P(s)} = 1 + K\frac{(s+z_1)(s+z_2)\cdots(s+z_j)}{s^n(s+p_1)(s+p_2)\cdots(s+p_l)} = 0 \text{ where } n+l=i \qquad (2\text{-}3\text{A-}2)$$

where K is the parameter of interest, $-z_1, -z_2, \ldots$ are the (open-loop) zeros and $-p_1, -p_2, \ldots$ are the (open-loop) poles. K is independent of s and must not appear in the polynomials $Z(s)$ and $P(s)$. The form of $KZ(s)/P(s)$ is important; these poles and zeros may be real or complex conjugates. Note in Eq. (2-3A-2) that the coefficient of s is always set equal to unity for root locus operations.

A *zero* is a value of s that makes $Z(s)$ equal to zero and is given the symbol ∘. Do not automatically assume that this zero is also a closed-loop zero that makes $N(s)$ equal to zero in the system (closed-loop) transfer function; it may be, but is not necessarily so. A *pole* is a value of s that makes $P(s)$ equal to zero and is given the symbol ×. The s^n term represents n poles, all equal to zero and located at the origin of the s plane. A root of the characteristic equation has previously been defined as a value of s that makes $D(s)$ equal to zero.

Since s is a complex variable and the poles and zeros may be complex, $KZ(s)/P(s)$ is a complex function and may, therefore, be handled as a vector having a magnitude and an associated angle or argument. Each of the factors on the right side of Eq. (2-3A-2) can also be treated as a vector with an individual magnitude and associated angle, as shown in Fig. 2-3A-1. Notice

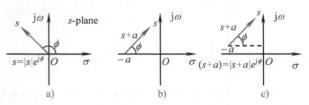

Fig. 2-3A-1 The angle and magnitude of root locus

that the angle ϕ is measured from the horizontal and is positive in the counterclockwise direction.

If we express each factor in polar form, then

$$K\frac{Z(s)}{P(s)} = \frac{K|s+z_1|e^{j\phi_{z1}}|s+z_2|e^{j\phi_{z2}}\cdots}{|s^n|e^{jn\phi_n}|s+p_1|e^{j\phi_{p1}}|s+p_2|e^{j\phi_{p2}}\cdots}$$

If we now collect the magnitudes together and multiply the exponentials together, we can write

$$K\frac{Z(s)}{P(s)} = \frac{K|s+z_1||s+z_2|\cdots}{|s^n||s+p_1||s+p_2|\cdots}e^{j\Sigma\phi} \text{ where } \Sigma\phi = \phi_{z1}+\phi_{z2}+\cdots-n\phi_n-\phi_{p1}-\phi_{p1}-\cdots \quad (2\text{-}3\text{A-}3)$$

Returning to the characteristic equation of Eq. (2-3A-2) and solving for $KZ(s)/P(s)$ yields

$$K\frac{Z(s)}{P(s)} = -1 = 1e^{j(2k+1)180°} \quad k=0,1,2,\cdots \quad (2\text{-}3\text{A-}4)$$

since -1 can be represented by a vector of unity magnitude and an angle that is an odd multiple of 180°. According to Eq. (2-3A-3) and Eq. (2-3A-4), we can find two criteria that make the characteristic function $D(s)$ equal to zero, i.e., there are two criteria which can find system (close-loop) poles as K is increased from 0 to ∞.

Magnitude criterion: $K\dfrac{|s+z_1||s+z_2|\cdots}{|s^n||s+p_1||s+p_2|\cdots} = 1$ or $K = \dfrac{\text{product of the pole distances}}{\text{product of the zero istances}}$

Angle criterion: $\Sigma\phi = \Sigma$ angles of the zero $-\Sigma$ angles of the poles $=(2k+1)180°$ $k=0,1,2,\cdots$

Rules for Root Locus Plotting

Applying angle and magnitude criterion, the root loci can obviously be plotted by a computer, however, we'll introduce rapid sketching techniques. The following guides are provided to facilitate the plotting of root loci:

1. For $K=0$ the close-loop poles coincide with the open-loop poles.

2. For $K\to\infty$ close-loop poles approach the open-loop zero.

3. There are as many as locus branches as there are open-loop poles. A branch starts, for $K=0$, at each open-loop pole. As K is increased, the closed-loop pole positions trace out loci, which end, for $K\to\infty$, at the open-loop zeros.

4. If there are fewer open-loop zeros than poles ($j<i$), those branches for which there are no

open-loop zeros left to go to tend to infinity along asymptotes. The number of asymptotes is $(i-j)$.

5. The directions of the asymptotes are found from the angle condition. The vectors from all m open-loop zeros and n open-loop poles to s have the same angle noted α. Hence the asymptote angles α must satisfy $\alpha = \dfrac{\pm(2k+1)180°}{i-j}$ (k = any integer). The angles are uniformly distributed.

6. All asymptotes intersect the real axis at a single point, at a distance ρ_0 to the origin:

$$\rho_0 = \frac{(\text{sum of o. l. poles}) - (\text{sum of o. l. zeros})}{(\text{number } n \text{ of o. l. poles}) - (\text{number } m \text{ of o. l. zeros})}$$

o. l. : open-loop

7. Loci are symmetrical about the real axis since complex open-loop poles and zeros occur in conjugate pairs.

8. Sections of the real axis to the left of an odd total number of open-loop poles and zeros on this axis form part of the loci, because any trial point on such sections satisfies the angle condition.

9. If the part of the real axis between two o. l. poles (o. l. zeros) belongs to the loci, there must be a point of breakaway from, or arrival at, the real axis. If no other poles and zeros are close by, the breakaway point will be halfway. In Fig. 2-3A-2d, adding the pole p_3 pushes the breakaway point away; a zero at the position of p_3 would similarly attract the breakaway point.

10. The angle of departure of loci from complex o. l. poles (or of arrival at complex o. l. zeros) is a final significant feature. Apply the angle condition to a trial point very close to p_1 in Fig. 2-3A-3. Then the vector angles from other poles and zeros are the same as those to p_1. The angle from p_1 to this point must satisfy the follows: *Departure angle*: $\phi_p = \mp 180° + \Sigma\phi$, similarly, the *arrival angle*: $\phi_z = \pm 180° + \Sigma\phi$.

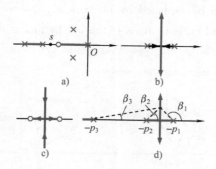

Fig. 2-3A-2 Root locus plot

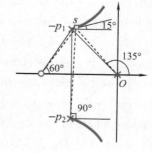

Fig. 2-3A-3 The departure angle of root locus

Root Loci for System Design and Compensation

Root loci are used for design to the extent of choosing the gain to obtain a specified damping ratio or time constant. A P (Proportion) control design does not change the shape of the loci. But if dynamic compensation is needed, a series compensator will add poles and zeros to the open-loop pole-zero pattern, in order to change the shape of the loci in a desirable direction.

As indicated in Fig. 2-3A-4, adding a pole pushes the loci away from that pole, and adding a zero pulls the loci toward that zero. These effects increase in strength with decreasing distance.[2] A zero can improve relative stability because it can pull the loci, or parts thereof, away from the imaginary axis, deeper into the left-half plane.

In analog control system, passive and active electrical networks are usually applied to realize these very important forms of compensation. Including a gain, the transfer functions can be written as follows:

$$G(s) = \frac{K(s+z)}{s+p}$$

For a phase-lead case, $z < p$, and for a phase-lag case, $z > p$. Fig. 2-3A-5 shows the pole-zero patterns. Phase-lead compensation is an approximation to PD (Proportion-Differential) control, and is often preferable to reduce the effect of signal noise and then to improve stability. Phase-lag compensation is commonly used, like PI (Proportion-integral) control, to improve accuracy. However, phase lead may also improve accuracy, and phase lag may improve stability.

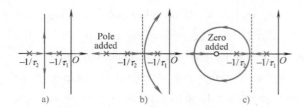

Fig. 2-3A-4　The effect of adding a pole or a zero　　Fig. 2-3A-5　Phase-lead and phase-lag patterns

Cases of phase-lead and phase-lag compensation:

In the system of Fig. 2-3A-6a, the P control is to be replaced by phase lead, intended to "pull" the locus branches for P control, shown as dashed curves, back to the left by means of the added zero. Ignoring for the weaker effect of the added pole, which is often placed at 10 times the distance to the origin,[3] the zero is chosen to satisfy the need for compensation.

In the case of phase-lag compensation, similarly, a pole-zero pair is employed. However, it is added close to the origin, much closer than pictured in Fig. 2-3A-6b to enable the shape of the loci near the origin to be indicated. As suggested by the vectors to a point on the dashed locus, for such a pair the net contribution to the

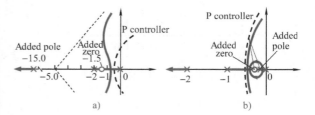

Fig. 2-3A-6　Phase compensations
a) Phase-lead compensation　b) Phase-lag compensation

vector angles at the point is small. Therefore, the main branches often change only little. The picture near the origin is of the type shown in Fig. 2-3A-4c. Although the main branches of system are changed little, the interest gain factor z/p contains in the loop gain function, which can be increased to improve the steady-errors.

WORDS AND TERMS

factored *adj.* 可分解的	intersect *v.* 相交
depict *v.* 描述	real axis 实轴
conjugate *adj.* 共轭的	symmetrical *adj.* 对称的
vector *n.* 矢量	breakaway point 分离点
argument *n.* 辐角，相位	arrival point 汇合点
counterclockwise *adj.* 逆时针的	departure angle 出射角
odd multiple 奇数倍	arrival angle 入射角
even multiple 偶数倍	thereof *adv.* 将它（们）
plot *v.* 绘图；*n.* 曲线图	imaginary axis 虚轴
sketch *v.*, *n.* （绘）草图，素描	passive *adj.* 被动的，无源的
facilitate *v.* 使容易，促进	active *adj.* 主动的，有源的
coincide *v.* 一致	network *n.* 网络，电路
asymptote *n.* 渐进线	phase-lead *n.* 相位超前
integer *n.* 整数	phase-lag *n.* 相位滞后

NOTES

［1］... as any single parameter, such as a gain or time constant, is varied from zero to infinity.

……当任意单一参数，如增益或时间常数，从零变到无穷时。

［2］These effects increase in strength with decreasing distance.

随着到原点距离的减小，它们的作用强度会增加。

此处 distance 是指零（极）点到原点的距离。

［3］Ignoring for the weaker effect of the added pole, which is often placed at 10 times the distance to the origin, the zero...

忽略常被置于 10 倍于零点到原点距离处的附加极点的微弱作用，零点……

B The Frequency Response Methods: Nyquist Diagrams

Introduction

There are times when it is necessary or advantageous to work in the frequency domain rather than in the Laplace domain of the root locus. For system analysis, the root locus method requires a transfer function, which may be difficult or even impossible to obtain for certain components, subsystems, or even systems. In many of these cases, the frequency response can be determined experimentally for sinusoidal test inputs of known frequency and amplitude.

The nature of the input also influences the choice of techniques to be used for system analysis and design. Many command inputs merely instruct a system to move from one steady-state condition to a second steady-state condition. This type of input can be described adequately by suitable steps in position, velocity, and acceleration, and the Laplace domain is appropriate for this purpose. If, however, the interval between such step inputs is decreased so that the system never has time to reach the corresponding steady state, the step representation and Laplace domain are no longer adequate. Such rapidly varying command inputs (or disturbance) may be periodic, random, or a combination thereof. The wind loading of a tracking radar antenna, for example, results from a mean velocity component that varies with time plus superimposed random gusts. [1] If the frequency distribution of these inputs can be calculated, measured, or even estimated, the frequency response can be used to determine their effects upon the system output.

The frequency response is a steady-state response. Although some information can be obtained about the transient response, it is only approximate and is subject to misinterpretation. [2]

The Frequency Transfer Function

It is necessary to develop an input-output relationship that can be used in the frequency domain, i.e., a frequency transfer function. Consider a linear system with a known transfer function $G(s)$ and apply the sinusoidal input

$$r(t) = r_0 \sin\omega_0 t \text{ or } R(s) = \frac{\omega_0 r_0}{s^2 + \omega_0^2}$$

where r_0 is the amplitude and ω_0 the input or forcing frequency. The transformed output is

$$C(s) = G(s)\frac{\omega_0 r_0}{s^2 + \omega_0^2}$$

The partial fraction expansion of $C(s)$ yields

$$C(s) = \frac{C_1}{s - j\omega_0} + \frac{C_2}{s + j\omega_0} + \frac{C_3}{s + r_1} + \frac{C_4}{s + r_2} + \cdots$$

where $-r_1, -r_2, \ldots$ are the roots of the characteristic equation of the transfer function. The inverse transform is

$$c(t) = C_1 e^{j\omega_0 t} + C_2 e^{-j\omega_0 t} + C_3 e^{-r_1 t} + C_4 e^{-r_2 t} + \cdots$$

where the first two terms represent an undamped oscillation resulting from the sinusoidal input, and the remaining terms are the transient response. If the system is stable, the transient response will disappear with time, leaving as the steady-state response

$$c_{ss} = C_1 e^{j\omega_0 t} + C_2 e^{-j\omega_0 t} \qquad (2\text{-}3\text{B-}1)$$

The coefficients C_1 and C_2 are evaluated by the Heaviside expansion theorem as

$$C_1 = \left[\frac{(s - j\omega_0)G(s)\omega_0 r_0}{s^2 + \omega_0^2}\right]_{s = +j\omega_0} = \frac{G(j\omega_0)r_0}{2j}; \quad C_2 = \left[\frac{(s - j\omega_0)G(s)\omega_0 r_0}{s^2 + \omega_0^2}\right]_{s = -j\omega_0} = \frac{-G(-j\omega_0)r_0}{2j}$$

With these values for C_1 and C_2, Eq. (2-3B-1) becomes

$$c_{ss} = \frac{r_0}{2j}G(j\omega_0)e^{j\omega_0 t} - \frac{r_0}{2j}G(-j\omega_0)e^{-j\omega_0 t} \qquad (2\text{-}3\text{B-}2)$$

Since they are complex functions,

$$G(j\omega_0) = \text{Re}G + j\text{Im}G = |G(j\omega_0)|e^{j\phi}; \quad G(-j\omega_0) = \text{Re}G - j\text{Im}G = |G(j\omega_0)|e^{-j\phi}$$

where the angle ϕ is the argument of $G(j\omega_0)$ and is equal to $\text{arctg}(\text{Im}G/\text{Re}G)$. Eq. (2-3B-2) can now be written as

$$c_{ss} = r_0|G(j\omega_0)|\left(\frac{e^{j\omega_0 t} - e^{-j\omega_0 t}}{2j}\right)$$

Since the bracketed terms are equal to $\sin(\omega_0 t + \phi)$, the steady-state response can be written as

$$c_{ss}(j\omega_0) = c_0\sin(\omega_0 t + \phi) \qquad \text{where } c_0 = |G(j\omega_0)|r_0$$

From these equations we see that sinusoidal input to a linear stable system produces a steady-state response that is also sinusoidal, having the same frequency as the input but displaced through a phase angle ϕ and having an amplitude that may be different. This steady-state sinusoidal response is called the *frequency response* of the system. Since the *phase angle* is the angle associated with the complex function $G(j\omega_0)$ and the *amplitude ratio* (c_0/r_0) is the magnitude of $G(j\omega_0)$, knowledge of $G(j\omega_0)$ specifies the steady-state input-output relationship in the frequency domain. $G(j\omega_0)$ is called the *frequency transfer function* and can be obtained from the transfer function $G(s)$ by replacing the Laplace variable s by $j\omega_0$. Consequently, if $G(j\omega_0)$ can be determined from experimental data, $G(s)$ can also be found by replacing $j\omega_0$ by s.

For a given system, the frequency response is completely specified if the amplitude ratio and phase angle are known for the range of input frequencies from 0 to $+\infty$ radians per unit time. Consider the stable first-order system of Fig. 2-3B-1 with a transfer function $G(s) = 1/(\tau s + 1)$, the frequency transfer function is $G(j\omega) = 1/(j\omega\tau + 1)$, where ω can be arbitrary frequency. The amplitude ratio is

$$M(j\omega) = \frac{c_0}{r_0} = |G(j\omega)| = \frac{1}{\sqrt{(\omega\tau)^2 + 1}}$$

and the phase angle is

$$\phi(j\omega) = \angle G(j\omega) = \angle 1 - \angle(j\omega\tau + 1) = -\arctan\omega\tau$$

As input frequency ω is increased from 0 to $+\infty$, we can draw the plot of M and ϕ, and a *polar plot* that traces the tip of the vector representing the frequency transfer function. Polar plots and M and ϕ versus ω plots are used to represent different types of complex functions in the frequency domain. Notice that the constant term in each factor is set equal to unity when working in the frequency domain for convenience, whereas in the Laplace domain the coefficient of the highest power of s is set equal to unity.

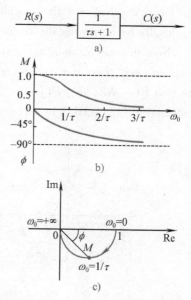

Fig. 2-3B-1 M, ϕ and polar plots of a first-order system

The Nyquist Stability Criterion

In the frequency domain, the theory of residues can be used to detect any roots in the right half of a plane. As with the root locus method, the characteristic function in the form $1 + KZ(s)/P(s)$

is used, where again the function $KZ(s)/P(s)$ may or may not be the open-loop transfer function. To develop the Nyquist criterion, the characteristic function itself is written as a ratio of polynomials so that

$$D(s) = 1 + K\frac{Z(s)}{P(s)} = \frac{P(s) + KZ(s)}{P(s)} = K'\frac{(s+r_1)(s+r_2)\cdots}{(s+p_1)(s+p_2)\cdots} = 0 \quad (2\text{-}3B\text{-}3)$$

Comparing the identities of Eq. (2-3A-2), we see that $-r_1, -r_2, \ldots$ are the roots of the characteristic equation and that $-p_1, -p_2, \ldots$ are the poles of both the characteristic function and $KZ(s)/P(s)$. Poles and roots at the origin have been omitted in the interests of simplicity. In many cases, however, it is difficult to factor the denominator polynomial of close-loop transfer function $D(s)$ to find the location of poles in the s-plane.

To prove stability for $D(s)$, it is necessary and sufficient to show that no zeros (for the closed-loop transfer function is poles) $-r_i$ are inside the right half of the s-plane. we introduce the Nyquist contour D shown in Fig. 2-3B-2, which encloses the entire right half of the s-plane. D consists of the imaginary axis from $-j\infty$ to $+j\infty$ and a semicircle of radius $R \to \infty$. In principle, stability analysis is based on plotting $[1 + KZ(s)/P(s)]$ in a complex plane as s travels once clockwise around the closed contour D. The factors $(s+r_i)$ and $(s+p_i)$ are vectors from $-r_i$ and $-p_i$ to s, and for any value of s the magnitude and phase of $[1 + KZ(s)/P(s)]$ can be determined graphically by measuring the vector lengths and angles in Fig. 2-3B-2, if the r_i were known.

Note that on the imaginary axis $s = j\omega$. The plot of $[1 + KZ(s)/P(s)]$ for s traveling up the imaginary axis from $\omega = 0+$ to $\omega \to \infty$ is in effect just the polar plot of the frequency response function $[1 + KZ(j\omega)/P(j\omega)]$. Hence frequency response function can be found graphically, as indicated in Fig. 2-3B-3 by measurement from the pole-zero pattern.

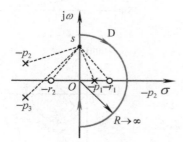
Fig. 2-3B-2 Nyquist contour D

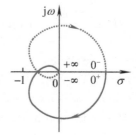
Fig. 2-3B-3 Nyquist plot

Fig. 2-3B-2 shows that if s moves once clockwise around D, vectors $(s+r_i)$ and $(s+p_i)$ rotate 360° clockwise for each pole and zero inside D, and undergo no net rotation for poles and zeros outside D. If the vector $(s+r_1)$ in the numerator rotates 360° clockwise, this will contribute a 360° clockwise rotation of the vector $[1 + KZ(s)/P(s)]$ in the complex plane in which it is plotted. If vector $(s+p_1)$ in the denominator rotates 360° clockwise, this will contribute a 360° counter clockwise revolution of $[1 + KZ(s)/P(s)]$. Poles and zeros outside D do not contribute any net rotation. The result can be expressed as follows:

Principle of the argument: If $[1 + KZ(s)/P(s)]$ has R zeros and P poles inside the Nyquist contour D, a plot of $[1 + KZ(s)/P(s)]$ as s travels once clockwise around D will encircle the origin of the complex plane in which it is plotted $N = R - P$ times in clockwise direction.

The encirclements of a plot of $[1 + KZ(s)/P(s)]$ around the origin equal the encirclements of a plot of $KZ(s)/P(s)$ around the -1 point, on the negative real axis. With this, the following has been proved.

Nyquist stability criterion: A feedback system is stable if and only if the number of counterclockwise encirclements of a plot of the loop gain function $KZ(s)/P(s)$ about the -1 point is equal to the number of poles of $KZ(s)/P(s)$ inside the right-half plane, called open-loop unstable poles.

In the marginal case where $KZ(s)/P(s)$ has poles on the imaginary axis, these will be excluded from the Nyquist contour by semicircular indentations of infinitesimal radius around them. This is shown in Fig. 2-3B-4 for the common case of a pole at the origin. The

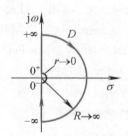

Fig. 2-3B-4 The case of a pole at the origin

plot of $KZ(s)/P(s)$ as s travels once around D is called a Nyquist diagram and is needed to use the criterion.

Gain Margin and Phase Margin

Most practical systems are not open-loop unstable, so that stability requires zero encirclements of the -1 point. To determine this, it is in fact not necessary to plot the complete Nyquist diagram; the polar plot, for ω increasing from $0+$ to $+\infty$, is sufficient.

Simplified Nyquist criterion: If $KZ(s)/P(s)$ does not have poles in the right-half s-plane, the closed-loop system is stable if and only if the -1 point lies to the left of the polar plot when moving along this plot in the direction of increasing ω.

For example, the polar plot of a loop gain function shown in Fig. 2-3B-5 indicates a stable system. If the curve passes through -1, the system is on the verge of instability. For adequate relative stability it is reasonable that the curve should not come too close to the -1 point. Gain margin and phase margin are two common design criteria, which specify the distance of a selected point of the polar plot to -1. Both are defined in Fig. 2-3B-5:

1. Gain margin $= 1/OC$.
2. Phase margin $\phi_m = 180°$ plus the phase angle of $KZ(s)/P(s)$ at the crossover frequency ω_c at which $|KZ(s)/P(s)| = 1$. It is also the negative phase shift (i.e., clockwise rotation) of $KZ(s)/P(s)$ which will make the curve pass through -1.[3]

Fig. 2-3B-5 Gain margin and phase margin

Each specifies the distance to -1 of only one point on the curve, so misleading indications are possible. Phase margin is used very extensively in practice.

WORDS AND TERMS

periodic   *adj.* 周期性的
random   *adj.* 随机的
misinterpretation   *n.* 曲解，误译
develop   *v.* 导出，引入
forcing frequency   强制频率
partial fraction expansion   部分分式展开式
bracket   *v.* 加括号
arbitrary   *adj.* 任意的
polar plot   极坐标图
tip   *n.* 顶端
versus   *prep.* ……对……
the theory of residues   余数定理
identity   *n.* 一致性，等式

omit   *v.* 省略
simplicity   *n.* 简单
contour   *n.* 轮廓，外形
enclose   *v.* 围绕
semicircle   *n.* 半圆形
radius   *n.* 半径
undergo   *v.* 经历
net   *n.* 净值；*adj.* 净值的
revolution   *n.* 旋转
encircle   *v.* 环绕
indentation   *n.* 缺口
infinitesimal   *adj.* 无限小的
misleading indication   导致错误的读数

NOTES

[1] The wind loading of a tracking radar antenna, for example, results from a mean velocity component that varies with time plus superimposed random gusts.

例如，跟踪雷达天线的风力负载是由一个随时间变化的平均速度成分与叠加的随机阵风组成的。

[2] ..., it is only approximate and is subject to misinterpretation.

……只是近似的而且容易判断错误。

[3] It is also the negative phase shift (i.e., clockwise rotation) of $KZ(s)/P(s)$ which will make the curve pass through -1.

它也是能使曲线通过 -1 点的 $KZ(s)/P(s)$ 的负相位移动（即顺时针旋转）的角度。

C   动词的翻译

同名词一样，英文中动词也存在一词多义的特点。例如，run 的所有词义共计有 200 种以上，每种含义之间既有相近之处，也有一定差别。动词在英文中的使用比较灵活，除了做谓语成分外，还有动名词、现在分词、宾语补足语等用法，很多动词本身还具有名词词性，因此在动词的翻译处理上要从三方面着手：语法成分、语言环境和专业基础，结合汉语表达习惯和词汇特点，对原文进行透彻理解和灵活表达。

例如，work 这个词做动词用常常仅被狭义地理解和翻译为"工作"，从而导致了译文缺乏文采，生硬难懂。而下列例句中则采取了灵活处理的方式，根据动作的执行主体和客体，对 work 的词义进行引申后，选择汉语中恰当的词来表达。

1. This method works well.

这种方法很有效。work well 译为"有效"。

2. My watch doesn't work.

我的表不走了。work 按汉语习惯译为"走";也可根据从上下文中获得的推断信息,把 doesn't work 合译为"坏了"或"停了"。

3. The machine works smoothly.

这台机器运转正常。work 译为"运转"。

4. Vibration has worked some connection loose.

震动使一些接线松了。work 译为"使"。

5. The instrument is not working well.

这台仪器失灵了。is not working well 合译为"失灵了"。

6. The movement of the spring is made to work a pointer on a dial, so weight is recorded.

弹簧伸缩带动刻度盘指针,从而记录下重量。is made to work 合译为"带动"。另外,句中的 spring 容易被误解为"春天";而把 movement 译成"伸缩"比"运动"更为具体形象和符合汉语习惯且便于理解。

7. Working with numerals, a computer is similar in many respects to an automatic language translator.

用数字进行运算的计算机在很多方面和自动语言翻译机很相似。working 译为"进行运算"较为符合汉语习惯。

8. To make a marriage work, the couple should realize the importance of respect.

为使婚姻幸福,夫妇双方应认识到尊重的重要性。work 引申意译为"幸福"。

9. The mine has been long worked.

这个矿已经开采很久了。根据主语"矿"选择 work 的含义为"开采"。

10. The interference worked much unstability in the system.

干扰给系统造成了很大的不稳定性。work 译为"造成"符合上下文语境和专业术语的搭配。

work 在其他情况下还可做名词使用,如:idle work power 无功功率;work frame 框架,机壳;cable works 电缆厂;the works of Marx 马克思的著作等,在翻译时要注意区分词性,避免混淆而造成译文的混乱。

UNIT 4

A   The Frequency Response Methods: Bode Plots

Bode Plots

The frequency transfer function of a system or of its $KZ(j\omega)/P(j\omega)$ function can be represented either by the single Nyquist diagram (a polar plot) or by plots of the amplitude ratio and the phase angle against the input (forcing) frequency. It is customary to plot the amplitude ratio in decibels and the phase angle in degrees against the common logarithm of the input frequency. In this form, the two plots are known as *Bode plots* (after H. W. Bode). There are exact Bode plots, which are best prepared with a computer, and straight-line asymptotic plots, which can be quickly and easily sketched or plotted by hand using the techniques to be developed and discussed in this article.

Bode plots of the system transfer function are used to determine the effects of various inputs (including a step) upon the steady-state response of the system. Since the frequency response is a steady-state response, the system must be stable and its stability must be determined before the system Bode plots can be used.

Bode plots are most commonly used with the frequency function $KZ(j\omega)/P(j\omega)$ to examine the stability of a system. When the function has no pole or zero inside the right-half s-plane, i.e., the function is *minimum phase*, the Bode plots can be sketched rather rapidly with a knowledge of the four *elementary factors* that appear in the function. These terms are: ① frequency-invariant terms K; ② zeros and poles at the origin $(j\omega)^{\pm n}$; ③ first-order terms or real poles and zeros $(j\omega\tau + 1)^{\pm n}$; ④ second-order poles and zeros $[(j\omega)^2 + j2\omega\zeta\omega_n + \omega_n^2]^{\pm 1}$.

For a product $KZ(s)/P(s) = M_1 e^{j\phi_1} M_2 e^{j\phi_2} \cdots = Me^{j\phi}$, $M = M_1 M_2 \cdots$ and $\phi = \phi_1 + \phi_2 + \cdots$. The phase angle ϕ is expressed as a sum. The magnitude M will also be expressed as a sum, by using decibels (dB) as units:

$$M \text{ in dB} = M_{dB} = 20\lg M = 20\lg M_1 + 20\lg M_2 + \cdots$$

In Bode plots, the magnitude M in dB and the phase angle ϕ in degrees are plotted against ω on semilog paper. The development has shown the following: Bode magnitude and phase-angle plots of $KZ(j\omega)/P(j\omega)$ are obtained by summing those of its elementary factors. These plots are much easier to make than polar plots or Nyquist diagrams, and can readily be interpreted in terms of different aspects of system performance.

Gain $K > 0$

$M_{dB} = 20\lg K$, $\phi = 0$, both independent of ω. Fig. 2-4A-1a.

Integrators $1/(j\omega)^n$ (poles at the origin)

$M_{dB} = 20\lg |j\omega|^{-n} = -20n\lg\omega$, at $\omega = 1$, $M_{dB} = 0$, and at $\omega = 10$, one decade (dec) away from $\omega = 1$, $M_{dB} = -20n$. Hence on a log scale the magnitude plot is a straight line at a slope of

$-20n$ dB/dec. Phase angle $\phi = -n\,90°$ and is independent of frequency.

Differentiators $(j\omega)^n$ (zeros at the origin): the plots are the mirror images of the corresponding integrator relative to the 0dB and 0° axes. This is also true for the *leads* corresponding to the simple and quadratic lag below. [1]

Simple lag $1/(j\omega\tau+1)$

$$\left.\begin{array}{l}M_{dB} = 20\lg[1/\sqrt{1+(\omega\tau)^2}] \\ \phi = -\arctan\omega\tau\end{array}\right\} \quad (2\text{-}4A\text{-}1)$$

The plot indicated in Fig. 2-4A-1c is the asymptotic approximation. The asymptotes meet at the break frequency or corner frequency given by $\omega\tau = 1$ (or $\omega = 1/\tau$) on the normalized plot.

closer to $\omega\tau = 1$, the actual values can be calculated from Eq. (2-4A-1). At $\omega\tau = 1$, the deviation is -3dB and the phase $-45°$.

Quadratic lag

$$\frac{1}{(j\omega/\omega_n)^2 + 2\zeta(j\omega/\omega_n) + 1}$$

Fig. 2-4A-1 a) Gain factor
b) Integrator factors c) Simple lag factor

$$M_{dB} = 20\lg\left[\left(1-\frac{\omega^2}{\omega_n^2}\right)+\frac{2\zeta\omega}{\omega_n}\right]^{-\frac{1}{2}}, \quad \phi = -\arctan\frac{2\zeta\omega/\omega_n}{1-\omega^2/\omega_n^2} \quad (2\text{-}4A\text{-}2)$$

where ω_n is undamped natural frequency, ζ is damping ratio. The Bode plot of qudratic lag for the low-frequency asymptote is the axis, the high-frequency asymptote crosses the 0dB axis at $\omega/\omega_n = 1$, at a slope of -40dB/dec. Closer to $\omega/\omega_n = 1$, Fig. (2-4A-2) gives the actual curves. Smaller damping ratios ζ cause more severe peaking of M_{dB} and more abrupt change of ϕ near $\omega/\omega_n = 1$.

Bode plots can be obtained by summing the magnitudes and angles of the elementary factors present respectively.

In Bode plot, the phase margin ϕ_m is the sum of 180° and the phase angle at the frequency where $|KZ(s)/P(s)| = 1$ (i. e., 0dB). Hence, as shown by the partial plots in Fig. 2-4A-2, the phase margin ϕ_m is the distance of the phase-angle curve above $-180°$ at the crossover frequency ω_c, where the magnitude plot crosses the 0dB axis. Similarly, the gain margin equals 1 divided by the magnitude at the frequency where the phase angle is $-180°$. GM_{dB}, the gain margin in dB, is therefore the distance of the magnitude below 0dB at this frequency, as shown in Fig. 2-4A-2.

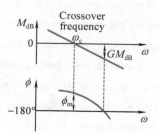

Fig. 2-4A-2 The phase margin and gain margin

Performance Specifications on the Bode Plot

An important reason for the widespread use of Bode is the ease of interpreting performance specifications on the asymptotic magnitude plot.

Relative stability The plots of the loop transfer function must have adequate length of not more than a -20dB/dec slope at or near crossover frequency ω_c. If not, it is immediately evident that ϕ_m will be inadequate.

Steady-state accuracy To improve steady, the low-frequency asymptote must be raised or its slope changed. The low-frequency asymptote is $K/(j\omega)^n$. For $n = 0$ (type 0) the steady-state error after a unit step is $1/(1+K)$, so reduces if the zero slope low frequency asymptote $20\lg K$ is raised. For zero steady-state error after a step, the system must be at least type 1 ($n = 1$), so the low-frequency asymptote must be at least -20dB/dec.

Accuracy in the operating range To ensure specified accuracy over a normal range of frequencies, the plot may not fall below a given level over this range. To improve accuracy, this level must be raised.

Crossover frequency and bandwidth Crossover frequency ω_c is a measure of bandwidth ω_b, so of speed of response, according to $\omega_c = 0.63\omega_b$.

Noise rejection To ensure a specified attenuation (reduction) of noise components in the input above a certain frequency should be below a certain level.[2]

These criteria show how different aspects of performance are reflected in individual features of the plot and permit specifications to be translated into requirements on the Bode plot. The task of system design is to derive the compensators that will meet these requirements.

Bode Plot Design

Design of a feedback control system using Bode techniques entails shaping and reshaping the Bode magnitude and phase angle plots until the system specifications are satisfied. These specifications are most conveniently expressed in terms of frequency-domain figures of merit such as gain and phase margin for the transient performance and the error constants for the steady-state time-domain response. And shaping the asymptotic Bode plots of continuous-time systems by cascade or feedback compensation is a relatively simple procedure.

Gain factor compensation It is possible in some cases to satisfy all system specifications by simple adjusting the open-loop gain factor K. As indicated in Eq. (2-3A-3), adjusting of the gain factor K does not affect the phase angle plot. It only shifts the magnitude plot up or down to correspond to the increase or decrease in K.

Lead compensation The addition of a cascade lead compensator to a system lowers the overall phase angle curve in the low-to-mid-frequency region. Lead compensation is normally used to increase the gain and/or phase margins of a system or increase its bandwidth. An additional modification of the Bode gain K_B is usually required with lead networks.

Lag compensation The lag compensation is employed in some cases to decrease the bandwidth of the system, and it is also used to improve the relative stability for a given value of error constant, or to reject the noise of high-frequency.

Lag-lead compensation It is sometimes desirable to simultaneously employ both lead and lag compensation. Although one each of these two networks can be connected in series to achieve the desired effect, it is usually more convenient to mechanize the combined lag-lead compensator.

WORDS AND TERMS

decibel   *n.* 分贝
common logarithm   常对数
minimum phase   最小相位
product   *n.* 乘积
semilog paper   半对数坐标（纸）
interpret   *v.* 解释，解析
slop   *n.* 斜率
quadratic   *adj.* 二次的；*n.* 二次方程

break frequency   转折频率
crossover frequency   穿越频率
bandwidth   *n.* 带宽
entail   *v.* 引起，使成为必要；*n.* 负担，需要
merit   *n.* 优点；指标，准则
procedure   *n.* 程序，过程
medchanize   *v.* 使机械化

NOTES

［1］This is also true for the leads corresponding to the simple and quadratic lag below.
对应于后面的一次和二次滞后的超前环节也是这样。

［2］To ensure a specified attenuation (reduction) of noise components in the input above a certain frequency should be below a certain level.
为保证对输入中高于一定频率的噪声成分指定的衰减，（伯德图中 *M*）应低于某一水平。

B   Nonlinear Control System

Introduction

In practice, most systems are nonlinear for large enough variations about the operating point, and linearization is based on the assumption that these variations are sufficiently small. But this cannot be satisfied, for example, for systems that include relays, which can switch position for very small changes. Startup and shutdown also frequently require the consideration of nonlinear effects, because of the size of the transients.

A differential equation $Ax'' + Bx' + Cx = f(t)$ is nonlinear if one or more of A, B, or C is a function of the dependent variable x or its derivatives. If A, B, or C of the system are also functions of the independent variable t, it will be a nolinear time-varying problem, which is not discussed here.

The principle of superposition does not apply to nonlinear systems. This has serious consequences. In fact, the analysis and design techniques discussed so far, including the use of transfer functions and Laplace transforms, are no longer valid. Worse, there is no general equivalent technique to replace them. Instead, a number of techniques exist, each of limited purpose and limited applicability.[1] We will only introduce the well-known phase plane and describing function methods.

Nonlinear Behavior and Common Nonlinearities

As a minimum, it is important to be aware of the main characteristics of nonlinear behavior, if only to permit recognition if these are encountered experimentally or in system simulations:

1. The nature of the response depends on input and initial conditions. For example, a nonlinear system can change from stable to unstable, or vice versa, if the size of a step input is doubled.

2. Instability shows itself frequently in the form of *limit cycles*. These are oscillations of fixed amplitude and frequency which can be sustained in the feedback loop even if the system input is zero. In linear systems an unstable transient grows theoretically to infinite amplitude, but nonlinear effects limit this growth.

3. The steady-state response to a sinusoidal input can contain harmonics and subharmonics of the input frequency.

4. The *jump phenomenon* is illustrated by the frequency response plot in Fig. 2-4B-1. If the frequency of the input is reduced from high values, the amplitude of the response drops suddenly at the vertical tangent point C to the value at D.

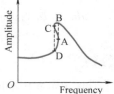

Fig. 2-4B-1   Jump phenomenon

Fig. 2-4B-2 shows common types of nonlinearities, with x as input and y as output.

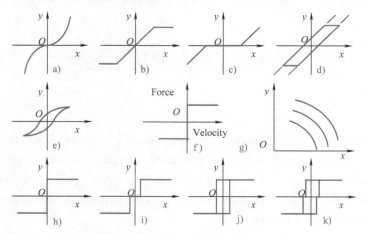

Fig. 2-4B-2   Common types of nonlinearities
a) Nonlinear gain   b) Saturation   c) Deadband   d) Backlash   e) Hysteresis
f) Coulomb friction   g) Characteristic curves   h) Ideal relay   i) Relay with deadband
j) Relay with hysteresis   k) Relay with hysteresis and deadband

Nonlinear gain very common (e.g., valve flow versus pressure drop or valve opening, or force versus deflection for rubber springs).

Saturation The output levels off to a constant limit beyond a certain value of the input. Amplifiers saturate, and valve flow cannot rise beyond pump capacity.

Deadband An insensitive zone, for example, in instruments or relays, or due to overlap of the lands on a hydraulic control valve spool over the ports to the cylinder. [2]

Backlash Due to play in mechanical connections.

Hysteresis In electromagnetic circuits; in materials.

Coulomb friction, or dry friction The friction force depends only on the direction of velocity.

Nonlinear characteristic curves The torque-speed curves of motors or the flow-pressure curves of valves.

Relays, with various imperfections A very important class of nonlinearities.

Phase-Plane Method

The phase-plane method (Fig. 2-4B-3) is a graphical method for finding the transient response of first or second order systems to initial conditions or simple inputs. Despite these restrictions, it is useful because of the insight it provides and because many systems approximate second order responses. Consider the nonlinear equation

$$x'' + g(x, x')x' + h(x, x')x = 0$$

Substitute into this $\quad x' = y \quad x'' = y' = \dfrac{dy}{dx}x' = y\dfrac{dy}{dx}$

Then the equation reduces to one of the first order

$$y\left(\frac{dy}{dx}\right) + g(x, y)y + h(x, y)x = 0$$

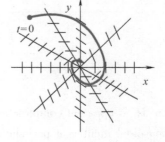

Fig. 2-4B-3 Phase-plane method

Rearranging this yields the *phase-plane equation*

$$\frac{dy}{dx} = \frac{-g(x, y)y - h(x, y)x}{y}$$

The phase plane is a plot of y versus x. At each point (x', y), dy/dx is the slope of the *phase-plane trajectory* through that point.

Isoclines are loci of constant trajectory slope. The isocline equation for $dy/dx = m$ is

$$y = \frac{-h(x, y)x}{g(x, y) + m}$$

This graphical technique is called the isocline method, and is useful to sketch the nature of the phase plane portrait. For numerical work, this and alternative graphical techniques have been largely replaced by computer methods.

Describing Function

The describing function technique is a response method, and its main use is in stability analysis (i.e., the prediction of limit cycles). In Fig. 2-4B-4, where G_1 and G_2 represent linear parts of the system and N a nonlinear element, the question is whether a limit cycle exists, that is, whether an oscillation can maintain itself around the loop for $R = 0$.

Limit cycles for second-order systems can also be constructed by phase plane methods. As illustrated in Fig. 2-4B-5, they are represented by closed curves in the phase plane. But limit cycles are distinguished from other possible closed curves in that the phase-plane trajectories tend toward or away from them asymptotically. A stable limit cycle is one that is approached by trajectories from both sides. Even the slightest disturbance causes trajectories to depart from unstable limit cycles. As indicated in the first of Fig. 2-4B-5, an initial condition or input outside the limit cycle leads to an unstable transient growth while transients following a sufficiently small disturbance decay to zero.[3] Thus it is also necessary to determine the type of limit cycle.

The model for N in Fig. 2-4B-4 used in this analysis is based on the following assumption: The input x to the nonlinearity is a sinusoidal

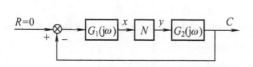

Fig. 2-4B-4 A nonlinear system

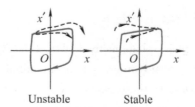

Fig. 2-4B-5 System transient for a small disturbance decaying to zero

$$x = A\sin\omega t \tag{2-4B-1}$$

There is an apparent contradiction here. In Fig. 2-4B-6 the square-wave output of an ideal relay for a sinusoidal input is a periodic function, so that it can be represented by a Fourier series of the general form

$$y(t) = b_0 + \sum_{n=1}^{\infty}(a_n\sin n\omega t + b_n\cos n\omega t) \tag{2-4B-2}$$

Thus y contains harmonics ($n > 1$) in addition to the fundamental Fourier component

$$y_f = a_1\sin\omega t + b_1\cos\omega t \tag{2-4B-3}$$

The harmonics would pass around the loop via G_2 and G_1 back to x, contradicting the assumption Eq. (2-4B-1). But for most practical systems, y_f is considerably larger than the harmonics, and G_2 acts as a low-pass filter which attenuates the harmonics much more strongly than y_f.

The combination of these effects usually justifies Eq. (2-4B-1), and also implies that only y_f is required. Hence the definition: The *describing function* (*DF*) N of a nonlinearity is the ratio $N = y_f/x$. From the theory of Fourier series, the Fourier coefficients a_1 and b_1 for y_f in Eq. (2-4B-3) are

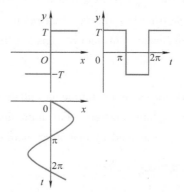

Fig. 2-4B-6 Describing function of an ideal relay

$$a_1 = \frac{\omega}{\pi}\int_0^{2\pi/\omega} y\sin\omega t\, dt, \quad b_1 = \frac{\omega}{\pi}\int_0^{2\pi/\omega} y\cos\omega t\, dt \qquad (2\text{-}4\text{B-}4)$$

If y can be extended into an odd function of time, as shown in Fig. 2-4B-6, then $b_1 = 0$ since $\cos\omega t$ is an even function of time, and the DF becomes

$$N = \frac{a_1}{A}$$

The DF is an equivalent linear gain which depends on the amplitude A, and sometimes also the frequency ω, of the input x. Example: DF of an ideal relay (Fig. 2-4B-6)

Since the DF is independent of ω, $\omega = 1$ can be assumed for simplicity, so that $x = A\sin t$.

$$a_1 = \frac{1}{\pi}\int_0^{2\pi} y\sin t\, dt = \frac{2}{\pi}\int_0^{\pi} T\sin t\, dt = \frac{4T}{\pi}$$

So the describing function is

$$N = \frac{4T}{\pi A}$$

As expected, this shows that the equivalent gain decreases as the input A increases, since the output is constant.

WORDS AND TERMS

linearazation   n. 线性化
assumption   n. 假设
relay   n. 继电器
startup   n. 启动
shutdown   n. 关闭
dependent variable   应变量
independent variable   自变量
valid   adj. 有效力的
equivalent   adj. 等价的；n. 等价
recognition   n. 认识
encounter   v. 遇到
vice versa   反之亦然
limit cycle   极限环
sustain   v. 维持
theoretically   adv. 理论上
harmonic   n. 谐波
subharmonic   n. 次谐波
tangent   adj. 切线的，正切的；n. 切线，正切
valve   n. 阀门

opening   n. 开度
deflection   n. 偏（离，差）
saturation   n. 饱和
deadband   n. 死区
insensitive   adj. 不敏感的
instrument   n. 仪器，工具
overlap   v., n. 重叠
hydraulic   adj. 水力的，液压的
spool   v. 绕；n. 卷筒，线圈，阀柱
backlash   n. 齿隙游移
hysteresis   n. 滞回线
coulomb friction   库仑摩擦
dry friction   干性摩擦
imperfection   n. 不完全，不足，缺点
phase-plane equation   相平面方程
isocline   n. 等倾线
portrait   n. 肖像，描写，型式
numerical   adj. 数值（字）的
prediction   n. 预测
contradiction   n. 矛盾

Fourier series 傅里叶级数
via *prep.* 经由
attenuate *v.* 减弱

justify *v.* 证明
imply *v.* 包含

NOTES

［1］Instead, a number of techniques exist, each of limited purpose and limited applicability.
有几种技术存在，但有各自限定的目的和限定的适用范围。

［2］…, or due to overlap of the lands on a hydraulic control valve spool over the ports to the cylinder.
……或由于液压控制阀柱上的挡圈与通向液压缸的出口的重叠。

［3］…, an initial condition or input outside the limit cycle leads to an unstable transient growth while transients following a sufficiently small disturbance decay to zero.
……当暂态响应跟随一个足够小的扰动向零衰减时，一个在极限环外的初始条件或输入会导致一种不稳定的暂态增长。

C 形容词的翻译

形容词在英文句中一般做表语或修饰名词的定语，因而在翻译时要把形容词与其所修饰的客体统一考虑，对二者的割裂会导致牛头不对马嘴的错误。例如：heavy 的基本词义是"重"，但在翻译时结合修饰的对象则译法各异：heavy current 强电流；heavy crop 大丰收；heavy traffic 交通拥挤；heavy industry 重工业。英语词义对上下文的依赖性是很大的。一个孤立的词，其词义通常是游移不定的，但当词处于特定的联立关系中时，其含义就受到相关词的制约而明朗和稳定了。因此，根据词的联立关系确定词义是词义辨析的重要且最为可行的手段。如 universal 有"宇宙的，世界的，普遍的，一般的，通用的"等含义，具体用法举例如下：

universal meter 万用表
universal motor 交直流两用电动机
universal agent 全权代理人
universal peace 世界和平
universal truth 普遍真理
universal valve 万向阀

universal constant 通用常数
universal use 普遍应用
universal class 全类
universal rules 一般法则
universal travel 环球旅行
universal dividing head 多用分度表头

一般的常用专业词组搭配都可以在英汉技术词典中查到，所以对含义不清楚的词组应借助词典来理解，"闭门造车"式的翻译只能导致主观臆断的错误。

对形容词的翻译还可采取词性转换的技巧，使对原文的表达更通顺和易于理解。例如：
1. 当形容词在句中做表语时，某些情况下可根据汉语习惯，将其和系词合译成动词形式。例如：

Internet is different from *Intranet* in many aspects though their spelling is alike.
虽然拼写相像，Internet 仍在很多方面不同于 Intranet。句中形容词译为"不同"和"相

像"。

If low-cost power becomes available from nuclear power plants, the electricity crisis would be solved.

如能从核电站获得低成本电力,电力紧张问题就会解决。becomes available 译为"获得"。

2. 某些表示事物特征的形容词,可在其后加上"性""度"等而转译为名词。例如:

IPC is more reliable than common computer.

工控机的可靠性比普通计算机高。

Experiment indicates that the new chip is about 1.5 times as integrative as that of the old ones.

试验表明,新型芯片的集成度是旧型号的1.5倍。

3. 有时形容词也可根据需要译成副词。例如:

The same principles of low internal resistance also apply to milliammeters.

低内阻原理也同样适用于毫安表。same 译为"同样"。

The modern world is experiencing rapid development of information technique.

当今世界的信息技术正在迅速地发展。development 名词动译,修饰它的形容词 rapid 也随之按副词翻译为"迅速地"。

词性转换技巧同样适用于其他词类,拘泥于原文的词性进行完全对应的翻译往往会给翻译带来极大的障碍,同时使得译文生涩而难以理解。翻译应从整体着眼,采取适当的词性转换、句子成分转换或句型转换的翻译技巧,可使译文既忠实于原意又通顺可读。

UNIT 5

A Introduction to Modern Control Theory

Introduction

Classical control theory is based on input-output relationships, principally the transfer function. When differential equations are encountered, they are linearized and subjected to whatever constraints are necessary to establish useful input-output relationships. Modern control theory, on the other hand, is based on direct use of the differential equations themselves. Although its techniques are powerful and relatively simple, classical control theory does have limitations and shortcomings that multiply as plants and control systems become more complex—thus the advent of modern control theory.

Several factors provided the stimulus for the development of modern control theory: ① the necessity of dealing with more realistic models of systems; ②the shift in emphasis towards optimal control and optimal system design; ③the continuing developments in digital computer technology; ④the shortcomings of previous approaches; ⑤ a recognition of the applicability of well-known methods in other fields of knowledge.

The transition from simple approximate models, which are easy to work with, to more realistic models produces two effects.[1] First, a larger number of variables must be included in the model. Second, a more realistic model is more likely to contain nonlinearities and time-varying parameters. Previously ignored aspects of the system, such as interactions with and feedback through the environment, are more likely to be included.

With an advancing technological society, there is a trend towards more ambitious goals. This also means dealing with complex systems with a larger number of interacting components. The need for greater accuracy and efficiency has changed the emphasis on control system performance. The classical specifications in terms of percent overshoot, settling time, bandwidth, etc., have in many cases given way to optimal criteria such as minimum energy, minimum cost, and minimum time operation. Optimization of these criteria makes it even more difficult to avoid dealing with unpleasant nonlinearities. Optimal control theory often dictates that nonlinear time-varying control laws be used, even if the basic system is linear and time-invariant.

The continuing advances in computer technology have had important effects on the controls field. The fast processing ability and large memory capacity of digital computers make it possible to solve practical problems with complex high-order differential equations. Classical control theory was dominated by graphical methods because at the time that was the only way to solve certain problems. Now every control designer has easy access to powerful computer packages for systems analysis and design. Although a computer can be used to carry out the classical transform-inverse transform methods, it is usually more efficient to be used to find (calculate) the numerical solutions of a differential equation directly. With the development of very large scale integrated circuits (VLSI), the cost, size, and reliability of microcomputers have made it possible to use them routinely in many systems. Computers are now so commonly used as just another component in the control system.

Modern control theory is well suited to the above trends because its time-domain techniques and its mathematical language (matrices, linear vector spaces, etc.) are ideal when dealing with a computer. Computers are a major reason for the existence of state variable methods.

Most classical control techniques were developed for linear constant coefficient system with one input and one output (perhaps a few inputs and outputs). The language of classical techniques is the Laplace or z-transform and transfer functions. When nonlinearities and time variations are present, the very basis for these classical techniques is removed. Some successful techniques such as phase-plane methods, describing functions, and other ad hoc methods, have been developed to alleviate this shortcoming. However, the greatest success has been limited to low-order systems. The state variable approach of modern control theory provides a uniform and powerful method of representing systems of arbitrary order, linear or nonlinear, with time-varying or constant coefficients. It provides an ideal formulation for computer implementation and is responsible for much of the progress in optimization theory.[2]

The Concept of State

The concept of state occupies a central position in modern control theory. However, it appears in many other technical and nontechnical contexts as well. In thermodynamics the equations of state are prominently used. Binary sequential networks are normally analyzed in terms of their *states*. In everyday life, monthly financial *statements* are commonplace. The President's State of the Union message is another familiar example.

In all of these examples the concept of state is essentially the same. It is a complete summary of the status of the system at a particular point in time. Knowledge of the state at some initial time t_0, plus knowledge of the system inputs after t_0, allows the determination of the state at a later time t_1. As far as the state at t_1 is concerned, it makes no difference how the initial state was attained.[3] Thus the state at t_0 constitutes a complete history of the system behavior prior to t_0, insofar as that history affects future behavior.[4] Knowledge of the present state allows a sharp separation between the past and the future.

At any fixed time the state of a system can be described by the values of a set of variables x_i, called *state variables*. One of the state variables of a thermodynamic system is temperature and its value can range over the continuum of real numbers R. In a binary network state variables can take on only two discrete values, 0 or 1. Note that the state of your checking account at the end of the month can be represented by a single number, the balance. The state of the Union can be represented by such things as gross national product, percent unemployment, the balance of trade deficit, ect. For the systems considered in this article the state variables may take on any scalar value, real or complex. That is, $x_i \in C$. Although some systems require an infinite number of state variables, only systems which can be described by a finite number n of state variables will be considered here. Then the state can be represented by an n component *state vector* $\boldsymbol{x} = \begin{bmatrix} x_1 & x_2 \cdots x_n \end{bmatrix}^T$. It belongs to an n-dimensional vector space defined over the field C.

For *Continuous-time* systems, the state is defined for all times in some interval, for example, a continually varying temperature or voltage. *Discrete-time* systems have their state defined only at discrete times, as with the monthly financial statement or the annual State of the Union message. Continuous-time and discrete-time systems can be discussed simultaneously by defining the times of interest as T. For continuous-time systems T consists of the set of all real numbers $t \in [t_0, t_f]$. For discrete-time systems T consists of a discrete set of times $\{t_0, t_1, t_2, \cdots, t_k\}$. In either case the initial time could be $-\infty$ and the final time could be ∞ in some circumstances.

The state vector $\boldsymbol{x}(t)$ is defined only for those $t \in T$. At any given t, it is simply an ordered set of n numbers. However, the character of a system could change with time, causing the *number of required state variables* (and not just the values) to change. If the dimension of the state space varies with time, the notation Σ_t could be used. It is assumed here that Σ is the same n-dimensional state space at all $t \in T$.

WORDS AND TERMS

constraint   *n.* 强迫，约束
multiply   *v.* 加倍，倍增
stimulus   *n.* 刺激，鼓励
optimal control   最优控制
dominate   *v.* 支配，使服从
package   *n.* 包
very large scale integrated circuits (VLSI)   超大规模集成电路
numerical   *adj.* 数字的
matrix   *n.* 矩阵   *pl.* matrices
linear vector space   线性矢量空间
ad hoc   尤其，特定地
alleviate   *v.* 减轻，缓和

arbitrary   *adj.* 任意的
formulation   *n.* 公式化（表达）
implementation   *n.* 实现，履行
attain   *v.* 达到，实现
constitute   *v.* 构造，组织
insofar as   到这样的程度，在……范围内
continuum   *n.* 连续
gross national product   国民生产总值
trade deficit   贸易赤字
scalar   *adj.* 数量的，标量的；*n.* 数量，标量
n-dimensional   *adj.* n 维的
circumstance   *n.* 状况，环境

NOTES

[1] The transition from simple approximate models, which are easy to work with, to more realistic models produces two effects.

从易于处理的简单近似模型到实际一些的模型的转换存在两方面问题。

[2] It provides an ideal formulation for computer implementation and is responsible for much of the progress in optimization theory.

它（现代控制理论）为计算机实现提供了理想的公式化方法，并对最优化理论的发展起了重要作用。

[3] As far as the state at t_1 is concerned, it makes no difference how the initial state was attained.

就 t_1 时刻的状态而言，它与初始状态是怎样实现的无关。

as far as...be concerned 意为"就……而言"；make no difference 意为"无论……都没有区别（没关系）"。

[4] Thus the state at t_0 constitutes a complete history of the system behavior prior to t_0, insofar as that history affects future behavior.

因此，t_0 时刻的状态构成了 t_0 以前系统活动状态的历史，这个历史状态在一定范围内影响（系统）未来的行为。

B State Equations

State Space Models

A state space model is a description in terms of a set of first-order differential equations which are written compactly in a matrix form. This standard form has permitted the development of general computer programs, which can be used for the analysis and design of even very large systems.

The derivation of state space models is no different from that of transfer functions in that the differential equations describing the system dynamics are written first. In transfer function models these equations are transformed and variables are eliminated between them to find the relation between selected input and output variables.[1] For state models, instead, the equations are arranged into a set of first-order differential equations in the terms of selected state variables, and the outputs are expressed in these same state variables. Because the elimination of variables between equations is not an inherent part of this process, state models are often easier to obtain. Two examples are given for illustration and to relate state models to the transfer functions used thus far.

A transfer function without zeros:

$$\frac{C}{R} = \frac{b}{s^3 + a_3 s^2 + a_2 s + a_1} \quad \text{or} \quad \dddot{c} + a_3 \ddot{c} + a_2 \dot{c} + a_1 c = br$$

A state model for the system described by this transfer function or the equivalent differential equation is not unique but depends on the choice of a set of state variables $x_1(t)$, $x_2(t)$, and $x_3(t)$. One possible choice is the following:

$x_1 = c, \ x_2 = \dot{c}, \ x_3 = \ddot{c}$ thus $\dot{x}_1 = x_2, \ \dot{x}_2 = x_3, \ \dot{x}_3 = -a_1 x_1 - a_2 x_2 - a_3 x_3 + br$

In matrix form, and with the output c expressed also in terms of all state variables,

$$\boldsymbol{x} = \begin{bmatrix} x_1 \\ x_2 \\ x_3 \end{bmatrix} \quad \dot{\boldsymbol{x}} = \begin{bmatrix} 0 & 1 & 0 \\ 0 & 0 & 1 \\ -a_1 & -a_2 & -a_3 \end{bmatrix} \boldsymbol{x} + \begin{bmatrix} 0 \\ 0 \\ b \end{bmatrix} r \quad c = \begin{bmatrix} 1 & 0 & 0 \end{bmatrix} \boldsymbol{x}$$

The general form of a state-space model is as follows:

$$\dot{x} = Ax + Bu \quad (\text{state equation})$$
$$y = Cx + Du \quad (\text{output equation}) \tag{2-5B-1}$$

where A is known as the *plant matrix* or *system matrix* and B as the *control matrix*. The unnamed matrices C and D relate the output variables to the state and control variables. The vector x of the state variables is the state vector. The controls vector u is the scalar function r in this example, and the output vector y the scalar function c, and $D = 0$.

A transfer function with zeros:

$$\frac{C}{R} = \frac{b_3 s^2 + b_2 s + b_1}{s^3 + a_3 s^2 + a_2 s + a_1} \quad \text{or} \quad \dddot{c} + a_3 \ddot{c} + a_2 \dot{c} + a_1 c = b_3 \ddot{r} + b_2 \dot{r} + b_1 r$$

First consider only the denominator:

$$\frac{V}{R} = \frac{1}{s^3 + a_3 s^2 + a_2 s + a_1} \quad \dddot{v} + a_3 \ddot{v} + a_2 \dot{v} + a_1 v = r$$

As in the case of no zero

$$x = \begin{bmatrix} v \\ \dot{v} \\ \ddot{v} \end{bmatrix} \quad \dot{x} = \begin{bmatrix} 0 & 1 & 0 \\ 0 & 0 & 1 \\ -a_1 & -a_2 & -a_3 \end{bmatrix} x + \begin{bmatrix} 0 \\ 0 \\ 1 \end{bmatrix} r$$

But $\quad C = (b_3 s^2 + b_2 s + b_1) V \quad$ or $\quad c = b_3 \ddot{v} + b_2 \dot{v} + b_1 v = \begin{bmatrix} b_1 & b_2 & b_3 \end{bmatrix} x$

Hence the output equation, with $y = c$, is

$$y = Cx \quad C = \begin{bmatrix} b_1 & b_2 & b_3 \end{bmatrix}$$

Thus the output equation represents the effect of system zeros, or derivatives of the input.

Comments on the State Space Representation

The selection of state variables is *not* a unique process. Various sets of state variables can be used. It is usually advantageous to use variables which have physical significance and, if possible, can be measured. There are several applicable methods of selecting state variables. The form of the information available for a system will often dictate which method should be used. In some cases, for example, a transfer function is obtained experimentally and must be used as the starting point.

In any case the advantages of the state space method include the following:

1. It provides a convenient, compact notation and allows the application of the powerful vector-matrix theory.

2. The uniform notation for all systems makes possible a uniform set of solution techniques.

3. The state space representation is in an ideal format for computer solution, either analog or digital. This is important because computers are frequently required for the analysis of complex systems.

4. The state space method gives a more complete description of a system than does the input-output approach. This will become more evident when the concepts of controllability and observability are considered later.

The state space method is the central theme of modern control theory. However, the ideas have been with us for some time. The generalized coordinates and momenta of Hamiltonian and Lagrangian mechanics are, in fact, a set of state variables.[2] For those familiar with these subjects, an alternative method of state variable selection is available.

Transfer Function Matrices and Stability

With a state-model description of the system dynamics, the first question to be answered is how stability may be determined. To derive the stability criterion, the generalizaion of the concept of a transfer function is considered first, by finding the transfer function matrix that corresponds to the state model. This requires Laplace transformation of the state-model equations. The Laplace transform of a vector is the vector of the Laplace transforms of its elements, so that the transforms of x and $\dot{x}$ are as follows:

$$X(s) = L[x(t)] = [L[x_1] \quad \cdots \quad L[x_n]]^T = [X_1(s) \quad \cdots \quad X_n(s)]^T$$

$$L[\dot{x}] = [L[\dot{x}_1] \quad \cdots \quad L[\dot{x}_n]]$$

where $L[\dot{x}_i] = sX_i(s) - x_i(0)$. Hence the transform of $\dot{x} = Ax + Bu$ is $sX(s) - x_0 = AX(s) + BU(s)$, or $(sI - A)X(s) = BU(s) + x_0$, where I is the unit matrix. Note that $sX(s) = sIX(s)$ and that $(s - A)$ is incorrect since s is a scalar and A is not. Then if the system output $y = Cx$, so $Y(s) = CX(s)$:

$$Y(s) = G(s)U(s) + C(sI - A)^{-1}x_0 \qquad (2\text{-}5\text{B-}2)$$

where

$$G(s) = C(sI - A)^{-1}B \qquad (2\text{-}5\text{B-}3)$$

Eq. (2-5B-2) shows the total response as a superposition of two separate components. The first term gives the input-output response, for $x_0 = 0$, and the second the output response to initial conditions, with $u = 0$. With $C = I$ the state response $X(s)$ is obtained.

$G(s)$ is called the *transfer function matrix* because it relates the transforms of the input and output vectors for zero initial conditions. As illustrated in Fig. 2-5B-1, $G(s)$ generalizes the concept of a transfer function. It is a matrix of ordinary transfer functions. For a system with r inputs and m outputs, the equation $Y = GU$ can be written out as

$$\begin{bmatrix} Y_1(s) \\ \vdots \\ Y_m(s) \end{bmatrix} = \begin{bmatrix} g_{11}(s) & \cdots & g_{1r}(s) \\ \vdots & \ddots & \vdots \\ g_{m1}(s) & \cdots & g_{mr}(s) \end{bmatrix} \begin{bmatrix} U_1(s) \\ \vdots \\ U_r(s) \end{bmatrix}$$

$$Y_i(s) = Y_{i1}(s) + \cdots + Y_{ir}(s) \qquad Y_{ij}(s) = g_{ij}(s)U_j(s)$$

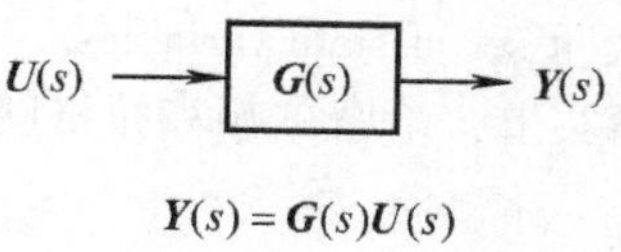

$U(s) \longrightarrow \boxed{G(s)} \longrightarrow Y(s)$

$Y(s) = G(s)U(s)$

Fig. 2-5B-1 Transfer function matrix

The element $g_{ij}(s)$ is an ordinary transfer function which gives the part Y_{ij} of Y_i that is due to input U_j.

For stability, all poles of all these transfer functions must lie in the left-half s-plane. The inverse in Eq. (2-5B-3) equals the adjoint divided by the determinant, so that the transfer function matrix becomes

$$G(s) = \frac{C \operatorname{adj}(sI - A) B}{|sI - A|}$$

The numerator is a matrix of polynomials, and all elements of **G** have the same denominator. This denominator is the polynomial $|s\boldsymbol{I} - \boldsymbol{A}|$. Hence

Stability Theorem: The system described by the state model (2-5B-1) is stable if and only if the *eigenvalues* of the system matrix **A**, that is, the roots of the system

$$Characteristic\ equation\quad |s\boldsymbol{I} - \boldsymbol{A}| = 0$$

all lie in the left-half *s*-plane.

The ability to exploit standard computer routines available to determine the eigenvalues of even very large matrices **A** is a major advantage of the state space formulation.

WORDS AND TERMS

derivation   *n.* 起源，得来
eliminate   *v.* 消除
inherent   *adj.* 固有的
equivalent   *adj.* 同等的，等效的；*n.* 同等，等效
dictate   *v.* 命令，要求
uniform   *adj.* 一致的
theme   *n.* 题目，主题，论文
generalize   *v.* 概括，一般化，普及
coordinate   *n.* 坐标，同等的人或物

momenta   *n.* 动量，冲量
Hamiltonian   哈密尔顿的
Lagrangian   拉格朗日的
superposition   *n.* 叠加
inverse   *n.* 反，逆，倒数
adjoint   *n.*, *adj.* 伴随（的），共轭（的）
determinant   *n.* 行列式
eigenvalue   *n.* 特征值（eigen- 特征）
exploit   *v.* 开发
routines   *n.* 程序

NOTES

[1] In transfer function models these equations are transformed and variables are eliminated between them to find the relation between selected input and output variables.

在传递函数模型中，这些方程经过（拉普拉斯）变换，并消去中间变量，以求得选定的输入输出变量间的关系。

[2] The generalized coordinates and momenta of Hamiltonian and Lagrangian mechanics are, in fact, a set of state variables.

事实上，哈密尔顿和拉格朗日力学的广义坐标和动量就是一组状态变量。

C 词性的转换

英语和汉语在语言结构和表达方式上有很多不同的特点，科技英语句子中的名词化结构较多，而谓语动词则只有一个；相对而言，汉语的一个句子中往往包含有几个动词；此外汉语中没有英语中的冠词、关系代词、关系副词、分词、不定式、动名词等语法成分，因而翻译时就需要做适当的词性转换处理。

1. 英语名词化结构及动名词译成汉语的动词。例如：

Integrated circuits are fairly recent development.

集成电路是近年来才发展起来的。development 译为"发展起来"。

In the dynamo, mechanical energy is used for rotating the armature in the field.

在直流发电机中机械能用来使电枢在磁场中转动。rotating 译为"使…转动"。

2. 介词在很多情况下，尤其是在构成状语时，往往可译成汉语的动词。例如：

Noise figure is minimized by a parameter amplifier.

采用参数放大器，即可将噪声系数减至最低。by 译为"采用"。

Power is needed to stall the armature against inertia.

为使电枢克服惯性而制动，就需要一定的能量。against 译为"克服"。

3. 如副词做表语，有时也可同系词一起合译成汉语的动词。例如：

In this case the temperature in the furnace is up.

在这种情况下，炉温就会升高。is up 合译为"升高"。

When the switch is off, the circuit is open and electricity doesn't go through.

开关断开时，电路开路，电流不能流过。is off 译为"断开"，is open 译为"开路"。

4. 某些英语动词的概念难以直接用汉语动词表达，翻译时可转换为汉语的名词形式。例如：The design aims at automatic operation, simple maintenance and high productivity.

设计的目的为自动操作，维护简单，生产率高。aims 译为"目的"。

5. 某些副词可用汉语的名词表达。例如：

This instrument is used to determine how fully the batteries are charged.

这种仪表用来测定电池充电的程度。how fully 译为"程度"。

This device is shown schematically in Fig. 3-11.

图 3-11 所示为这种装置的简图。schematically 转译为"简图"，而且全句在翻译时做了句型转换的处理，使译文表达更接近汉语习惯。

6. 某些副词可转换为汉语的形容词表达。例如：

An current varies directly as the voltage force and inversely as the resistance.

电流的大小与电压成正比，与电阻成反比。directly 译为"正"；inversely 译为"反"。

Java is chiefly characterized by its simplicity of operation and compatibility with almost all OSs.

Java 主要的特点是：操作简单，并和几乎所有操作系统兼容。副词 chiefly 译为形容词"主要的"；被动动词结构 is characterized 译为名词"特点"。

UNIT 6

A Controllability, Observability, and Stability

Controllability and Observability

A plant (or system) is said to be *completely state-controllable* if it is possible to find an unconstrained control vector $u(t)$ that will transfer any initial state $x(t_0)$ to any other state $x(t)$ in a finite time interval. Since complete state controllability does not necessarily mean complete control of the output, and vice versa, *complete output controllability* is separately defined in the same manner.[1] A plant is said to be completely observable if the state $x(t)$ can be determined from a knowledge of the output $c(t)$ over a finite time interval.

The dual concepts of controllability and observability are fundamental to the control of multivariable plants, particularly with regard to optimal control. Complete controllability ensures the existence of an unconstrained control vector and thus the existence of a possible controller. However, it does not tell how to design the controller, nor does it guarantee either a realistic control vector or a practical controller. Complete observability ensures a knowledge of the state or internal behavior of the plant from a knowledge of the output. It does not, however, guarantee that the output variables are physically measurable.

The significance of these two concepts can be illustrated by consideration of a generalized nth-order plant, which will have n state variables and thus n transient (dynamic) modes. The number of control variables will be designated by m, and the number of output variables by p. In a practical control system we expect m and p to be less than n and would like them both to be small in number. If the plant is not completely controllable, there will be modes (state variables) that can not be controlled in any way by one or more of the control variables; these modes are decoupled from the control vector. If the plant is not completely observable, there will be modes whose behavior cannot be determined; these modes are decoupled from the output vector.

A plant can be divided into four subsystems, as shown in Fig. 2-6A-1. Since only the first subsystem A, which is both controllable and observable, has an input-output relationship, it is the only subsystem that can be represented by a transfer function or a transfer function matrix. Conversely, a transfer function or matrix representation of this plant reveals nothing about the dynamic behavior of subsystems B and D and provides no control

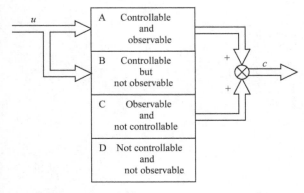

Fig. 2-6A-1   Four subsystems of a plant

over the behavior of subsystems C and D. For example, if the modes comprising subsystem B reacted violently to any of the control variables, the output variables would give no indication of such behavior. Undesirable transients in subsystem C would affect the output, but nothing could be done to modify them. This plant can be made completely controllable by appropriately adding control variables. The task of making the plant completely observable, however, is more difficult and will not be discussed further.

Stability

Leaving the concepts of controllability and observability, we need to reexamine the concepts and definitions of stability with regard to continuous-variable systems in general. The stability of stationary linear systems is relatively straight-forward in that it is a property of the system characteristics only, being independent of the initial state and of the magnitude and type of inputs. There is one finite equilibrium (singular) state, and we call the system stable if it returns to that state if disturbed. Stability is determined by the location of the eigenvalues (roots of the characteristic equation), and there are various techniques for locating the eigenvalues.

For nonstationary linear system and particularly for nonlinear system, stability is no longer dependent only upon the system properties but is also dependent upon the initial state and the type and magnitude of any input. Furthermore, there may well be more than one equilibrium state. To discuss stability for these systems, additional definitions and criteria are necessary. We shall limit ourselves to autonomous systems since stability theory for arbitrary inputs is still undeveloped.

A system is said to be stable if trajectories leaving an initial state return to and remain within a specified region surrounding an equilibrium state. This general definition of stability is often referred to as stability in the sense of Liapunov and permits limit cycles and vortices. If the trajectories of a system that is stable in the sense of Liapunov eventually converge to the equilibrium state, the system is said to be asymptotically stable. If the system is stable only for initial states within a bounded region of state space, it is said to be locally stable or stable in the small. If it is stable for all initial states within the entire state space, it is said to be globally stable or stable in the large.

We should like our control systems to have asymptotic stability, preferably global; if not global, then the region of asymptotic stability should be large enough to include any anticipated disturbances. The stability of classical control theory is asymptotic. It may appear at first glance to be global, but in reality it is local since no system is truly linear. Only local asymptotic stability with respect to the established equilibrium state can be guaranteed for linear analyses.[2]

There are three basic methods for determining the stability of nonlinear autonomous systems. One method is to approximate the actual system by a second-order system, plot many trajectories in the phase plane, and examine the resulting phase portrait for regions of stability and instability. The describing function method can be used in conjunction with the phase plane to search for and identify limit cycles. Another method is known as the first, or indirect, method of Liapunov. It consists of linearizing the nonlinear vector equations about each equilibrium state by means of the Jacobian matrix and then examining the corresponding eigenvalues for local stability only. The two methods just mentioned are sometimes lumped together as Liapunov's first method.

The third technique is the second, or direct, method of Liapunov, so called because it does not require solution of the differential equations. It is applicable to all types of differential equations of any order, provides some answers to global as well as local stability, and is widely used.

In using the second method of Liapunov, the equilibrium state being investigated is translated to the origin of the state space so that the autonomous system can be represented by the equation

$$\dot{x} = f(x) \qquad (2\text{-}6\text{A-}1)$$

with the equilibrium state $x_{eq} = 0$. The asymptotic stability theorem of Liapunov is the essence of this direct method. This theorem states that the system of Eq. (2-6A-1) is asymptotically stable within the closed region R surrounding the origin if there exists a positive definite scalar function $V(x)$ which will vanish with time along all trajectories originating within the region R. If the region R includes all of the state space, the system is globally stable; if not, the system is locally stable within the finite region R. The scalar function $V(x)$ is known as a Liapunov function. It must be continuous within the region R, as must its first partial derivatives. The requirement that it be positive definite means that $V(x)$ must be greater than zero for all nonzero values of the state variables and that $V(0)$ be equal to zero. In order for $V(x)$ to vanish along all trajectories starting within R, $dV(x)/dt$ must be less than zero, that is, be negative definite. If $\dot{V}(x) \leq 0$, it is negative semidefinite and the system is guaranteed to be stable only in the sense of Liapunov; if $\dot{V}(x) < 0$ along a trajectory, the system is asymptotically stable. Finally, if $\dot{V}(x)$ is indefinite, nothing has been proved with respect to the stability of the system, and we must try different $V(x)$ functions until the system has been shown to be either stable or unstable. Incidentally, for stable system the size of the region of guaranteed stability can vary with the choice of the Liapunov function.

WORDS AND TERMS

state-controllable *adj.* 状态可控（制）的
observable *adj.* 可观测的
dual *adj.* 双的，对偶的，孪生的
fundamental *n.* 基本原理
multivariable *adj.* 多变量的
guarantee *v., n.* 保证，担保
generalize *v.* 一般化，普及
decouple *v.* 解耦，退耦
reveal *v.* 显现，揭示
comprise *v.* 包含
violently *adv.* 激烈地
straight-forward *adj.* 直截了当的，简单的

eigenvalue *n.* 特征根
autonomous *adj.* 自治的，自激的
trajectory *n.* 轨迹
Liapunov 李亚普诺夫
vortices *n.* vortex 的复数，旋转体（面）
converge *v.* 集中，汇聚，收敛
asymptotically stable 渐近稳定
bound *v.* 限制
locally stable 局域稳定
globally stable 全局稳定
portrait *n.* 描述
conjunction *n.* 结合

identify   v. 确认，识别，辨识
Jacobian matrix   雅戈比矩阵
positive definite   正定
incidentally   adv. 偶然地

NOTES

[1] Since complete state controllability does not necessarily mean complete control of the output, and vice versa, complete output controllability is separately defined in the same manner.

因为状态完全能控性不一定意味着输出的完全可控，而且反之亦然，所以输出完全能控性以类似的方式单独定义。

[2] Only local asymptotic stability with respect to the established equilibrium state can be guaranteed for linear analyses.

只有相对于（系统）建立的平衡状态的局域渐近稳定才能保证线性分析（可以应用）。

B Optimum Control Systems

In recent years much attention has been focused upon optimizing the behavior of systems. A particular problem may concern maximizing the range of a rocket, maximizing the profit of a business, minimizing the error in estimation of position of an object, minimizing the energy or cost required to achieve some required terminal state, or any of a vast variety of similar statements. The search for the control which attains the desired objective while minimizing (or maximizing) a defined system criterion constitutes the fundamental problem of optimization theory.

The design sequence for an optimal control system has five basic steps:

1) Modeling of plant;
2) Establishment of constraints;
3) Selection of the performance index;
4) Minimization of the performance index;
5) Determination of the controller configuration.

For continuous, deterministic, and lumped parameter plant the end results of the first step are the state and output equations

$$\dot{x}(t) = f(x, u, t)$$
$$c(t) = g(x, u, t)$$

Obviously, these equations must adequately describe the plant. Plant modeling is not a trivial task, nor is the selection of the best state, control, and output variables. Complete controllability in the mathematical sense is a necessary but not sufficient condition for the existence of an optimal control.[1] In addition, if control is to be feedback, the plant must be completely observable. Remember that observability does not guarantee physical measurability.

The constraints of the second step are physical constraints imposed on the state and control variables as well as any other physical constraints that might affect the performance of the plant. The lack of proper constraints leads to physically unrealistic and ridiculous solutions. State constraints may be equality constraints whereby the initial and/or final states are specified or inequality constraints restricting the range of permissible values of specific state variables. Control and other constraints are generally inequality constraints; e.g. the maximum acceleration of the plant or fuel used must be less than a specified value. State trajectories and controls that satisfy all the constraints are called admissible trajectories and admissible controls and are candidates for further investigation. Those trajectories and controls that do not satisfy the constraints are termed inadmissible and are rejected.

The formulation of the performance index (PI) may well be the most critical and difficult step of all. The performance index is an attempt to express quantitatively the deviations in plant performance from an ideal express performance. The performance index is written as the functional:

$$J = J_1[\boldsymbol{x}(t_f)] + \int_{t_0}^{t} J_2[\boldsymbol{x}, \boldsymbol{u}, t] dt$$

where t_0 and t_f are the initial and final times. J_1 is evaluated at the final state and is not necessarily specified. J_2, the cost or loss function, is evaluated over the entire control interval ($t_0 \sim t_f$). Weighting factors are used to assign relative importance to various terms in J_2 that describe the deviations from the ideal performance. Each admissible control will yield a single value for a given performance index. This value of J can be used as a figure of merit in comparing competing controls; the smaller the value of J, the better the control, mathematically that is.[2] The admissible control that produces an admissible trajectory and minimizes the value of the performance index is called the *optimal control* and given the symbol $\boldsymbol{u}^*$. the trajectory is called the *optimal trajectory* and symbolized by $\boldsymbol{x}^*$.

If an optimal control, which is a function of time, is specified only for a particular initial state

$$\boldsymbol{u}^*(t) = f_1[\boldsymbol{x}(t_0), t]$$

then the control is open-loop. If, however, the optimal control is a function of both time and the state

$$\boldsymbol{u}^*(t) = f_2[\boldsymbol{x}(t), t]$$

then the control is closed loop with state-variable feedback and we call $\boldsymbol{u}^*$ the *optimal control law*. For example,

$$\boldsymbol{u}^*(t) = \boldsymbol{E}\boldsymbol{x}(t)$$

where $\boldsymbol{E}$ is a constant matrix, the optimal control law is a stationary linear feedback of state variables.

An optimal control need not exist for a given performance index nor be unique if it does exist. In addition, changing the performance index will result in a different optimal control and trajectory. The designer must be capable of interpreting and choosing from several optimal controls in terms of physical and practical considerations.

With the exception of some special cases, minimization of the performance index to obtain an optimal control does not yield analytical solutions and the computational effort is high. The two basic methods for minimization are dynamic programming and a calculus of variations approach known as Pontryagin's minimum principle.

Dynamic programming is a multistage decision process that searches directly for minimum of the performance index, which is written as a recurrence equation. The distinguishing characteristic of dynamic programming is the use the principle of optimality to reduce the area of search sufficiently to make direct search feasible. The reduction of the search area by principle of optimality and the state and control constraints is illustrated in Fig. 2-6B-1. As applied to the optimal control problem the principle of optimality states that a control that is optimal over a complete interval must be optimal over every subinterval. The computational procedure is to start at a final state and work backward in stages to an initial state, finding the optimal control for each stage in turn. If the optimal control is found for N stages, it contains the optimal control for any lesser number of stages ending at the same final stage. This is known as the principle of embedding.

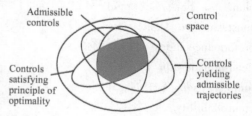

Fig. 2-6B-1 Illustration of optimum control

Dynamic programming yields an optimal control law, but not in analytical form. If analytical approximation is not possible, the tabulated control values must be stored and be accessible when needed. Dynamic programming basically uses difference equations; they may be an approximation of the differential equations of a continuous plant or may represent an actual sampled-data plant. The major disadvantage of dynamic programming is the requirement for rapid-access storage with a large capacity that increases rapidly as the dimensionality of the plant increases.

With the exception of some linear plants, the variational approach of Pontryagin's minimum principle leads to a nonlinear two-point boundary-value problem that must be solved by numerical methods.[3] Three applicable techniques are the method of steepest descent, the variation of extremals, and quasi linearization. A fourth numerical technique known as gradient projection minimizes a function of several variables subject to constraints. In contrast to dynamic programming the inequality constraints complicate the solution using the variational approach. Furthermore, the numerical solutions for the optimal control are in open-loop form.

Open-loop optimal control may be acceptable, particularly when unknown or unpredictable disturbance are absent or small. If closed-loop control is needed or desired, the open-loop optimal solution obtained by the variational approach can be used to limit the area to be searched with dynamic programming. This combination of the two approaches makes it possible to obtain an optimal control law with an appreciable reduction in the requirements for computer storage and in computation time.

When the designer has selected the best optimal control, he still faces the problem of constructing a practical controller to generate this particular control vector. As a direct synthesis approach, optimal control system has some drawbacks, such as, the limited number of analytical solutions, difficult implementing with hardware. So sometimes the designer may be willing to accept a low quality of performance in exchange for a simple and less expensive controller, which is called suboptimal controller.

WORDS AND TERMS

optimal control   最优控制
plant   *n.* 机器，设备被控对象
constraint   *n.* 约束条件
performance index   性能指标
deterministic   *adj.* 确定的
lumped   *adj.* 集中的
controllabillity   *n.* 能控性
observability   *n.* 能观性
admissible   *adj.* 可采纳的，允许的
trajectory   *n.* 轨迹，弹道
constant matrix   常数矩阵
multistage   *adj.* 多级的，多步的
quasi   *adj.* 近似的
linearization   *n.* 线性化
variational   *adj.* 变化的，变种的
suboptimal   *adj.* 次优的

NOTES

[1] Complete controllabillity in the mathematical sense is a necessary but not sufficient condition for the existence of an optimal control.

在数学意义上，完全能控是最优控制存在的必要条件但非充分条件。

[2] ... mathematically that is.

……数学上是如此的。

[3] With the exception of some linear plants, the variational approach of Pontryagin's minimum principle leads to a nonlinear two-point boundary-value problem that must be solved by numerical methods.

除去一些线性控制对象外，庞垂根最小原理的推论引出了一个必须用数字方法求解的非线性两点边界值问题。

C 语法成分的转换

在翻译中，虽然总体上可以保持译文与原文中的主语、谓语、宾语等句子成分的对应，但由于英汉两种语系在思维方式和表达习惯上的差别，很多情况下要对原句做语法成分的必要转换，以克服翻译中遇到的表达上的障碍。不论是前几节中介绍的意译、转译或词性转换的方法，还是下面将讨论的语法成分转换的方法，都是建立在对原文透彻理解的基础上对其含义的灵活表达，而意译、词性转换在多数情况下必然引起语法成分的变化。

1. 在被动句中，主语作为动作的接受者而常常被转换成宾语来翻译，而宾语则按主语翻译。例如：

Considerable use has been made of these data.

这些资料得到了充分的利用。

2. 当 care，need，attention，improvement，emphasis 等名词或名词化结构做主语时，可采取把主语译为谓语的方式。例如：

Care should be taken to protect the instrument from dust and dump.

应当注意保护仪器免受灰尘和潮湿。

3. 某些英语谓语动词如 act，characterize，feature，behave，relate，conduct 等在翻译成

汉语时往往要将词性转换为名词而在句中做主语。例如：

Fuzzy control acts differently from conventional PID control.
模糊控制的作用不同于传统的比例积分微分控制。
Copper conducts electricity very well.
铜的导电性能良好。

4. 为了突出原句中的定语，可在翻译时将其转换为谓语或表语。例如：

There is a large amount of energy wasted due to the fraction of commutator.
换向器引起的摩擦损耗了大量的能量。
Many factors enter into equipment reliability.
涉及设备可靠性的因素很多。定语形容词 many 转换成表语。

5. 对于英语中有 to have a voltage of..., to have a height of... 等结构时，如照字面直译为汉语，会十分生硬或绕口，因此在翻译时通常把"电压""高度"等译作主语，同时对句中其他语法成分做相应的转换。例如：

The output of transformer has a voltage of 10 kilovoltages.
变压器的输出电压为 10kV。
Solid state disk provides a relatively high access speed than traditional hard disk.
固态硬盘的存取速度比传统硬盘相对要快。
USB flash disk has a much more data volume than that of CD.
闪存盘的数据存储量比光盘大得多。

原文的写作是为了表达某一思路或阐述某一事物的性质、过程，而翻译原文的目的同样是表达这一思路或阐述这一事物的性质、过程，因此不必过分拘泥于原文的表达方式，片面追求忠实原文而忽视英汉两种语言的内在差异；换句话说，根据实际情况需要，采取意译、转译、词性转换、语法成分转换的处理是极为必要的。

UNIT 7

A Conventional and Intelligent Control

The term "conventional (or traditional) control" is used here to refer to the theories and methods that were developed in the past decades to control dynamical systems, the behavior of which is primarily described by differential and difference equations.[1] Note that this mathematical framework may not be general enough in certain cases. In fact it is well known that there are control problems that cannot be adequately described in a differential/difference equations framework. Examples include discrete event manufacturing and communication systems, the study of which has led to the use of automata and queuing theories in the control of systems.

In the minds of many people, particularly outside the control area, the term "intelligent control" has come to mean some form of control using fuzzy and/or neural network methodologies. This perception has been reinforced by a number of articles and interviews mainly in the nonscientific literature. However intelligent control does not restrict itself only to those methodologies. In fact, according to some definitions of intelligent control not all neural/fuzzy controllers would be considered intelligent. The fact is that there are problems of control which cannot be formulated and studied in the conventional differential/difference equation mathematical framework. To address these problems in a systematic way, a number of methods have been developed that are collectively known as intelligent control methodologies.

There are significant differences between conventional and intelligent control and some of them are described below. It is worth remembering at this point that intelligent control uses conventional control methods to solve "lower level" control problems and that conventional control is included in the area of intelligent control. Intelligent control attempts to build upon and enhance the conventional control methodologies to solve new challenging control problems.

The word control in "intelligent control" has different, more general meaning than the word control in "conventional control". First, the processes of interest are more general and may be described, for example by either discrete event system models or differential/difference equation models or both. This has led to the development of theories for hybrid control systems, that study the control of continuous-state dynamic processes by discrete-state sequential machines. In addition to the more general processes considered in intelligent control, the control objectives can also be more general. For example, "replace part A in satellite" can be the general task for the controller of a space robot arm; this is then decomposed into a number of subtasks, several of which may include for instance "follow a particular trajectory", which may be a problem that can be solved by conventional control methodologies. To attain such control goals for complex systems over a period of time, the controller has to cope with significant uncertainty that fixed feedback robust controllers or adaptive controllers cannot deal with. Since the goals are to be attained under large uncertainty, fault diagnosis and control reconfiguration, adaptation and learning are important considerations in intelligent controllers. It is also clear that task planning is an important area in intelligent control design. So the control problem in intelligent control is an enhanced version of the problem in conventional control. It is much more ambitious and general. It is not surprising then that these increased control demands require methods that are not typically used in conventional control. The area of intelligent control is in fact interdisciplinary, and it attempts to combine and extend theories and methods from areas such as control, computer science and operations research to attain demanding control goals in complex systems.

Note that the theories and methodologies from the areas of operations research and computer science cannot, in general be used directly to solve control problems, as they were developed to address different needs; they must first be enhanced and new methodologies need to be developed in combination with conventional control methodologies, before controllers for very complex dynamical systems can be designed in systematic ways. Also traditional control concepts such as stability may have to be redefined when, for example, the process to be controlled is described by discrete event system models; and this issue is being addressed in the literature. Concepts such as reachability and deadlock developed in operations research and computer science are useful in intelligent control, when studying planning systems. Rigorous mathematical frameworks, based for example on predicate calculus are being used to study such questions. However, in order to address control issues, these mathematical frameworks may not be convenient and they must be enhanced or new ones must be developed to appropriately address these problems. This is not surprising as the techniques from computer science and operations research are primarily analysis tools developed for nondynamic systems, while in control, synthesis techniques to design real-time feedback control laws for dynamic systems are mainly of interest. In view of this discussion, it should be clear that intelligent control research, which is mainly driven by applications has a very important and challenging theoretical component. Significant theoretical strides must be made to address the open questions and control theorists are invited to address these problems. The problems are nontrivial, but the pay-off is very high indeed.

As it was mentioned above, the word control in intelligent control has a more general meaning than in conventional control; in fact it is closer to the way the term control is used in everyday language. Because intelligent control addresses more general control problems that also include the problems addressed by conventional control, it is rather difficult to come up with meaningful bench mark examples.[2] Intelligent control can address control problems that cannot be formulated in the language of conventional control. To illustrate, in a rolling steel mill, for example, while conventional controllers may include the speed (rpm) regulators of the steel rollers, in the intelligent control framework one may include in addition, fault diagnosis and alarm systems; and perhaps the problem of deciding on the set points of the regulators, that are based on the sequence of orders processed, selected based on economic decisions, maintenance schedules, availability of machines etc. All these factors have to be considered as they play a role in controlling the whole production process which is really the overall goal.

Another difference between intelligent and conventional control is in the separation between controller and the system to be controlled. In conventional control the system to be controlled, called the plant, typically is separate and distinct from the controller. The controller is designed by the control designer, while the plant is in general given and cannot be changed; note that recently attempts to coordinate system design and control have been reported in areas such as space structures and chemical processes, as many times certain design changes lead to systems that are much easier to control. In intelligent control problems there may not be a clear separation of the plant and the controller; the control laws may be imbedded and be part of the system to be controlled. This opens new opportunities and challenges as it may be possible to affect the design of processes in a more systematic way.

Research areas relevant to intelligent control, in addition to conventional control include areas such as planning, learning, search algorithms, hybrid systems, fault diagnosis and reconfiguration, automata, Petri nets, neural nets and fuzzy logic. In addition, in order to control complex systems, one has to deal effectively with the computational complexity issue; this has been in the periphery of the interests of the researchers in conventional control, but now it is clear that computational complexity is a central issue, whenever one attempts to control complex systems.

It is appropriate at this point to briefly comment on the meaning of the word intelligent in "intelligent control". Note that the precise definition of "intelligence" has been eluding mankind for thousands of years.[3] More recently, this issue has been addressed by disciplines such as psychology, philosophy, biology and of course by artificial intelligence (AI); note that AI is defined to be the study of mental faculties through the use of computational models. No consensus has emerged as yet of what constitutes intelligence. The controversy surrounding the widely used IQ tests also points to the fact that we are well away from having understood these issues. In this report we do not even attempt to give general definitions of intelligence. Instead we introduce and discuss several characterizations of intelligent systems that appear to be useful when attempting to address some of the complex control problems mentioned above.

Some comments on the term "intelligent control" are now in order. Intelligent controllers are envisioned emulating human mental faculties such as adaptation and learning, planning under large uncertainty, coping with large amounts of data etc. in order to effectively control complex processes; and this is the justification for the use of the term intelligent in intelligent control, since these mental faculties are considered to be important attributes of human intelligence. Certainly the term intelligent control has been abused and misused in recent years by some, and this is of course unfortunate. Note however that this is not the first time, nor the last that terminology is used to serve one's purpose. Intelligent control is certainly a catchy term and it is used (and misused) with the same or greater abundance by some, as for example the term optimal has been used (or misused) by others; of course some of the most serious offenses involve the word "democracy"! For better or worse, the term intelligent control is used by many. An alternative term is "autonomous (intelligent) control". It emphasizes the fact that an intelligent controller typically aims to attain higher degrees of autonomy in accomplishing and even setting control goals, rather than stressing the (intelligent) methodology that achieves those goals. On the other hand, "intelligent control" is only a name that appears to be useful today. In the same way the "modern control" of the 60's has now become "conventional (or traditional) control", as it has become part of the mainstream, what is called intelligent control today may be called just "control" in the not so distant future. What is more important than the terminology used are the concepts and the methodology, and whether or not the control area and intelligent control will be able to meet the ever increasing control needs of our technological society. This is the true challenge.

I would like to finish this brief outline with an optimistic note; and there are many reasons for being optimistic. This is an excellent time indeed to be in the control area. We are currently expanding our horizons, we are setting ambitious goals, opening new vistas, introducing new challenges and we are having a glimpse of the future that looks exciting and very promising.

WORDS AND TERMS

framework n. 构架，结构	rigorous adj. 严密的，精确的
discrete adj. 离散的	synthesis n. 综合
fuzzy adj. 模糊的	periphery n. 外围
hybrid adj. 混合的	terminology n. 术语学
decompose v. 分解	mainstream n. 主流
trajectory n. 轨迹，轨道	vista n. 展望
diagnosis n. 诊断	queuing theory 排队论
interdisciplinary adj. 跨学科的	bench mark 基准点
reachability n. 能达到性	neural network 神经网络
deadlock n. 死锁，僵局	artificial intelligence 人工智能

NOTES

[1] The term "conventional (or traditional) control" is used here to refer to the theories and methods that were developed in the past decades to control dynamical systems, the behavior of which is primarily described by differential and difference equations.

"传统控制"这个术语是指在过去的几十年里发展起来的用于控制以微分和差分方程表述的动态系统的理论和方法。

[2] Because intelligent control addresses more general control problems that also include the problems addressed by conventional control, it is rather difficult to come up with meaningful bench mark examples.

由于智能控制解决了包含传统控制解决的问题在内的更多、更广泛的问题，所以提出有代表性的例子相当困难。

[3] Note that the precise definition of "intelligence" has been eluding mankind for thousands of years.

我们注意到"智能"的精确定义已经数千年不为人类所知了。

B Artificial Neural Networks

Artificial Neural Network is a system loosely modeled on the human brain. The field goes by many names, such as connectionism, parallel distributed processing, neuro-computing, natural intelligent systems, machine learning algorithms, and artificial neural networks. It is an attempt to simulate within specialized hardware or sophisticated software, the multiple layers of simple processing elements called neurons. Each neuron is linked to certain of its neighbors with varying coefficients of connectivity that represent the strengths of these connections. Learning is accomplished by adjusting these strengths to cause the overall network to output appropriate results.

Neuron

The most basic components of neural networks are modeled after the structure of the brain. Some neural network structures are not closely to the brain and some does not have a biological counterpart in the brain. However, neural networks have a strong similarity to the biological brain and therefore a great deal of the terminology is borrowed from neuroscience.

The most basic element of the human brain is a specific type of cell, which provides us with the abilities to remember, think, and apply previous experiences to our every action. These cells are known as neurons, each of these neurons can connect with up to 200,000 other neurons. The power of the brain comes from the numbers of these basic components and the multiple connections between them.

All natural neurons have four basic components, which are dendrites, soma, axon, and synapses. Basically, a biological neuron receives inputs from other sources, combines them in some way, performs a generally nonlinear operation on the result, and then output the final result. The Fig. 2-7B-1 shows a simplified biological neuron and the relationship of its four components.

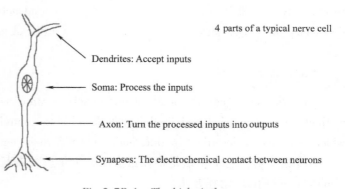

Fig. 2-7B-1 The biological neuron

The basic unit of neural networks, the artificial neurons, simulates the four basic functions of natural neurons. Artificial neurons are much simpler than the biological neuron; the Fig. 2-7B-2 shows the basic structure of an artificial neuron.

Note that various inputs to the network are represented by the mathematical symbol, $x(n)$. Each of these inputs are multiplied by a connection weight, these weights are represented by $w(n)$. In the simplest case, these products are simply summed, fed through a transfer function to generate a result, and then output.

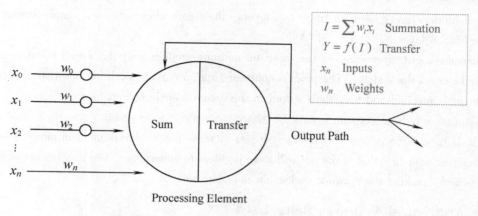

Fig. 2-7B-2 The basic structure of an artificial neuron

Even though all artificial neural networks are constructed from this basic building block, the fundamentals may vary in these building blocks and there are differences.

Layer

Biologically, neural networks are constructed in a three dimensional way from microscopic components. These neurons seem capable of nearly unrestricted interconnections. This is not true in any man-made network. Artificial neural networks are the simple clustering of the primitive artificial neurons. This clustering occurs by creating layers, which are then connected to one another. How these layers connect may also vary. Basically, all artificial neural networks have a similar structure of topology. Some of the neurons interface the real world to receive its inputs and other neurons provide the real world with the network's outputs. All the rest of the neurons are hidden from view.

As the Fig. 2-7B-3 shows, the neurons are grouped into layers. The input layer consists of neurons that receive input from the external environment. The output layer consists of neurons that communicate the output of the system to the user or external environment. There are usually a

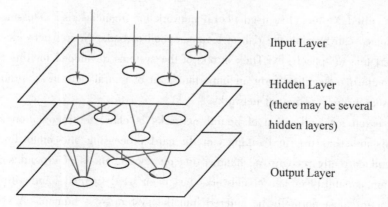

Fig. 2-7B-3 Three layers of neural networks

number of hidden layers between these two layers; the figure above shows a simple structure with only one hidden layer.

When the input layer receives the input its neurons produce output, which becomes input to the other layers of the system. The process continues until a certain condition is satisfied or until the output layer is invoked and fires their output to the external environment.

The number of hidden neurons must be determined so that the network performs its best, one of the methods used often is trial and error.[1] If you increase the hidden number of neurons too much you will get an over fit, that is the net will have problem to generalize. The training set of data will be memorized, making the network useless on new data sets.

Where Are Neural Networks Being Used?

Neural networks are performing successfully where other methods do not, recognizing and matching complicated, vague, or incomplete patterns.[2] Neural networks have been applied in solving a wide variety of problems.

The most common use for neural networks is to project what will most likely happen. There are many areas where prediction can help in setting priorities. For example, the emergency room at a hospital can be a hectic place, to know who needs the most critical help can enable a more successful operation. Basically, all organizations must establish priorities, which govern the allocation of their resources. Neural networks have been used as a mechanism of knowledge acquisition for expert system in stock market forecasting with astonishingly accurate results. Neural networks have also been used for bankruptcy prediction for credit card institutions.

Although one may apply neural network systems for interpretation, prediction, diagnosis, planning, monitoring, debugging, repair, instruction, and control, the most successful applications of neural networks are in categorization and pattern recognition. Such a system classifies the object under investigation (e. g. an illness, a pattern, a picture, a chemical compound, a word, the financial profile of a customer) as one of numerous possible categories that, in return, may trigger the recommendation of an action (such as a treatment plan or a financial plan).

A company called Nestor, has used neural network for financial risk assessment for mortgage insurance decisions, categorizing the risk of loans as good or bad. Neural networks have also been applied to convert text to speech; NETtalk is one of the systems developed for this purpose. Image processing and pattern recognition form an important area of neural networks, probably one of the most actively research areas of neural networks.

Another research for application of neural networks is character recognition and handwriting recognition. This area has use in banking, credit card processing and other financial services, where reading and correctly recognizing handwriting on documents is of crucial significance. The pattern recognition capability of neural networks has been used to read handwriting in processing checks. The amount must normally be entered into the system by a human. A system that could automate this task would expedite check processing and reduce errors. One such system has been developed by HNC (Hecht-Nielsen Co.) for BankTec.

One of the best known applications is the bomb detector installed in some U.S. airports. This device called SNOOPE, determines the presence of certain compounds from the chemical configurations of their components.

In a document from International Joint Conference, one can find reports on using neural networks in areas ranging from robotics, speech, signal processing, vision, character recognition to musical composition, detection of heart malfunction and epilepsy, fish detection and classification, optimization, and scheduling. One may take under consideration that most of the reported applications are still in research stage.

Basically, most applications of neural networks fall into the following five categories:

Prediction

Use input values to predict some output, e.g. pick the best stocks in the market, predict weather, identify people with cancer risk.

Classification

Use input values to determine the classification, e.g. the input the letter A, is the blob of the video data a plane and what kind of plane it is.

Data Association

Like classification but it also recognizes data that contain errors. e.g. not only identify the characters that are scanned but identify when the scanner is not working properly.

Data Conceptualization

Analyze the inputs so that grouping relationships can be inferred, e.g. extract from a database of the names of those who are most likely to by a particular product.

Data Filtering

Smooth an input signal. e.g. take the noise out of a telephone signal.

WORDS AND TERMS

sophisticated   *adj.* 非常复杂、精密或尖端的
neuron   *n.* 神经元
dendrite   *n.* 树突
soma   *n.* 体细胞
axon   *n.* 轴突
synapse   *n.* 神经键
topology   *n.* 拓扑

bi-directional   *adj.* 双向的
hierarchical   *adj.* 分级的
resonance   *n.* 共振,共鸣
recurrent   *adj.* 再发生的,循环的
vague   *adj.* 含糊的,不清楚的
debugging   *n.* 调试
data filtering   数字滤波

NOTES

[1] The number of hidden neurons must be determined so that the network performs its best, one of the methods used often is trial and error.

必须确定隐含神经元的数量以使网络性能最优,一种常用的方法是试凑法。

[2] Neural networks are performing successfully where other methods do not, recognizing and matching complicated, vague, or incomplete patterns.

神经网络在识别和匹配复杂、含糊和不完整图案的问题上得到了成功的应用，而其他方法对此问题无能为力。

C 增 词 译 法

增词译法是指：为了使译文通顺达意而易于理解，在翻译时根据意义上、语法上、搭配上的需要而增补某些必要的词。翻译中，死守原文，一一对应查字典式的机器翻译会导致译文生硬而缺乏文采，妨碍了对原文的清晰表达。当然增词译法是有一定原则限制的。

1. 翻译英语省略句，可根据需要在译文中补充省略部分。例如：

The letter I stands for the current in amperes, E the electromotive force in voltages, and R the resistance in ohms.

字母 I 代表电流的安培数，E 代表电动势的伏特数，R 代表电阻的欧姆数。

2. 英语中有些被动句没有行为主体，翻译时可增添适当的主语使译文通顺流畅，如"人们""有人""大家""我们"等。例如：

Rubber is found a good material for the insulation of cable.

人们发现，橡胶是一种用于电缆绝缘的理想材料。

3. 原文中虽然没有某些关联词语，而翻译成汉语时可从逻辑出发增添上这些关联词语。例如：

If the system gain is increased beyond a certain limit, the system will become unstable.

如果系统增益超过了一定界限，系统就会不稳定。增译了关联词"就"。

The pointer of ampere meter moves from zero to three and goes back to one.

安培表的指针先从 0 转到 3，然后又回到 1。增译了关联词"先……然后又"。

4. 英语可数名词没有汉语中的量词修饰，翻译时应根据汉语搭配而增加量词；复数名词前可增补"许多""一些""各种"等词，使其复数意义更明确。例如：

The three computers in this section are infected by virus.

这个部门有 3 台计算机感染了病毒。增译了量词"台"。

In spite of difficulties, the problem was overcome eventually.

尽管有各种困难，问题最终还是解决了。增译了表复数的词"各种"。

5. 根据需要适当增加一些概括性词。例如：

Based upon the relationship between magnetism and electricity are motors and generators.

电动机和发电机就是以磁和电这二者之间的关系为基础的。增译了"这二者"。

6. 当英语中某些词单独译出意思不明确时，可增加适当的说明性词语以使译文意思明确。例如：

A new kind of computer—cheap, small, light is attracting increasing attention.

一种新型计算机正在引起人们越来越多的关注——它（造价）低，（体积）小，（重量）轻。

Speed and reliability are the chief advantages of the IPC.

速度（快）、可靠性（高）是工控机的主要优点。

7. 当含有动作意义的抽象名词表示具体概念时，通常可通过增加说明性词缀使译文具体化，常用的说明性词缀主要有：作用、现象、效应、方法、情况、性、过程等。例如：

Oxidation will make ground-wire of lightning arrester rusty.

氧化（作用）会使避雷装置的接地线生锈。

Integration can get rid of the static error in closed-loop system.

积分（作用）可以消除闭环系统中的静差（现象）。

PART 3

Computer Control Technology

UNIT 1

A Computer Structure and Function

This section introduces the internal architecture of a computer, describes how instructions are stored and interpreted, and explains how the instruction execution cycle is broken down into its various components.[1]

At the most basic level, a computer simply executes binary-coded instructions stored in memory. These instructions act upon binary-coded data to produce binary-coded results. For a general-purpose programmable computer, four necessary elements are the memory, central processing unit (CPU, or simply processor), an external processor bus, and an input/output system as indicated in Fig. 3-1A-1.

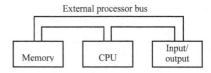

Fig. 3-1A-1 Basic elements of a computer

The memory stores instructions and data.

The CPU reads and interprets the instructions, reads the data required by each instruction, executes the action required by the instruction, and stores the results back in memory. One of the actions that is required of the CPU is to read data from or write data to an external device. This is carried out using the input/output system.

The external processor bus is a set of electric conductors that carries data, address, and control information between the other computer elements.

The Memory

The memory of a computer consists of a set of sequentially numbered locations. Each location is a register in which binary information can be stored. The "number" of a location is called its address. The lowest address is 0. The manufacturer defines a word length for the processor that is an integral number of locations long. In each word the bits can represent either data or instructions. For the Intel 8086/87 and Motorola MC68000 microprocessors, a word is 16 bits long, but each memory location has only 8 bits and thus two 8-bit locations must be accessed to obtain each data word.

In order to use the contents of memory, the processor must fetch the contents of the right location. To carry out a fetch, the processor places (enables) the binary-coded address of the desired location onto the address lines of the external processor bus. The memory then allows the contents of the addressed memory location to be read by the processor. The process of fetching the contents of a memory location does not alter the contents of that location.

Instructions in Memory Instructions stored in memory are fetched by the CPU and unless program branches occur, they are executed in the sequence they appear in memory. An instruction written as a binary pattern is called a machine-language instruction. One way to achieve meaningful patterns is to divide up the bits into fields as indicated in Fig. 3-1A-2, with each field containing a code for a different type of information.[2]

Each instruction in our simple computer can be divided up into four fields of 4 bits each. Each instruction can contain *operation code* (or opcode, each instruction has a unique opcode), *operand address*, *immediate operands*, *Branch address*.

In a real instruction set there are many more instructions. There is also a much larger number of memory locations in which to store instructions and data. In order to increase the number of memory locations, the address fields and hence the instructions must be longer than 16 bits if we use the same approach. There are a number of ways to increase the addressing range of the microprocessor without increasing the instruction length: variable instruction fields, multiword instructions, multiple addressing modes, variable instruction length. We will not discuss them in detail.

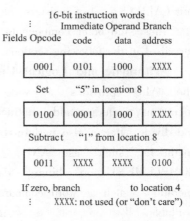

Fig. 3-1A-2 Arrangement of program and data in memory

Data in Memory data is information that is represented in memory as a code. For efficient use of the memory space and processing time, most computers provide the capability of manipulating data of different lengths and representations in memory. The various different representations recognized by the processor are called its data types. The data types normally used are: bit, binary-coded decimal digit (4-bit nibble, BCD), byte (8 bits), word (2 bytes), double word (4 bytes).

Some processors provide instructions that manipulate other data types such as single-precision floating-point data types (32 bits) and double-precision floating-point data types (64 bits). There is another type of data—character data. It is also usually represented in 8 bits. Each computer terminal key and key combination (such as shift and control functions) on a standard terminal keyboard has a 7-bit code defined by the American Standard Code for Information Interchange (ASCII).

Types of Memory In the applications of digital control system, we also concerned with the characteristics of different memory techniques. For primary memory, we need it to be stored information temporarily and to be written and got information from successive or from widely different locations. This type memory is called random-access memory (RAM). In some case we do not want the information in memory to be lost. So we are willing to use special techniques to write into memory. If writing is accomplished only once by physically changing connections, the memory is called a read-only memory (ROM). If the interconnection pattern can be programmed to be set, the memory is called a programmable read-only memory (PROM). If rewriting can be accomplished when it is necessary, we have an erasable programmable read-only memory (EPROM). An electronically erasable PROM is abbreviated EEPROM.

The CPU

The CPU's job is to fetch instructions from memory and execute these instructions. The structure of the CPU is shown in Fig. 3-1A-3. It has four main components: an arithmetic and logical unit (ALU), a set of registers, an internal processor bus and controller.

These and other components of the CPU and their participation in the instruction cycle are described in the following sections.

Arithmetic and Logical Unit (ALU) The ALU provides a wide arithmetic operations, including add, subtract, multiply, and divide. It can also perform boolean logic operations such as AND, OR, and COMPLEMENT on binary data. Other operations, such as word compares, are also available. The majority of computer tasks involve the ALU, but a great amount of data movement is required in order to make use of the ALU instructions. [3]

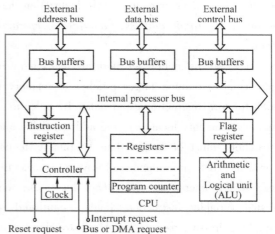

Fig. 3-1A-3 The structure of the CPU

Registers A set of registers inside the CPU is used to store information.

Instruction register When an instruction is fetched, it is copied into the instruction register, where it is decoded. Decoding means that the operation code is examined and used to determine the steps of the execution sequence.

Programmer's model of the CPU The collection of registers that can be examined or modified by a programmer is called the programmer's model of the CPU. The only registers that can be manipulated by the instruction set, or are visibly affected by hardware inputs or the results of operations upon data, are the registers represented in the model.

Flag register The execution sequence is determined not only by the instruction but also by the results of the previous instructions. For example, if an addition is carried out in the ALU, data on the result of the addition (whether the result is positive, negative, or zero, for example) is stored in what is known as a flag register, status register, or condition register. If the next instruction is a conditional branch instruction, the flag word is tested in that instruction to determine if a branch is required.

Program counter (instruction pointer) The address of the next instruction is located in a register called the program counter.

Data registers When an instruction uses the registers to store data, the reference to the register in the instruction is called register addressing. The reasons of making use of the internal registers to store data are that they can make the instructions shorter and make execution faster.

Address registers The internal registers can also be used for the storage of addresses of data in memory data. In such a case, the instruction word contains a register number (i.e. a register address). In the register is contained the address of memory data to be used in the instruction. This form of addressing is called register indirect addressing. The contents of the register are said to point to the data in memory.

Internal Processor Bus The internal processor bus moves data between internal registers. A bus is a set of closely grouped electric conductors that transfers data, address, and control information between functional blocks of the CPU. Data from a source register can be passed to a destination register when both are enabled onto (connected to) the bus.

Controller The controller provides the proper sequence of control signals for each instruction in a program cycle to be fetched from memory. A total program cycle comprises many instruction cycles, each instruction cycle can be divided up into its component machine cycles, and each machine cycle comprises a number of clock cycles.

In order to fetch an instruction, for example illustrated in Fig. 3-1A-4, the address in the program counter is placed on the address lines of the external bus (AB) at the onset of clock cycle C_1. Simultaneously, using a code on the control lines of the bus (CB), the CPU informs all devices attached to the bus that an "opcode fetch" machine cycle is being executed by the CPU. The memory allows the memory address to select the memory location

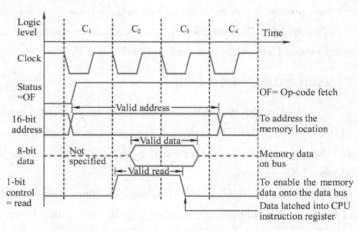

Fig. 3-1A-4 A timing diagram for "operation-code fetch"

containing the instruction. At C_2 the controller places a "read" command onto the control bus which allows the memory data to be placed onto the data bus. The controller then gates the data into the instruction register and removes the read command from the control bus in C_3. At C_4, The controller removes the address from the address bus and begins to decode the operation-code portion of the instruction to see what steps are required for execution. The decoding operation may take several more clock cycles at the end of which the "opcode fetch" machine cycle.

External attention requests It is often necessary to stop the normal instruction processing sequence. One type of external attention request is the *reset request*. In the case of an unrecoverable error, a computer system may be required to reset itself. This would have the effect of initializing all important registers in the system and starting instruction execution from a standard memory location-usually location 0.

An input that is more commonly activated during the normal course of events is the *interrupt request*. An interrupt request signal from an external device can cause the CPU to immediately execute a service subroutine which carries out the necessary actions. After completing the service subroutine, the processor will continue the task from which it was originally interrupted.

The third type of input is the *bus request*, or *direct memory access* (DMA) *request*. It is possible to have a terminal interface that stores up all the characters in a line of text until it receives a "carriage return." Then the interface requests the use of the system bus, at which time the complete line of data is transferred to memory as fast as possible. In this way the processor simply becomes inactive until the transfer is completed.

Buses

The bus is the most important communication system in a computer system. Under control of the CPU, a data source device and a data destination device are "enabled" onto (equivalent to being connected to) the bus wires for a short transmission.

External processor bus   The internal processor bus described in Sec. is connected to the external processor bus by a set of bus buffers located on the microprocessor integrated circuit.

System bus   The microcomputer board can communicate with other boards by connecting its bus to an external system bus through a connector.

Computer Input and Output

A set of registers external to the CPU is associated with what is known as the input/output (I/O) system. The I/O system is connected to the external processor bus using control, address, and data buses through an I/O registers in an interface. There are basically two ways that are used to address I/O registers.

In the first method, called I/O-mapped input/output, the operation code itself has special I/O instructions that address a numbered register in the interface called an I/O port.

The second method of addressing I/O registers gives the I/O ports addresses that lie within the memory address range of the CPU. This is called memory-mapped I/O. Of course there must not be any memory locations at the same address as I/O locations.

One of the benefits of the memory-mapped approach is that the full range of memory addressing modes is available to the addressing of I/O registers.

WORDS AND TERMS

architecture   *n.* 体系结构
instruction set   指令集
binary-coded   *adj.* 二进制编码的
central processing unit (CPU)   中央处理器
processor   *n.* 处理器
location   *n.* （存储）单元
word length   字长
access   *v.* 存取，接近

fetch   *v.*, *n.* 取来
field   *n.* 域，字段
opcode   *n.* 操作码
operand *n.* 操作数
address *n.* 寻址
single-precision   *adj.* 单精度的
floating-point   *adj.* 浮点的
terminal   *n.* 终端

complement   *v.* 补充，求补
decode   *v.* 解码，译码
request   *n.* 请求
inactive   *n.* 不活动，停止
I/O-mapped   *adj.* 输入/输出映射的（单独编址）
memory-mapped   *adj.* 存储器映射的（统一编址）

NOTES

[1] ... how the instruction execution cycle is broken down into its various components.
……指令执行周期怎样分解成不同的部分。

[2] One way to achieve meaningful patterns is to divide up the bits into fields...
一种得到（指令）有效形式的方法是将（这些）位分成段……

[3] The majority of computer tasks involve the ALU, but a great amount of data movement is required in order to make use of the ALU instructions.
计算机的大多数工作涉及 ALU（逻辑运算单元），但为了使用 ALU 指令，需要传送大量的数据。

B Fundamentals of Computers and Networks

Organization of Computer System

A computer is a fast and accurate symbol manipulating system that is organized to accept, store, and process data and produce output results under the direction of a stored program of instructions. This section explains why a computer is a system and how a computer system is organized. Key elements in a computer system include input, processing, and output devices. Let's examine each component of the system in more detail.

Input Devices Computer systems use many devices for input purpose. Some INPUT DEVICES allow direct human/machine communication, while some first require data to be recorded on an input medium such as a magnetizable material.[1] Devices that read data magnetically recorded on specially coated plastic tapes or flexible or floppy plastic disks are popular. The keyboard of a workstation connected directly to (or ONLINE to) a computer is an example of a direct input device.[2] Additional direct input devices include the mouse, input pen, touch screen, and microphone. Regardless of the type of device used, all are components for interpretation and communication between people and computer systems.

Central Processing Unit The heart of any computer system is the central processing unit (CPU). There are three main sections found in the CPU of a typical personal computer system: the primary storage section, the arithmetic-logic section, and the control section. But these three sections aren't unique to personal computers. They are found in CPUs of all sizes.

Output Devices Like input units, output devices are instruments of interpretation and communication between humans and computer systems of all sizes. These devices take output results from the CPU in machine-coded form and convert them into a form that can be used (a) by people (e.g. a printed and/or displayed report) or (b) as machine input in another processing cycle.

In personal computer systems, display screen and desktop printers are popular output devices. Larger and faster printers, many online workstations, and magnetic tape drives are commonly found in larger systems.

The input/output and secondary storage units are sometimes called peripheral devices (or just peripherals). This terminology refers to the fact that although these devices are not a part of the CPU, they are often located near it. Besides, a computer system also includes buses, ROM(read only memory), RAM(random access memory), parallel port and serial port, hard disk, floppies and CD(compact disk) drive, and so on.

Operating System

Operating systems have developed over the past thirty years for two main purposes. First, they provide a convenient environment for the development and execution of programs. Second, operating systems attempt to schedule computational activities to ensure good performance of the computing system.

The operating system must ensure correct operation of the computer system. To prevent user programs from interfering with the proper operation of the system, the hardware was modified to create two modes: user mode and monitor mode. Various instructions (such as I/O instructions and halt instructions) are privileged and can only be executed in monitor mode. The memory in which the monitor resides must also be protected from modification by the user. A timer prevents infinite loops. Once these changes (dual mode, privileged instructions, memory protection, timer interrupt) have been made to the basic computer architecture, it is possible to write a correct operating system.

As we have stated, operating systems are normally unique to their manufacturers and the hardware in which they are run. Generally, when a new computer system is installed, operational software suitable to that hardware is purchased. Users want reliable operational software that can effectively support their processing activities.

Though operational software varies between manufacturers, it has similar characteristics. Modern hardware, because of its sophistication, requires that operating systems meet certain specific standards. For example, considering the present state of the field, an operating system must support some form of online processing. Functions normally associated with operational software are:

1) Job management;

2) Resource management;

3) Control of I/O operations;

4) Error recovery;
5) Memory management.

Networks

Communication between distributed communities of computers is required for many reasons. At a national level, for example, computers located in different parts of the country use public communication services to exchange electronic messages (mail) and to transfer files of information from one computer to another. Similarly, at a local level within, say, a single building,[3] distributed communities of computer-based workstations use local communication networks to access expensive shared resources—for example, printers and disks tapes and copiers, etc.—that are also managed by computers. Clearly, as the range of computer-based products and associated public and local communication networks proliferate, computer-to-computer communication will expand rapidly and ultimately dominate the field of distributed systems.

Although the physical separation of the communicating computers may vary considerably from one type of application to another, or, at the heart of any computer communication network is the data communication facility which, may be a PSDN, a private LAN or perhaps a number of such networks interconnected together. However, irrespective of the type of data communication facility, an amount of hardware and software is required within each attached computer to handle the appropriate network-dependent protocols. Typically, these are concerned with the establishment of a communication channel access the network and with the control of the flow of messages across this channel. The provision of such facilities is only part of the network requirements, however, since in many applications the communicating computers may be of different types. This means that they may use different programming languages and, more importantly, different forms of data representation interface between user (application) programs, normally referred to as application processes or APs, and the underlying communication services may be different. For example, one computer may be a small single-user computer, while another may be a large multi-user system.

WORDS AND TERMS

stored program 存储程序
input device 输入设备
output device 输出设备
primary storage (memory) 主存储器
secondary storage (memory) 辅助存储器
arithmetic-logic unit 算术逻辑部件
workstation n. 工作站
magnetic tape drive 磁带机
display screen 显示屏
peripheral n. 外围设备，外设

schedule v. 调度
electronic message (mail) 电子信息，电子邮件
local communication network 局域网
shared resource 共享资源
proliferate v. 激增
PSDN 公共交换数据网
private LAN 专用局域网
irrespective adj. 不考虑的
attached adj. 附加的

protocol   n. 协议   underlying   adj. 根本的

NOTES

[1] Some INPUT DEVICES allow direct human/machine communication, while some first require data to be recorded on an input medium such as a magnetizable material.

有些输入设备允许直接的人机对话，而有些输入设备则需要先将数据存储在诸如磁性材料等输入介质上。

由 while 连接的并列复合句，有转折的意思，译成"而""可是"等。

[2] ... connected directly to (or ONLINE to) a computer...

online 与"直接连接"的概念不同，前者是"联机"，此处是指工作站键盘与主机连在一起工作，后者是指"物理连接"。

[3] Similarly, at a local level within, say, a single building...

句中的 say 是"比如说"的意思。

C 减 词 译 法

上节介绍了增词译法在英译汉时的使用，同样减词译法在很多情况下也是必要的。首先，汉语不像英语那样大量地使用代词，英译汉时往往将代词略去不译；其次，冠词是汉语中没有的词类，在很多情况下都省略不译；另外，两种语言中很多其他的差异也需要在翻译时进行必要的词量增减处理。翻译中，过分受原文制约，不考虑汉语习惯而一个单词不落地全盘汉化，常常会使得译文不够精练，甚至言语啰唆。

1. 冠词的省译。例如：

The controlled output is the process quantity being controlled.

被控输出量是指被控的过程变量。定冠词"the"被省译。

A transistor is a device controlling the flow of electricity in a circuit.

晶体管是控制电路中的电流的器件。三个不定冠词"a"和一个定冠词"the"被省译。

2. 对代词省译或对指代的对象省译，以使译文简练清晰。例如：

Different metals differ in their conductivity.

不同的金属具有不同的导电性能。省译了物主代词"their"。

The current will blow the fuses when it reaches certain limit.

当电流达到一定界限时会使熔丝熔断。对"it"指代的对象"current"省译。

3. 对做形式主语或形式宾语的"it"省译。例如：

It seems that the result of this test will be very important to us.

看来，这个试验结果对我们非常重要。

The development of IC made it possible for electronic devices to become smaller and smaller.

集成电路的发展使电子器件可以做得越来越小。

4. 英语中引出从句的关系代词有时也可以省译。例如：

The energy that has to be supplied by the generator in order to overcome the opposition is transformed into heat within the conductor.

发电机克服电阻所必须损耗的能量在导体内转变成了热量。"that"省译。

The rate at which the electricity flows is measured in amperes.

电流的大小用安培度量。"which"省译。

5. 英语中的系动词 be, become 等可根据具体情况在翻译中省译。例如：

The electric power industry is primarily concerned with energy conversion and distribution.

电力工业同能源的转化和分配密切相关。

The importance of Mobile Internet becomes known to all day by day.

移动互联网的重要性日渐被所有人认识。

6. 英文中有些介词在汉语中没必要使用或用法不同，可根据汉语习惯表达。例如：

The notebook computer carries with it the storage battery of itself.

笔记本电脑自带电池。"with"和"of"省译。

USB flash disk is a good storage medium with a volume of several hundred gigabytes.

闪存盘是一种很好的存储介质，其存储量有几百吉字节。"with"省译。

UNIT 2

A Interfaces to External Signals and Devices

Introduction

Signals on a system bus occur in a very orderly sequence. Each signal is initiated by a particular sequence of events and it is terminated either after a fixed time or by another sequence of events. Bus events are directly initiated by the microprocessor in a system containing only memory and a microprocessor.

Autonomous external devices and signals having no bus-compatible signals and no temporal relationship with the system bus signals cannot be connected to the system bus directly. Communication with the system bus is accomplished via an input/output interface. The main purposes of an input/output interface are coordinating the transfer of data between the processor and independent outside devices, and converting data between a modality recognized by the processor. Other functions of interface may be to provide electric isolation, amplification, noise rejection, temporary data storage, or data format conversion.

There are several types of interfaces such as parallel input/output, digital-to-analog conversion, analog-to-digital conversion, real-time clock, direct memory access (DMA) are usually used. Here, we will limit our discussing to parallel input/output and A/D D/A.

Parallel Input Interfaces

The terms "input" and "output" usually refer to the CPU and output from the CPU. A simple standard dorm of interfacing is shown in Fig. 3-2A-1. The input interface is controlled only from the processor. Whenever the processor addresses the input port, data on the external data bus is "enabled" onto the processor data bus and that this data is available at the moment the data is enabled onto the bus by the CPU.

In this interface, there is a "tristate" buffer which when it is enabled will force the processor bus to have the same binary value as the external data lines.[1] When it is not enabled, the buffer output goes into a high impedance "third" state which effectively removes the data lines from the bus, allowing other data to use the system bus.

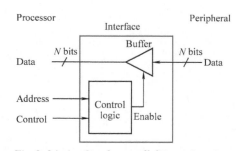

Fig. 3-2A-1 Simple parrallel input interface

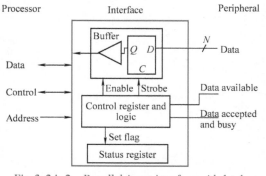

Fig. 3-2A-2 Parrallel input interface with latch

In some cases, the use of *handshaking* (sequence used for request, permission, and transfer) signals shown in Fig. 3-2A-2 is necessary. When the data is available, the peripheral negates the data-available line, and data is then strobed into the interface. At about the same time, a "ready" flag is set in the status register to indicate to the CPU that data is available. In order to know this, the CPU must be continually "polling" the interface (reading the status register), and finally latched the data.

The input task can also be carried out using DMA (direct memory access) I/O.

Parallel Output Interfaces

One difference between input and output interfaces is that an output interface must have a data register, since the processor data is constant for only a very short time on the CPU data bus. Fig. 3-2A-3 shows a simple form of interface that accepts data whenever the CPU issues the correct address of the interface data register and performs a "write" operation. The data is constantly available to the external world after latching.

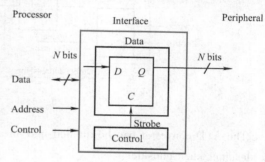

Fig. 3-2A-3 Simple output

Digital-to-Analog Conversion Interface

A digital-to-analog (D/A) converter as indicated in Fig. 3-2A-4 is used to produce an analog voltage or current that is proportional to a digital binary number over a given "full-scale" range.

Digital Representation of Analog Voltage
For a straight binary representation, the output voltage for any digital value can be found from

$$U_{OUT} = U_{FS}(b_7 2^{-1} + \cdots + b_0 2^{-8})$$

For the offset binary, A bipolar output can be obtained for voltages between, for example, $-1V$ and $+1V$.

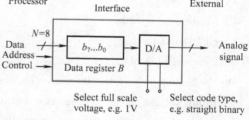

Fig. 3-2A-4 Digital-to-analog conversion

Twos complement Another way of representing at polar output is to use twos complement numbers. If the most-significant bit (MSB) is considered to be a sign, the most-positive number becomes 01111111, and the most-negative number becomes 10000000.

Other possible representations are binary-coded decimal and sign-plus magnitude. There may also be several jumper-selectable analog voltage ranges which the board can produce. The most common voltages are 5/10V, 5/12V, or 10/24V.

In addition to the digital representation, and the full-scale voltages available, the specifications of a D/A converter should at least give its Resolution; Slew rate; Settling time; Linearity; Temperature coefficients of gain and offset.

Analog-to-Aigital Conversion Interface

An analog signal acquisition system is shown in Fig. 3-2A-5. Its purpose is to create a digital representation for the voltage on one of the N analog input channels at a particular instant of time, called the sampling time. As in the D/A converter, the desired digital representation is a design parameter that may be different for different available A/D interfaces (binary offset binary, twos complement.

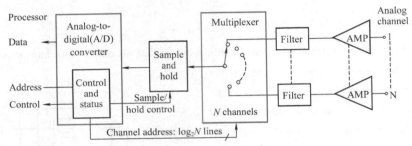

Fig. 3-2A-5 Analog signal acquisition system

The A/D converter can be divided into an analog side and a digital side. On the digital side, the designer must consider:

1) The integrated circuit technology;
2) Logic levels and tristate capability;
3) Resolution;
4) Conversion speed;
5) CPU handshaking;
6) External controls.

On the analog side of the interface the designer must be concerned about:

1) Input voltage range (difference between the analog voltage that produces the maximum digital value and the analog voltage that produces the maximum digital value);
2) The various error sources;
3) The equivalent input noise generated in the analog portions of the A/D.

Just as in the D/A converter, there is a possible gain and offset error and linearity errors. The temperature coefficients of these errors are also quite important, and their contribution to the total error should be calculated over the temperature range.

Successive Approximation A/D There is a wide variety of different techniques used in the construction of A/D converters. The most common is the successive approximation converter. It has the advantage of both moderate speed and moderate resolution.

It functions by first comparing the input voltage with a test voltage set equal to one-half the full scale A/D reference voltage. The test voltage is obtained by using a calibrated D/A converter. If the input voltage is greater than one-half full scale after the first comparison is made, the most-significant output bit is set. If the input voltage is less than one-half full scale, the one-half full-scale voltage is removed from the test voltage, otherwise the voltage remains the same.

Next, a one-fourth full-scale reference voltage is then added to the test voltage and the input voltage and test voltage are again compared. If the input voltage exceeds the test voltage, the next most-significant bit is set and the test voltage, the second bit is reset (set to 0), and the one-fourth full-scale reference component is removed.

The process is repeated with successively smaller binary-weighted voltages until the least-significant bit (LSB) has been tested.

Dual Slop A/D Toward the high-accuracy end of the performance spectrum lies the dual slope converter. In this converter, an unknown positive (constant) input voltage U_i is applied to an electronic integrated from zero volts for a fixed time T, producing a positive-going output voltage ramp proportional to U_iT (the integral of a constant voltage over the time T). U_i is then removed and a known negative constant reference voltage U is integrated which produces a ramp down.[2] This second ramp crossed zero at U_iT/U seconds from the time the reference was applied. The time is measured by a high-speed counter; and since T and U are constant, the counter holds a value that is proportional to the input voltage. If the input voltage, for example, is equal to the reference voltage U, the two integration times are equal and the counter would be set to reach its maximum. This type of converter is usually quite linear and converters up to 20 bits can be obtained, but the conversion time is relatively long.

Flash Converter At the high-speed end of the performance spectrum, the parallel (flash) converter can provide conversion rate greater than 100MHz. This is accomplished by providing internal voltage references for each of the $2^N - 1$ quanta into which the analog voltage range is divided. The analog signal is compared with all the reference voltages at once by a string of high-speed comparators whose outputs are used to generate the binary output. Because of the number and quality of the components required for this A/D, implementations are more common for 8 bit or less. Flash converters are usually expensive.

Sample-and-hold Circuit In front of the A/D converter in an analog conversion system is usually located a sample-and-hold circuit. A constant input is especially required in a successive approximation converter because the input is compared with a reference several times over the conversion period.

Multiplexer The multiplexer shown in Fig. 3-2A-5 is conceptually like a rotary switch that can be rotated to "address" any input channel under control of the processor. In a multiplexer with N channels, the effective per-channels. The multiplexer can be constructed from either mechanical switches or solid-state devices (such as CMOS switches). The multiplexing can be carried out in a variety of ways such as "single-ended" and "differential" connection. The single-ended connection is useful when a signal is referenced to ground. The differential multiplexer is useful when you are interested in the difference between two voltages, such as the two arms of a strain gauge bridge.

WORDS AND TERMS

autonomous   *adj.* 自治的
bus-compatible   *adj.* 总线兼容的
temporal   *adj.* 暂时的
handshaking   *n.* 握手
direct memory access（DMA）   直接存储器访问
strobe   *v.* 选通，发选通脉冲
poll   *v.* 登记，通信，定时询问
latch   *v.* 抓住，占有；*n.* 寄存器
full-scall   *adj.* 满量程的
most-significant bit（MSB）   最高有效位
least-significant bit（LSB）   最低有效位
resolution   *n.* 分辨率
slew rate   转换速度
sample   *v.* 采样
successive approximation   逐次逼近
dual slop   双积分
flash converter   闪速转换器
sample-and-hold   *n.* 采样保持
spectrum   *n.*（光）谱，领域，范围
multiplexer   *n.* 多路器（开关）

NOTES

[1] In this interface, there is a "tristate" buffer which when it is enabled will force the processor bus to have the same binary value as the external data lines.

在这种接口中，有一个三态缓冲器，它能迫使处理器总线与外部数据线有相同的二进制值。

[2] U_i is then removed and a known negative constant reference voltage U is integrated which produces a ramp down.

然后去掉 U_i，对一个已知的负常数参考电压 U 积分以产生一个下降的斜坡。

B   The Applications of Computers

The use of computers has been universal in nearly all walks of life in developed countries. Here are some examples for application of computers:

Scientific Calculations   The earliest computers were created to take the job of scientific calculations which involved complex and difficult mathematics or time-consuming, tedious and repetitive, numerical calculations. For example, calculating the trajectory of artillery shells requires resolving a set of differential equations in a few seconds or designing a large dam involves resolving sets of simultaneous algebraic equations which may have up to hundreds of variables may take mathematicians years, but can be done by a computer program in hours.

Data Processing   Computers have been widely used in data processing, for example, accounting, statistics, census. The operations involved are very simple—addition, subtraction, multiplication and division, but the amount of data is overwhelming, beyond human capability and patience. Database products, for instance, Lotus-1, 2, 3, provide the user with formal data structures for sorting, categorizing, storing, accessing, retrieving data. With database software equipped, the computer can handle data to the satisfaction of the user.

Automatic Control   Computers take over the jobs which used to be exclusive for humans with special skill and knowledge, for example, controlling productive processes, manipulating machines, inspecting product quality, managing production plans, administering inventory, etc., all on an automatic scale with high efficiency and accuracy. In NC (Numerical Control) systems, PID control systems, servo control systems, group control systems, optimal control and adaptive control systems, computers as a central control unit take all relevant calculations involved in the automatic control process and schedule all other working units in the system. The CIMS (Computer Integrated Manufacturing System) contains not only a production control system but also production planning and management systems aiming at the integration of factory automation (FA) and an office automation (OA), constituting a computer network for a whole company. The HIMS (Human Integrated Manufacturing System) is a form of high-level computer control. Virtual reality technology is used to create virtual space for human operators through the use of high-performance computers and specific software.

Computer Simulation Computer simulation as a powerful analytic tool widely used in scientific research and engineering design demonstrates unrivalled advantages.[1] With computer simulation, scientists and engineers do not have to build real hardware models or primary prototypes when they observe an unknown phenomenon, analyze a complex process, design a machine or a building. Computer simulation is particularly significant when the object under study and examination is costly or it is impossible to build it into a real model.

Virtually computer simulation is based on mathematical models representing the nature of the object under study or examination. The mathematical model comprises a series of equations that depict the inherent process of the object in mathematical terms. A computer simulation program includes algorithms that are derived from those equations. Many computer simulation systems have been developed and proved to be cost-effective, because using computer simulation programs, engineers can accomplish iterative process each time by inputting different schemes and parameters into their computer models rather than building many different real models.

Robotics The controller in a robot is mostly a computer—from microprocessors to minicomputers. NC (Numerical Control) and SC (Servo Control) are widely used. They are reprogrammable to produce sequences of instructions for all movements and actions to be taken by the robot, in accordance with the program. For example, a controller sends a series of pulses to a step motor in a joint of a robot arm to rotate it a certain angle exactly as the program requires. When all joints driven in the same way, the robot arm can reach the designed position and attitude, and its end effector performs its jobs as the controller instructs it. The accuracy of movement is determined by the controller itself.

CAD and CAM CAD (Computer-aided Design) is software that can help engineers in their designs for new products, buildings, print circuit boards, civil works like bridges and airports, relieving them from the tedious, back-breaking and time-consuming jobs like drafting and drawing. When embarking on their designs, engineers frequently make reference into various manuals listed in which are details of structures, parts, materials and auxiliary materials ready for designers' choice for their designs. CAD products incorporate the content of all these manuals into a series of libraries in the software product, providing the engineers with information, for instance, names, dimensions, functions, performance, specifications, shapes, color, manufacturers, prices of the machines, parts, components, tools, materials, etc.—all necessary for engineering designs.

CAM (Computer-aided Manufacture) is software helping engineers to analyze a product or a project, and give advice for manufacturing it or constructing it. Data, diagrams, tables, etc. showing its shape, dimension, structure, fabrication and the material it is made from are input as the software requires.[2] Then it will give out suggestions about its manufacturing, for instance, machining procedure, machine tools and facilities to be used, technical parameters like allowance for finish, machining accuracy, as well as special processings.

Management Management is one of the decisive factors that tell success or failure of any bank, corporation, firm, university, research institute in competition. Management is a comprehensive technique, involving every aspect of the unit—task (products, inventions, creations, patents), personnel (administrators, clerks, technicians, servants), finance, real estate, equipment, etc. Computerized management is software designed to provide tools for management in any profession, for example, regulation of personnel, accounting, sales, inventory, taxation, wages, etc. Every kind of software embodies the up-to-date theory and method in its profession, and is rather easy to learn. More and more management software products emerge, taking the place of human managers.

Computerized Communication Progress in this field began in the early 1960s, when we witnessed the problem of connecting remote user terminals to central computer facilities. The solution to that problem was then based on asynchronous low-speed lines organized in either a star topology with a line dedicated to each terminal, or a tree topology with multidrop lines.[3] By the end of the 60s, the field had made a major leap with the advent of distributed resource-sharing networks. The goal was to interconnect computers and their users at various geographically distributed sites in order to allow the sharing, by all users connected to the network, of hardware and software resources developed at any of the sites. The most prominent example of such a network is ARPANET, which began implementation in 1969.

A radio-based terminal-access network called the ALOHA system was built at the University of Hawaii in 1970. Satellite transponder in the INTERNET equipped with a global coverage antenna can transmit data using the ground stations, thus achieving full connectivity among them. Optics-fiber cables connecting remote user terminals to central computer facilities can transmit data, graphs, video and audio signals better than any existing networks.

Advanced Applications Artificial intelligence (AI) is a subdivision of computer science. Its purpose is to develop theory and method to create "intelligent" computer programs that work in a human-like way, rather than subjecting human users to the stereotyped computer-dominated working style. In a sense of analogy, human intelligence is added to computer programs which then exhibit more intelligent behavior and more extensive ability, for example, thinking and reasoning, acquiring knowledge and applying it to solving more complex and difficult problems that present computers can't.

Expert systems are the most successful example of AI. An expert system oriented to a specific profession works like a human expert of that field and provides advice to solve the problem proposed by the user. Expertise is derived and organized in its knowledge base ready for the user to retrieve. Today, many expert systems have been commercially available, and more are under development.

Knowledge Engineering is another subject of AI. Knowledge engineering is devoted to the study of mimicking the human mind by computer programs, particularly to the simulation of its ability to obtain and apply knowledge. In other words, knowledge engineering tends to create computers that can learn, that is, can enlarge its own storage of knowledge by itself.

Computer vision is another application for AI. Computer vision is the use of computers to analyze and evaluate visual information; in other words, computers that can see. A computer vision system can extract key features from visual information of, say, photographs, pictures, scenes, etc., to identify or distinguish objects that have been definitely categorized in computer programs. This system can work more efficiently than human eyes. For example, examining aerial photographs to identify specific objects, say, missiles, bombers, warships, can help commander-in-chief make his decisions at battlefield.

Education can be another application of AI. Unlike traditional CBT (Computer Based Training), artificial intelligent CBT can adjust its tutorial to the students' knowledge, experience, strengths, and weaknesses. As a result, artificial intelligent CBT is far more effective than conventional CBT.

WORDS AND TERMS

artillery shell　炮弹
census　*n.* 人口统计
overwhelming　*adj.* 压倒一切的
retrieve　*v.* 检索
inventory　*n.* 仓库管理
servo control system　伺服控制系统
group control system　群控系统
virtual reality　虚拟现实
computer simulation　计算机仿真
prototype　*n.* 原型（机）
cost-effective　*adj.* 性能价格比（高）的
iterative　*adj.* 重复的，反复的
interactive　*adj.* 交互式的
step motor　*n.* 步进电动机
end effector　终端执行机构
embark　*v.* 从事，着手
auxiliary material　辅助材料

specification　*n.* （复）规格
machine tool　机床
fabrication　*n.* 构成，组成，制作
allowance for finish　加工余量
comprehensive　*adj.* 综合（性）的
real estate　不动产
asynchronous　*adj.* 异步的
topology　*n.* 拓扑结构
transponder　*n.* 发射机应答器
terminal　*n.* 终端（机）
stereotyped　*adj.* 僵化的
AI　人工智能
reason　*v.* 推理
expertise　*n.* 专门知识
knowledge base　知识库
knowledge engineering　知识工程
mimic　*v.* 模仿，学样儿

tutorial n. 个人辅导 | expert system 专家系统

NOTES

[1] Computer simulation as a powerful analytic tool widely used in scientific research and engineering design demonstrates unrivalled advantages.

作为一种广泛应用于科学研究和工程设计的有力分析工具，计算机仿真显示出无与伦比的优点。

[2] Data, diagrams, tables, etc. showing its shape, dimension, structure, fabrication and the material it is made from are input as the software requires.

表示产品形状、尺寸、结构、组成和制造材料的数据、图形和表格等将按软件的需求输入。

[3] The solution to that problem was then based on asynchronous low-speed lines organized in either a star topology with a line dedicated to each terminal, or a tree topology with multidrop lines.

此问题的解决基于异步低速的传输线，其结构可以是每台终端单配一线的星状拓扑结构，也可以是多分支线的树状拓扑结构。

solution, answer, key, guide 等词在表示所属关系时用介词 to 而不用 of。

C 常用数学符号和公式的读法

自动控制领域大量应用了高等数学、线性代数、积分变换、复变函数等多门数学的基础知识，因而在相关的文献中会经常出现很多数学公式。这些公式的读法是在基础英语中所未曾涉及的，下面介绍一些常用数学符号和公式的读法：

$4/5$	four fifths
0.025	zero point zero two five
38.49	thirty-eight point four nine
2%	two per cent
5^2	the second power of five; five to the power two
$\sqrt{x}$	the square root of x
6×10^7	six times the seventh power of ten
$+$	plus; positive
$-$	minus; negative
$\times$	multiplied by; times
$\div$; $/$	divided by
$=$	is equal to; equals
$(\ )$	round brackets; parentheses
i ; j	imaginary unit
$a!$	factorial a
$\sin x$	sine of x

arcsin x	arc sine of x
$\prod$	the product of the terms indicated
$\sum$	the sum of the terms indicated
b'	b prime
b''	b second prime
b_2	b sub two
b''_m	b second prime sub m
dy/dx	the first derivative of y with respect to x
d^2y/dx^2	the second derivative of y with respect to x
$\int_a^b$	integral between limits a and b
$x \rightarrow \infty$	x approaches to infinity
$a + b = c$	a plus b is equal to c
$a - b = c$	a minus b equals c
$s = vt$	s equals v multiplied by t
$v = s/t$	v equals s divided by t
$(a + b - c \times d)/e = f$	a plus b minus c multiplied by d, all divided by e equals f
$C/R = G/(1 + GH)$	C over R equals G divided by the sum of one plus H times G

UNIT 3

A PLC Overview

Automation of many different processes, such as controlling machines or factory assembly lines, is done through the use of small computers called a programmable logic controller (PLC). Programmable logic controllers were first created to serve the automobile industry, and the first programmable logic controller project was developed in 1968 for General Motors to replace hardwired relay systems with an electronic controller. Since the advent of programmable logic controllers, the ability to centralize factory processes, especially in the automotive industry, has improved greatly.

PLC's Architecture

Programmable Logic Controllers (PLCs) are diskless compact computers including all the necessary software and hardware interfaces to the process. They are generally used for automation control application (e.g. closed loop control) either standalone or connected to distributed inputs/outputs, to other PLCs and/or to supervision PCs. The connections are established by means of fieldbuses such as WorldFIP, MPI, PROFIBUS or Ethernet. The Fig. 3-3A-1 shows a typical PLC system, which is established by fieldbuses of MPI and PROFIBUS.

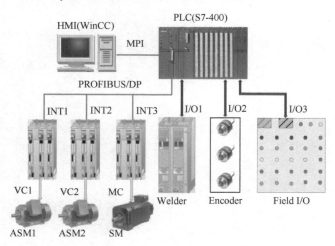

Fig. 3-3A-1 A typical PLC system

A typical PLC consists in:
- A power supply;
- A CPU where the user program runs;
- Input/output modules;
- Optional communication modules.

The available I/O modules support a wide range of electrical interfaces:
- Analog module (+/−10V, +/−1V, 4−20mA, resistor, etc.);
- Temperature measurement(pt100, Ni 100, etc);
- Digital module (±24V, 220V, etc.);
- TTL module (Beckhoff I/Omodule, etc.);
- RS 232 module;
- etc.

These modules are either connected to the internal bus of the PLC or to a bus coupler connected to a fieldbus segment (PROFIBUS, WorldFIP or CAN) shared with a PLC.

Custom hardware can hardly be connected directly the PLC internal bus. A solution consists in integrating the custom hardware with a standard fieldbus interface (e. g. PROFIBUS, CAN, and WorldFIP) by means of specific cards (e. g. HMS's Any Bus cards).

Nowadays, PLCs are provided with Ethernet-based communications. Although based on TCP/IP and IEEE 802.2, the PLC protocols are manufacturer specific. Therefore, PLCs of different manufacturers can not exchange, by default, data via Ethernet. However, Schneider PLC has a socket library and one can implement the RFC1006 used by the SIEMENS PLC. One can also implement a gateway by means of an OPC DX server, a SCADA application or a specific communication card such as the APPLICOM one.[1]

PLC-based solutions are well adapted to two-level control architectures where the front-end layer has to be autonomous and independent from the supervision layer. The process control (input/output readout, closed loop control, etc.) does not depend on the network neither on a remote computer, it is more secure.

PLCs have a long term availability and support: old generation of hardware or software are typically supported during more than 10 years by their respective manufacturers.

PLC Operation

PLCs are provided with manufacturer specific operating systems. The OS handles:
- The states of the CPU (start, stop, init, etc.);
- The calls to the user program;
- The interrupts;
- The memory;
- The communication with programming devices and other communicator partners.

PLCs operate in a polling mode with a precise execution cycle. This cycle consists basically in three steps (as shown in the Fig. 3-3A-2 SIEMENS PLC cycle) which are continuously executed:
- In the "Read Inputs" state, the PLC kernel reads all the input modules and copy the values into its internal input memory;
- In the "Operate user program" state, the PLC kernel executes the user program which has access to all the PLC memory areas. It stores the execution results in its output internal memory;
- In the "Write Outputs" state, the PLC kernel copies the internal output memory to the

output modules.

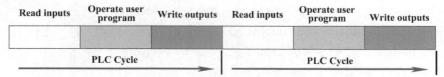

Fig. 3-3A-2 SIEMENS PLC cycle

In addition to the cycle, the OS manages the interrupts (hardware interrupts, user program errors, etc.). User program (control program) are produced with a development environment which is vendor dependent and then downloaded to the target CPU. It remains in the ROM memory of the CPU even if the power supply is off. In general, all the PLCs have two different operating modes: STOP and RUN selected by key switch or by software. At startup, the PLC goes in RUN or STOP mode depending on what was configured.

PLC Programming

The development of PLC programs is two folds. One shall define the hardware configuration of the PLC and produce the user program source code.

The hardware configuration describes the modules (I/O and communication) and the order in which they are installed in the PLC, the addresses of the I/O are automatically generated according to this order.

The source code is produced with Integrated Development Environments which are vendor specific. They consist, in general, in a set of tools for:
- Setting up and managing the applications;
- Configuring and assigning parameters to hardware;
- Configuring the fieldbus and the devices connected to it;
- Configuring communication links;
- Creating the user program for the PLC;
- Downloading the user program to the target and testing it.

The user program can be done in one of the five programming languages of IEC 1131-3 (International Electrotechnical Commission's standard). The IEC 1131-3 is a worldwide standard that has tried to merge PLC programming languages under one international standard. It harmonizes the way people look to industrial control by standardizing the programming interface. This includes the definition of the language Sequential Function Charts (SFC), used to structure the internal organization of a program, and four inter operable languages: Ladder Diagram (LD), Function Block Diagram (FBD), Structured Text (ST) and Instruction List (IL). The first three languages are graphical and the last two are textual. Each manufacturer can also have their own proprietary languages.

By using modularization and declaration of variables, each program is structured and simplified, increasing its reusability. In addition, IEC 1131-3 structures the way a control system is configured. There are, unfortunately, some differences in the languages, the source code is not 100% portable from one PLC to another. The major issues are in hardware addressing and the PLC kernel (how it operates, how it handles the interrupts, how it calls the user program, etc.)

Most of the PLC vendors belong to the PLC open organization which is a vendor and product independent worldwide association that brings greater value to users of Programmable Controllers through the pursuit of the IEC 1131-3 open software development standard.

Today's PLC

As PLC technology has advanced, so have programming languages and communications capabilities, along with many other important features. Today's PLCs offer faster scan times, space efficient high-density input/output systems, and special interfaces to allow non-traditional devices to be attached directly to the PLC. Not only can they communicate with other control systems, they can also perform reporting functions and diagnose their own failures, as well as the failure of a machine or process.

Size is typically used to categorize today's PLC, and is often an indication of the features and types of applications it will accommodate. Small, non-modular PLCs (also known as fixed I/O PLCs) generally have less memory and accommodate a small number of inputs and outputs in fixed configurations. Modular PLCs have bases or racks that allow installation of multiple I/O modules, and will accommodate more complex applications.

When you consider all of the advances PLCs have made and all the benefits they offer, it's easy to see how they've become a standard in the industry, and why they will most likely continue their success in the future.

WORDS AND TERMS

hard-wired   adj. 硬接线的
supervision   n. 监督，管理
socket   n. 插座
kernel   n. 内核
vendor   n. 卖主，供应商
harmonize   v. 协调

polling   n. 轮询
reusability   n. 可用性
CAN   控制器局域网，一种现场总线
accommodate   v. 容纳，使适应
rack   n. 架子，导轨

NOTES

[1] One can also implement a gateway by means of an OPC DX server, a SCADA application or a specific communication card such as the APPLICOM one.

可以使用 OPC DX 服务器、SCADA 应用，或者特定通信接口卡，如 APPLICOM one，作为网关。

句中的 OPC DX 即 OLE for Process Control Data Exchange。OLE (Object Linking and Embedding) 为对象的链接与嵌入，SCADA (Supervisory Control and Data Acquisition) 为数据

采集与监视控制系统。

B PACs for Industrial Control, the Future of Control

With a number of vendors producing Programmable Automation Controllers that combine the functionality of a PC and reliability of a PLC, PACs today are increasingly being incorporated into control systems. This white paper explores the origins of the PAC, how PACs differ from PLCs and PCs, and the future direction of industrial control with PACs.

Introduction of PACs

For the last decade a passionate debate has raged about the advantages and disadvantages of PLCs (programmable logic controllers) compared to PC-based control. As the technological differences between PC and PLC wane, with PLCs using commercial off the shelf (COTS) hardware and PC systems incorporating real-time operating systems, a new class of controllers, the PAC is emerging. PAC, a new acronym created by Automation Research Corporation (ARC), stands for Programmable Automation Controller and is used to describe a new generation of industrial controllers that combine the functionality of a PLC and a PC. The PAC acronym is being used both by traditional PLC vendors to describe their high end systems and by PC control companies to describe their industrial control platforms.

The "80-20" Rule

During the three decades following their introduction, PLCs have evolved to incorporate analog I/O, communication over networks, and new programming standards such as IEC 61131-3. However, engineers create 80 percent of industrial applications with digital I/O, a few analog I/O points, and simple programming techniques. Experts from ARC, Venture Development Corporation (VDC), and the online PLC training source PLCS.net estimate that:

- 77% of PLCs are used in small applications (less than 128 I/O);
- 72% of PLC I/O is digital;
- 80% of PLC application challenges are solved with a set of 20 ladder-logic instructions.

Because 80 percent of industrial applications are solved with traditional tools, there is strong demand for simple low-cost PLCs. This has spurred the growth of low-cost micro PLCs with digital I/O that use ladder logic. It has also created a discontinuity in controller technology, where 80 percent of applications require simple, low cost controllers and 20 percent relentlessly push the capabilities of traditional control systems. The applications that fall within the 20 percent are built by engineers who require higher loop rates, advanced control algorithms, more analog capabilities, and better integration with the enterprise network.

In the 1980s and 1990s, these "20 percenters" evaluated PCs for industrial control. The PC provided the software capabilities to perform advanced tasks, offered a graphical rich programming and user environment, and utilized COTS components allowing control engineers to take advantage of technologies developed for other applications. These technologies include floating point processors; high speed I/O busses, such as PCI and Ethernet; non-volatile data storage; and graphical development software tools. The PC also provided unparalleled flexibility, highly productive software, and advanced low-cost hardware.

However, PCs were still not ideal for control applications. Although many engineers used the PC when incorporating advanced functionality, such as analog control and simulation, database connectivity, web based functionality, and communication with third party devices, the PLC still ruled the control realm. The main problem with PC-based control was that standard PCs were not designed for rugged environments.

The PC presented three main challenges:

1) Stability: Often, the PC's general-purpose operating system was not stable enough for control. PC-controlled installations were forced to handle system crashes and unplanned rebooting.

2) Reliability: With rotating magnetic hard drives and non-industrially hardened components, such as power supplies, PCs were more prone to failure.

3) Unfamiliar Programming Environment: Plant operators need the ability to override a system for maintenance or troubleshooting. Using ladder logic, they can manually force a coil to a desired state, and quickly patch the affected code to quickly override a system. However, PC systems require operators learn new, more advanced tools.

Although some engineers use special industrial computers with rugged hardware and special operating systems, most engineers avoided PCs for control because of problems with PC reliability. In addition, the devices used within a PC for different automation tasks, such as I/O, communications, or motion, may have different development environments.

So the "twenty percenters" either lived without functionality not easily accomplished with a PLC or cobbled together a system that included a PLC for the control portion of the code and a PC for the more advanced functionality.[1] This is the reason many factory floors today have PLCs used in conjunction with PCs for data logging, connecting to bar code scanners, inserting information into databases, and publishing data to the Web. The big problem with this type of setup is that these systems are often difficult to construct, troubleshoot and maintain. The system engineer often is left with the unenviable task of incorporating hardware and software from multiple vendors, which poses a challenge because the equipment is not designed to work together.

Building a Better Controller

With no clear PC or PLC solution, engineers with complex applications worked closely with control vendors to develop new products. They requested the ability to combine the advanced software capabilities of the PC with the reliability of the PLC. These lead users helped guide product development for PLC and PC-based control companies.

The software capabilities required not only advanced software, but also an increase in the hardware capabilities of the controllers. With the decline in world-wide demand for PC components, many semiconductor vendors began to redesign their products for industrial applications. Control vendors today are incorporating industrial versions of floating point processors, DRAM, solid-state storage devices such as CompactFlash, and fast Ethernet chipsets into industrial control products. This enables vendors to develop more powerful software with the flexibility and usability of PC-based control systems that can run on real-time operating systems for reliability.

The resulting new controllers, designed to address the "20 percent" applications, combine the best PLC features with the best PC features. Industry analysts at ARC named these devices programmable automation controllers, or PACs. In their "Programmable Logic Controllers Worldwide Outlook" study, ARC identified five main PAC characteristics. These criteria characterize the functionality of the controller by defining the software capabilities:

"**Multi-domain functionality**, *at least two of the logic, motion, PID control, drives, and process on a single platform.*"[2] Except for some variations in I/O to address specific protocols like SERCOS; logic, motion, process, and PID are simply a function of the software. For instance, motion control is a software control loop which reads digital inputs from a quadrature encoder, performs analog control loops, and outputs an analog signal to control a drive.

"**Single multi-discipline development platform** *incorporating common tagging and a single database for access to all parameters and functions.*"[3] Because PACs are designed for more advanced applications such as multi-domain designs, they require more advanced software. In order to make system design efficient, the software must be a single integrated software package instead of disparate software tools which are not engineered to seamlessly work together.

"**Software tools that allow the design by process flow across several machines or process units**, *together with IEC61131-3, user guidance, and data management.*" Another component that simplifies system design is high level graphical development tools that make it easy to translate an engineer's concept of the process into code that actually controls the machine.

"**Open, modular architectures** *that mirror industry applications from machine layouts in factories to unit operations in process plants.*" Because all industrial applications require significant customization, the hardware must offer modularity so the engineer can pick and choose the appropriate components. The software must enable the engineer to add and remove modules to design the required system.

"**Employ de-facto standards for network interfaces, languages, etc.**, *such as TCP/IP, OPC & XML, and SQL queries.*" Communication with enterprise networks is critical for modern control systems. Although PACs include an Ethernet port, the software for communication is the key to trouble-free integration with the rest of the plant.

Two Approaches to Software

While software is the key difference between PACs and PLCs, vendors vary in their approach to providing the advanced software. They typically start with their existing control software and work to add the functionality, reliability, and ease-of-use required to program PACs. Generally, this creates two camps of PAC software providers: those with a background in PLC control and those with a background in PC control.

Software Based on PLC Philosophy Traditional PLC software vendors start with a reliable and easy-to-use scanning architecture and work to add new functionality. PLC software follows a general model of scanning inputs, running control code, updating outputs, and performing housekeeping functions. A control engineer is concerned only with the design of the control code because the input cycles, output cycles, and housekeeping cycles are all hidden. With much of the work done by the vendor, this strict control architecture makes it easier and faster to create control systems. The rigidity of these systems also eliminates the need for the control engineer to completely understand the low-level operation of the PLC to create reliable programs. However, the rigid scanning architecture which is the main strength of the PLC, can also make it inflexible. Most PLC vendors create PAC software by adding into the existing scanner architecture new functionality such as Ethernet communication, motion control, and advanced algorithms. However, they typically maintain the familiar look and feel of PLC programming and the inherent strengths in logic and control. The result is PAC software generally designed to fit specific types of applications such as logic, motion, and PID, but is less flexible for custom applications such as communication, data logging, or custom control algorithms.

Software Based on PC Philosophy Traditional PC software vendors start with a very flexible general-purpose programming language, which provides in-depth access to the inner workings of the hardware. This software also incorporates reliability, determinism, and default control architectures. Although engineers can create the scanner structure normally provided to the PLC programmer, they are not inherent to PC-based control software. This makes the PC software extremely flexible and well suited for complex applications that require advanced structures, programming techniques, or system level control but more difficult for simple applications.

The first step for these vendors is to provide reliability and determinism, which are often not available in a general-purpose operating system such as Windows. This is accomplished through real-time operating systems (RTOS) such as Phar Lap from Ardence (formerly Venturcom) or VxWorks from Wind River. These RTOSs provide the capability to control all aspects of the control system, from the I/O read and write rates to the priority of individual threads spawned on the controller.[4] These vendors then add abstractions and I/O read/write structures to make it simpler for engineers to build reliable control applications. The result is flexible software suited for custom control, data logging, and communication but lacking the familiar PLC programming architectures, making application development more demanding.

WORDS AND TERMS

white paper　白皮书
passionate　*adj.* 激烈的
commercial off the shelf (COTS)　商业现货
acronym　*n.* 首字母缩写词
functionality　*n.* 功能性

high end　高端
spur　*v.* 刺激
discontinuity　*n.* 不连续
relentlessly　*adv.* 无情地，残酷地
unparalleled　*adj.* 无比的，空前的

system crash 系统崩溃
reboot *n.* 重新启动
patch *v.* 修补
factory floor 工厂车间
bar code scanner 条码扫描仪
criteria *n.* 标准（criterion 的复数）
quadrature encoder 正交编码器
seamless *adj.* 无缝的
architecture *n.* 架构

mirror *v.* 镜像
de-facto *adj.* 事实上的，实际的
XML 可扩展标记语言
SQL 结构化查询语言
housekeeping *n.* 常规事务
rigidity *n.* 严格
data logging 数据记录
in-depth *adj.* 深入的
determinism *n.* 确定性

NOTES

[1] So the "twenty percenters" either lived without functionality not easily accomplished with a PLC or cobbled together a system that included a PLC for the control portion of the code and a PC for the more advanced functionality.

因此那些要开发"20%应用"的工程师们，要么使用无法轻松实现系统所需的功能的 PLC，要么采用既包含 PLC 又包含 PC 的混合系统，他们利用 PLC 来执行代码的控制部分，用 PC 来实现更高级的功能。

[2] Multi-domain functionality, at least two of the logic, motion, PID control, drives, and process on a single platform.

多功能性，是指在一个平台上有逻辑、运动、PID 控制、驱动和处理中的至少两种以上功能。

[3] Single multi-discipline development platform incorporating common tagging and a single database for access to all parameters and functions.

单一的多规程开发平台使用通用标签和单一的数据库来访问所有的参数和功能。

[4] These RTOSs provide the capability to control all aspects of the control system, from the I/O read and write rates to the priority of individual threads spawned on the controller.

这些 RTOS 能控制系统的各个层面，从 I/O 读写速率到控制器上各个线程的优先级。

C 科技论文的结构与写作

科技论文是科技人员介绍有关研究成果的文章，从结构上通常分为标题（Title）、作者（Author）、摘要（Abstract）、关键词（Keywords）、正文部分（通常由引言 Introduction、主体 Body 和结论 Conclusion 构成）以及致谢（Acknowledgment）和参考文献（Reference）几个部分。

本节主要介绍论文正文部分的写作。

文章的开头，即引言部分主要介绍与文章有关的背景知识以及现存的问题，从而引出论文的写作目的和主题。其常用的英语句型有：

The experiments (research) on... was carried by...

Recent experiments by... have suggested that...

The previous work on... has indicated that...

The paper is divided into five major sections as follows...

文章的主体部分篇幅大、内容多，是主题思想的展开和论述。选材上要围绕主题，段落划分既要结构严谨，又要保证全文的整体性和连贯性。作者可根据需要在文章中加小标题，将主体内容分为几个部分进行论述。英文写作通常把每段的主题句（Topic Sentence）放在段落的第一句，全段围绕主题句论述，必要的例子和对实验数据的分析通常是不可缺少的。常用句型有：

The problem is chiefly concerned with the nature (effect, activity) of... based on the study of...

The core of the problem is the interaction (origin, connection) of...

It is described as follows...

The present study was made with a view to show (demonstrate, determine)...

Studies of these effects cover various aspects of...

结论是对研究或实验结果（仿真结果）的分析，在其中要阐明作者的观点，指出争议的问题，得出最后的结论。常用的句型有：

On the basis of…, the following conclusion can be made...

From..., we now conclude (sum up) that...

We have demonstrated in this paper...

The results of the experiment (simulation) indicate (show)...

Finally, a summary is given of...

此外，为了保证论文内部应有的内在联系，文章中应适当使用一些连接词语，例如：

并列：and, also, in addition, as well as

转折：although, even though, but, on the other hand, otherwise, while, yet, in spite of

承接：first of all, secondly, thirdly, furthermore, moreover, besides, finally

条件：if, whether, unless, on condition that, so long as

举例：for example, for instance, as an illustration, such as

原因：for, because, for that reason, as, since, in order to

结果：as a result of, hence, so, consequently, thus, therefore, then

总结：in conclusion, to sum up, to summarize, to conclude

UNIT 4

A Fundamentals of Single-Chip Microcomputers

The single-chip microcomputer is the culmination of both the development of the digital computer and the integrated circuit, arguably the two most significant inventions of the 20th century.[1]

These two types of architecture are found in single-chip microcomputers. Some employ the split program/data memory of the Harvard architecture, as shown in Fig. 3-4A-1, others follow the philosophy, widely adopted for general-purpose computers and microprocessors, of making no logical distinction between program and data memory as in the Princeton architecture, as shown in Fig. 3-4A-2.

In general terms a single-chip microcomputer is characterized by the incorporation of all the units of a computer into a single device, as shown in Fig. 3-4A-3.

Read Only Memory (ROM) ROM is usually for the permanent, non-volatile storage of an applications program. Many microcomputers and microcontrollers are intended for high-volume applications and hence the economical manufacture of the devices requires that the contents of the program memory be committed permanently during the manufacture of the chips. Clearly, this implies a rigorous approach to ROM code development since changes cannot be made after manufacture. This development process may involve emulation using a sophisticated development system with a hardware emulation capability as well as the use of powerful software tools.

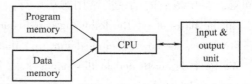

Fig. 3-4A-1 A Harvard type

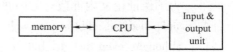

Fig. 3-4A-2 A conventional Princeton computer

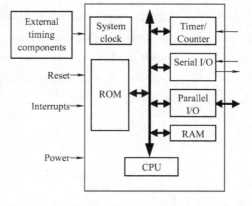

Fig. 3-4A-3 Principal features of a microcomputer

Some manufacturers provide additional ROM options by including in their range devices with (or intended for use with) user programmable memory. The simplest of these is usually a device which can operate in a microprocessor mode by using some of the input/output lines as an address and data bus for accessing external memory. This type of device can behave functionally as the single-chip microcomputer from which it is derived albeit with restricted I/

O and a modified external circuit. The use of these ROMless devices is common even in production circuits where the volume does not justify the development costs of custom on-chip ROM [2]; there can still be a significant saving in I/O and other chips compared to a conventional microprocessor based circuit. More exact replacements for ROM devices can be obtained in the form of variants with 'piggy-back' EPROM (Erasable Programmable ROM) sockets or devices with EPROM instead of ROM. These devices are naturally more expensive than the equivalent ROM device, but do provide complete circuit equivalents. EPROM based devices are also extremely attractive for low-volume applications where they provide the advantages of a single-chip device, in terms of on-chip I/O, etc., with the convenience of flexible user programmability.

Random Access Memory (RAM) RAM is for the storage of working variables and data used during program execution. The size of this memory varies with device type but it has the same characteristic width (4, 8, 16 bits, etc.) as the processor. Special function registers, such as a stack pointer or timer register are often logically incorporated into the RAM area. It is also common in Harvard type microcomputers to treat the RAM area as a collection of registers; it is unnecessary to make a distinction between RAM and processor register as is done in the case of a microprocessor system since RAM and registers are not usually physically separated in a microcomputer.

Central Processing Unit (CPU) The CPU is much like that of any microprocessor. Many applications of microcomputers and microcontrollers involve the handling of binary-coded decimal (BCD) data (for numerical displays, for example), hence it is common to find that the CPU is well adapted to handling this type of data. It is also common to find good facilities for testing, setting and resetting individual bits of memory or I/O since many controller applications involve the turning on and off of single output lines or the reading of a single line. These lines are readily interfaced to two-state devices such as switches, thermostats, solid-state relays, valves, motors, etc.

Parallel Input/Output Parallel input and output schemes vary somewhat in different microcomputers; in most a mechanism is provided to at least allow some flexibility of choosing which pins are outputs and which are inputs. This may apply to all or some of the ports. Some I/O lines are suitable for direct interfacing to, for example, fluorescent displays, or can provide sufficient current to make interfacing to other components straightforward. Some devices allow an I/O port to be configured as a system bus to allow off-chip memory and I/O expansion. This facility is potentially useful as a product range develops, since successive enhancements may become too big for on-chip memory and it is undesirable not to build on the existing software base.

Serial Input/Output Serial communication with terminal devices is a common means of providing a link using a small number of lines. This sort of communication can also be exploited for interfacing special function chips or linking several microcomputers together. Both the common asynchronous and synchronous communication schemes require protocols that provide framing (start and stop) information. This can be implemented as a hardware facility or U(S)ART (Universal (synchronous) asynchronous receiver/transmitter) relieving the processor (and the applications programmer) of this low-level, time-consuming, detail. It is merely necessary to select a baud-rate and possibly other options (number of stop bits, parity, etc.) and load (or read from) the serial transmitter (or receiver) buffer. Serialization of the data in the appropriate format is then handled by the hardware circuit.

Timer/Counter Facilities   Many applications of single-chip microcomputers require accurate evaluation of elapsed real time. This can be determined by careful assessment of the execution time of each branch in a program but this rapidly becomes inefficient for all but the simplest programs. The preferred approach is to use a timer circuit that can independently count precise time increments and generate an interrupt after a preset time has elapsed. This type of timer is usually arranged to be preloadable with the required count. The timer then decrements this value producing an interrupt or setting a flag when the counter reaches zero. Better timers then have the ability to automatically reload the initial count value. This relieves the programmer of the responsibility of reloading the counter and assessing the elapsed time before the timer is restarted, which otherwise would be necessary if continuous precisely timed interrupts were required (as in a clock, for example). Sometimes associated with a timer is an event counter. With this facility there is usually a special input pin, that can drive the counter directly.

Timing Components   The clock circuitry of most microcomputers requires only simple timing components. If maximum performance is required, a crystal must be used to ensure the maximum clock frequency is approached but not exceeded. Many clock circuits also work with a resistor and capacitor as low-cost timing components or can be driven from an external source. This latter arrangement is useful if external synchronization of the microcomputer is required.

WORDS AND TERMS

culmination   *n.* 顶点，极点
split   *adj.* 分离的
philosophy   *n.* 原理，原则
incorporation   *n.* 合并，结合
volatile   *adj.* 易变的
commit   *v.* 保证
emulation   *n.* 竞争
albeit   *conj.* 虽然
custom   *adj.* 定制的
variant   *adj.* 不同的，替换的
piggy-back   *adj.* 背负式的

erasable   *adj.* 可擦除的
socket   *n.* 插座
thermostat   *n.* 恒温器
protocol   *n.* 协议
consuming   *adj.* 控制的
baud   *n.* 波特
elapse   *v.* 经过
evaluation   *n.* 估计
preset   *adj.* 事先调整的
preloadable   *adj.* 预载的
decrement   *n.* 减少量

NOTES

[1] The single-chip microcomputer is the culmination of both the development of the digital computer and the integrated circuit, arguably the two most significant inventions of the 20th century.

单片机是数字计算机和集成电路发展中的一个顶峰，而这二者可以说是 20 世纪的两项最有意义的发明。

[2] on-chip ROM 译为"片内 ROM"。

B Understanding DSP and Its Uses

Digital Signal Processor (*noun*)

A Digital Signal Processor is a super-fast chip computer, which has been optimized for the detection, processing and generation of real world signals such as voice, video, music, etc, in real time. It is usually implemented in a single chip, or nowadays just part of an IC, about 0.5 cm^2 to 4 cm^2, costing between \$3 and \$300 and residing in every mobile phone, MP3 player, computer and most cars on our planet.

In contrast, a *microprocessor* (a term with which we are all probably familiar) is traditionally a much less powerful computer that performs the mundane "behind the scenes" tasks, often controlling other devices—e.g., keyboard entry, central heating, washing machine cycles, etc. Having said this, a Pentium III chip would normally be classed as a microprocessor rather than a DSP, yet, running at speeds in excess of 1 GHz, this can hardly be described as "much less powerful"!

Today, there is in fact a blurring between the roles of microprocessor and digital signal processor[1]—perhaps synergy is a better word. Many so-called DSPs now have micro-processor functionality on board. Many so-called micros now have DSP functionality on board. Distinguishing between DSP and micro by application is perhaps not the best way forward after all. In fact, the main difference lies deep inside the devices themselves, in the internal chip architecture, with DSP devices particularly optimized for high-speed, high-accuracy multiplication.

Digital Signal Processing (*verb*)

To "*digital signal process*" is to manipulate signals that have either originated in, or are to be exported to, the real world, where those signals are represented as digits (numbers). This means that an integral part of DSP applications is the conversion of the real world signals—analog voice, music, video, engine speed, ground vibration—to numerical values for processing by the DSP (*noun*)—a process termed *analog to digital conversion* (A/D). Going in the other direction—the conversion of numerical values generated within the DSP to real world signal—a process termed *digital to analog conversion* (D/A) is also involved. This sequence is illustrated in Fig. 3-4B-1.

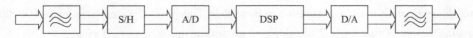

Fig. 3-4B-1 A digital signal processing system

Fundamental to DSP (*verb*) is the fact that any real world signal, e.g. music, can be accurately represented by samples of the signal taken at periodic intervals. (A man called *Nyquist* has devised a formula for this.). These samples can then be converted into numbers (e. g. representing the volume of the music at the sample point), and these numbers are expressed in binary form. We are now in the world of computing and software where manipulation of binary numbers is bread and butter stuff. [2]

If we simply scale all of the numbers representing our samples by an equal amount, and then convert the numbers back to voltage samples to drive our headphones or speakers, we have implemented a DSP volume control. If, on the other hand, we scale the numbers by a different amount each second, we have the beginnings of a simple audio FX processor.

In DSP (*verb*) terms, these are very simple tasks, and do not even begin to harness the *processing* power available. An example of a more challenging application is a graphic equalizer. Essentially this allows you to change the volume of components of the music source depending on their frequency. To implement a graphic equalizer requires the separation of the signal into frequency bands using *filters*, scaling each filtered segment independently, and then adding the segments back together for sending to the output.

Digital filtering is child's play for a DSP, but actually involves a lot of computation. Consecutive samples must be stored, each scaled independently and then summed to perform even the most simple of filter functions. However, the DSP (*noun*) and its instruction set are optimized for tasks such as this (hence the distinction between a DSP and a microprocessor), and this function can indeed be implemented using just three lines of code on most DSP devices.

DSP —A Very Short History Lesson

In the 1960s, DSP hardware used discrete components and consequently, because of the high cost and volume, its application could only be justified for very specialized requirements (or large-budget research programs!). In the 1970s, monolithic components for some of the DSP subsystems appeared, primarily dedicated digital multipliers and address generators, and DSP systems could be implemented using bit slice microprocessors. The breakthrough for mass exploitation of DSP techniques came in 1979 when Intel introduced the 2920, a completely self-contained signal processing device in a 40-pin DIP (Double In-line Package) package incorporating on-board program EPROM, data RAM, A/D and D/A converters, and an architecture and instruction set powerful enough to implement a full duplex 1200 bps (bits per second) modem, including transmit and receive filters. This breakthrough was followed up convincingly by Texas Instruments in 1982 with the launch of the TMS32010.

Since the Intel 2920 there have been five further generations of general-purpose signal processing devices, with a sixth generation announced. The latest devices have some 100,000 times the processing power of that early 2920 device, all within the space of 20 years.

Today, all of the major semiconductor device manufacturers have DSP products either already released or under development, reflecting their confidence in an enormous growth of demand in the late 1980s, 1990s and now into the 21st century, paralleling the spectacular success of microprocessors in the 1970s.

Speech analysis was the driving force behind initial attempts to process signals digitally. The exacting tolerances demanded of filters in speech processing systems simply could not be maintained over time with analog techniques, subject as they are to temperature drift, component tolerances, and aging[3]. Digital processing operations, in contrast, consisting of nothing more than sequences of binary multiplication, addition, subtraction, etc, make the outcome entirely predictable and thus reproducible. This feature, together with the ability to replicate multiple analog processing tasks within a single low-power device, is the reason why DSP is a thriving multi-billion dollar industry.

DSP —Where and How Is It Used

Table 3-4B-1 lists many of the applications for which DSP is now part of the staple diet.

Table 3-4B-1 Some typical applications of DSP

General-purpose DSP	Graphics/Imaging	Instrumentation
Digital filtering	3-D rotation	Spectrum analysis
Convolution	Robot vision	Function generation
Correlation	Image transmission/compression	Pattern matching
Hilbert transforms	Pattern recognition	Seismic processing
Fast Fourier transforms	Image enhancement	Transient analysis
Adaptive filtering	Homomorphic processing	Digital filtering
Windowing	Workstations	Phase-locked loops
Waveform generation	Animation/digital map	
Voice/Speech	**Control**	**Military**
Voice mail	Disk control	Secure communications
Speech vocoding	Servo control	Radar processing
Speech recognition	Robot control	Sonar processing
Speaker verification	Laser printer control	Image processing
Speech enhancement	Engine control	Navigation
Speech synthesis	Motor control	Missile Guidance
Text to speech		Radio frequency modems
Telecommunication	**Telecommunication**	**Automotive**
Echo cancellation	FAX	Engine control
ADPCM transcoders	Cellular telephones	Vibration analysis
Digital PBXs	Speaker phones	Antiskid brakes
Line repeaters	Digital speech interpolation(DSI)	Adaptive ride control
Channel multiplexing	X.25 Packet Switching	Global positioning
1200 to 19200 bps modems	Video conferencing	Navigation
Adaptive equalizers	Spread spectrum	Voice commands
DTMF encoding/decoding	Communications	Digital radio
Data encryption		Cellular telephones
Consumer	**Industrial**	**Medical**
Rader detectors	Robotics	Hearing aids
Power tools	Numeric control	Patient monitoring
Digital audio/TV	Security access	Ultrasound equipment

Music synthesizers   Power line monitors   Diagnostic tools
Educational toys   Prosthetics
                                                                   Fetal monitors

PC Modem Example

To briefly answer the question "How is DSP used?", consider the example of a PC modem. Fig. 3-4B-2 highlights the distinction between the analog world of the telephone cable, the DSP world of real-time signal manipulation, and the PC world of non-real-time data entry, display, and storage.

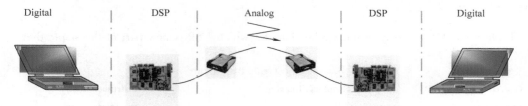

Fig. 3-4B-2 PC modem system

Taking the modulation (transmission) half of the modem first, we see that for this example, the data to be processed by the DSP is already in digital form—e. g. an e-mail file, a zipped data file, etc. The task of the DSP is to take the binary information and convert it into a form suitable for transmission over the telephone line. Unlike the cable connecting a PC to a printer, the telephone line cannot pass the 1's and 0's of binary data as they stand, not least because the information-bearing portion of the telephone line is ac-coupled (dc power is fed to the phone over the same pair of wires). The DSP thus translates groups of data bits into "symbols". These symbols are typically a set of discrete amplitude and phase values of a carrier signal at about 1800Hz. In the process of generating these symbols, special filtering is performed to shape the modulated signal, so that as much data can be packed into the limited bandwidth of the telephone channel as possible[4]. The digital sampled version of the now filtered symbol waveform is passed through a digital to analog converter and anti-alias filter, and onto the phone line.

On the receiving end of the line, analog to digital conversion provides the entry point to the DSP domain, where the waveform is further filtered to remove noise on the line. There is also usually a function known as equalization performed at this stage, which attempts to compensate for attenuation and variable delay over the telephone cables. Equalization takes the form of an adaptive filter, which alters its properties based on the measured response of the link. This function is almost impossible to do without DSP techniques.

The next task is to convert the symbols back into groups of data bits, probably with some error checking and correction in place before passing the data onto the PC data bus.

All of the modem processing can be implemented within a single DSP costing only $3. In fact, the latest DSP devices can realize up to 128 modems simultaneously in a single chip. These super

DSPs are used extensively in modem banks at telephone exchanges and with Internet Service Providers (ISPs).

 **WORDS AND TERMS**

mundane *adj.* 平凡的	convolution *n.* 卷积
entry *n.* 入口	Hilbert transform 希尔伯特变换
audio *adj.* 音频的	fast Fourier transform 快速傅里叶变换
harness *v.* 利用	pattern recognition 模式识别
graphic equalizer 图像均衡器	homomorphic processing 同态处理
consecutive *adj.* 连续的	phase-locked loop 锁相环
instruction set 指令集	data encryption 数字加密
monolithic *adj.* 单片的	Cellular telephone 蜂窝电话
address generator 地址发生器	anti-alias filter 抗混叠滤波器
duplex *adj.* 全双工的	attenuation *n.* 衰减
temperature drift 温度漂移	channel *n.* 信道

NOTES

[1] Today, there is in fact a blurring between the roles of microprocessor and digital signal processor.

当前,微处理器和数字信号处理器两者的任务界限已经变得模糊了。

[2] We are now in the world of computing and software where manipulation of binary numbers is bread and butter stuff.

我们现在生活在计算机和软件的世界中,在这个世界中,二进制数字的处理就好比是面包和黄油,不可缺少。

[3] The exacting tolerances demanded of filters in speech processing systems simply could not be maintained over time with analog techniques, subject as they are to temperature drift, component tolerances, and aging.

在语音处理系统里使用模拟技术,滤波器对误差的苛刻要求就得不到保证,这主要由于模拟技术存在温度漂移、元件误差和老化等问题。

[4] In the process of generating these symbols, special altering is performed to shape the modulated signal, so that as much data can be packed into the limited bandwidth of the telephone channel as possible.

在生成这些符号的过程中,用专门的滤波器完成已调信号的整形,以便让尽可能多的数据能够压缩进带宽有限的电话信道。

C　论文的标题和摘要

科技论文的标题和摘要是概括全文和吸引读者的重要部分。为了便于国际学术交流和文献索引,即使是在国内发表的论文往往也需要加上英文的标题和摘要。

科技论文的标题应该能隐含文章的主题与大意,便于索引。一般情况下,标题应力求简明扼要,没有主、谓、宾的固定结构,长度在10个词以内,最多不超过15个词,句式应避免问句格式,且不要使用不标准的缩略语。

作者和作者的单位放在标题的下方。

摘要是全文的缩影和总结。通过阅读摘要,读者能很快地了解全文的梗概,决定是否阅读全文。摘要的长度一般在100个单词左右或稍长一些,要保证重点突出,言简意赅,内容完整,结构严谨,避免过分简单和使用不标准的缩略语。此外,摘要中应不使用问句或感叹句。摘要中常用句型和例句有:

This paper treats (introduces) an important problem in...

A new technique (method, manner) on... is shown in this paper.

This paper provides (shows, develops, extends, describes) a new approach for...

The author describes the technique of...

It is suggested that some basic steps be taken in order to...

This paper presents a thorough study of the input/output stability of relay pulse sender.

The purpose of this article is to explore the relation between sampling period and stability.

关键词是论文中最重要且出现率最高的词或词组。列出关键词有助于读者对全文的理解,同时便于查阅和检索。关键词一般使用名词形式,词数为3~6个。

下面介绍一篇论文的标题和摘要的例子:

Wind Turbines Based on Doubly Fed Induction Generator under Asymmetrical Voltage Dips

Lopez, Jesus Gubia, Eugenio

(Univ Publ Navarra, Dept Elect & Elect Engn, Pamplona 31006, Spain.)

Abstract: Many large wind farms employ doubly fed induction generators (DFIGs). The main drawback of these machines is their large sensitivity to grid disturbances, especially to voltage dips. As the penetration of wind energy in the network increases, it becomes more relevant to understand the behavior of these machines under voltage dips. This paper analyzes the operation of the DFIG under asymmetrical voltage dips and explains why such kinds of dips are more harmful than are symmetrical dips. The influence of the dip type and instant of occurrence is also studied.

Keywords: doubly fed induction generator (DFIG), protection, wind power generation

中文摘要和英文摘要的写作原则基本相同。同一篇论文的中英文摘要必须保证内容上的一致,但在分句方式和表达方式上则可以适当灵活。

UNIT 5

A A First Look at Embedded Systems

Introduction

These are the days when the term like embedded is increasingly becoming more and more popular in the world. We are flooded with embedded systems that seem to be everywhere. Now the question is what basically embedded systems are. We can define an embedded system as "A microprocessor based system that does not look like a computer". Or we can say that it is "A combination of computer hardware and software, and perhaps additional mechanical or other parts, designed to perform a dedicated function. In some cases, embedded systems are part of a larger system or product, as is the case of an antilock braking system in a car".

Embedded systems include a variety of hardware and software components, which perform specific functions in host systems, for example, satellites, washing machines, robots, hand-held telephones and automobiles.

Modem, hard drive, floppy drive, and sound card—each of which is an embedded system. Each of these devices contains a processor and software and is designed to perform a specific function. For example, the modem is designed to send and receive digital data over an analog telephone line. That's it. And all of the other devices can be summarized in a single sentence as well. If an embedded system is designed well, the existence of the processor and software could be completely unnoticed by a user of the device. Such is the case for a microwave oven, VCR, or alarm clock. In some cases, it would even be possible to build an equivalent device that does not contain the processor and software.

Attributes of an Embedded System

As embedded systems started progressing, they started becoming more and more complex. Additionally, new attributes that got added to these systems were smart and intelligent. Not only were the embedded devices able to do their jobs but also were able to do them smartly. These attributes can be defined as:

Computational Power These devices have some amount of computing power. A very simple 8-bit controller or a high-end 64-bit microprocessor could provide this computation power.

Memory The next requirement is memory. These devices possess some amount of memory that can be used by the processor and also some to remember user data and preferences.

Realtime All the devices have to respond to user/environmental inputs within a specified period of time.

Communication The device must be able to receive inputs given by other devices in the environment, process it and provide some solid output to the other devices or users.

Dynamic decisions   The system should be able to change its next course of activity based on the change of input from its sensors or surroundings.

Applications of Embedded Systems

Embedded systems are the applications that fuel some of the microprocessors that play a hidden but crucial role in our everyday lives. These are the tiny, quick, and smart microprocessors that live inside printers, answering machines, elevators, cars, cash machines, refrigerators, thermostats, wristwatches, and even toasters. Embedded systems are on the cutting edge of consumer electronics, poised to revolutionize various technologies by making them "smarter."[1]

Embedded System Applications describes the latest techniques for embedded system design in a variety of applications. This also includes some of the latest software tools for embedded system design. Some of the examples of embedded systems are as follows:

1) Automatic teller machines;
2) Computer printers;
3) Disk drives;
4) Cellular telephones and telephone switches;
5) Inertial guidance systems for aircraft and missiles;
6) Medical equipments;
7) Video game consoles;
8) Industrial machinery use programmable logic controllers to handle automation and monitoring;
9) Engine control computers and antilock brake controllers for automobiles;
10) Wristwatches;
11) Household appliances, including microwave ovens, air conditioners, iron, washing machines, and television sets;
12) Home automation products, like thermostats, sprinkler, and security monitoring systems;
13) Network equipment, including routers and firewalls;
14) Traffic control (e.g. intelligent traffic lights);
15) Music systems;
16) Card reader.

Benefits of Embedded Systems

Taking reference of the above examples, we can well imagine how much such kinds of systems have served humanity. Whether these benefits are in concerned with security purposes or with human's comforts, in fact these intelligent systems helped humanity in every phase of life. They have helped man in their progress by developing the new technologies. Embedded microprocessors enable firms to compete on product and service innovation, by adding product and service features that customer's value, but which would be largely impossible without this technology.[2] According to market researchers, consumers love electronic equipment that can do "smart" things like transmit instructions to other devices wirelessly via infrared signals; be programmed to operate automatically; and connect to super-technologies, such as satellites, to bring remote power into their own hands. Such systems have made their life easy and comfortable.

Future Prospects

It is unavoidable that computer will continue to become cheaper, smaller and more powerful, and that eventually they will be inexpensive enough to put in nearly every product. In addition, nearly all computers equipped products will have some kind of access to either local networks, or the Internet.

Over the next decade, many common household items will be given embedded systems, reinventing them, and changing them for forever. Like desktop publishing, and later the Internet, embedded systems is a technology that will fundamentally, and permanently, change the way advertising and marketing works. It will also permanently change the kind of products that are made, and how they are made. The development of intelligent products, and intelligent product marketing, made possible by embedded systems, where these machines exist for the convenience of people. The rise of embedded systems marks a new phase in industrialization.

No doubt, the field of embedded system is getting more and more challenging, and issues in development of embedded software are becoming very attracting to a wide number of researchers both in industry and academia. In fact, lots of effort is being done in the development of embedded systems and many researchers are still working on it to make the systems more intelligent and smarter. With the fast pace of technological progress, that future may be right around the corner.

WORDS AND TERMS

embedded system　嵌入式系统
antilock braking system　防抱死系统
equivalent　*adj.* 相等的，相当的
attribute　*n.* 品质，特征
high-end　*n.* 高端
preferences　*n.* 参数选择
solid output　可靠输出
automatic teller machine　自动柜员机
answering machine　电话答录机

thermostat　*n.* 自动调温器
inertial guidance system　惯性导航系统
aircraft　*n.* 飞行器
missile　*n.* 导弹
sprinkler　*n.* 洒水车，洒水装置
infrared　*adj.* 红外线的
reinvent　*v.* 彻底改造
Video Cassette Recorder (VCR)　录像机

NOTES

[1] Embedded systems are on the cutting edge of consumer electronics, poised to revolutionize various technologies by making them "smarter."

嵌入式系统是电子消费中最先进的技术，它们的"智能化"将彻底改变各种技术。

[2] Embedded microprocessors enable firms to compete on product and service innovation, by

adding product and service features that customer's value, but which would be largely impossible without this technology.

嵌入式处理器通过增加产品和服务特色，为用户创造价值，使得公司在产品和服务上实现创新。如果没有这种技术，用户价值是不可能实现的。

B Embedded Systems Design

Fundamental Components of Embedded Systems

Usually all embedded systems have a lot in common in terms of their components and their requirements. Some of these requirements and components are:

Computational/Processing Power This is one of the primary requirements of an embedded system. The processing, however, can be done using microprocessor or a hydraulic circuit or a simple electrical/electronic circuit. This processing power is required to translate the request from the user, changes in the environment to the output as desired by the end user.

This processing logic that used to be "hardwired" in a chip or other electrical circuits grew up exponentially and is so complex nowadays that many functionalities are simply unimaginable without software. The usual practice is to hardwire 'mature' features in hardware and use software to implement developing features.

An embedded system can also take inputs from the environment, e.g., a refrigerator or an air conditioner has various functionalities like defrost, air circulation, temperature control etc. Some advanced refrigerators may have sensors to deodorize and detect inactivity.

To compute and regulate the various parameters a system may require various levels of computing power. The microcontroller can be chosen based on the required level of computing power.

Memory Memory is a very precious resource and is always found wanting in many embedded systems. As due to intense price wars, every resource must be handled with extreme care. In many systems, some space has to be allocated for future expansions. We cannot afford expansion slots as in PC for embedded system due to embedded-hardware design constraints.[1] So, memory should be handled very carefully. Algorithms that use a huge amount of memory or copying of huge data structures are ignored unless it is an absolute necessity.

Many embedded systems do not carry hard disk or floppy disk drives with them. The usage of secondary storage is not possible in most embedded systems. So these systems usually have some ROM and nonvolatile RAM where the code and user preferences are stored.

Some of the programs do not terminate and tend to run forever. In case of some special type of systems, when emergency strikes or when some irrecoverable error occurs, embedded systems implement watchdog timers which just reset the system.

Realtime We can define a system as a collection of subsystems or components that respond to the inputs from the user or the environment or from itself (e.g., timers). Typically, there is a time lapse between the time at which the input is given and the time at which the system responds. In any system, it is quite natural to expect some response within a specific time interval. But, there are systems where, very strict (not necessarily short) deadlines have to be met. These systems are called realtime systems. These systems are characterized as: A late answer is a wrong answer. [2] Realtime systems can be classified as:

- Hard Realtime Systems: A realtime system where missing a deadline could cause drastic results that could lead to loss of life/property. For example, aircrafts, nuclear reactors etc.
- Soft Realtime Systems: A realtime system where a few missed deadlines may not cause any significant inconvenience to the user. For example, televisions, DVD player or music system, multimedia streaming over Internet (where loss of some packets can be afforded).

The realtime systems can also be classified as fast and slow systems based on the time deadlines they operate with.

Closely associated with the concept of realtime is the concept of determinism. This is also a very important concept that differentiates realtime programming from normal application programming. A realtime system is the one that behaves predictably, as it responds within a particular amount of time. The time interval between the instant at which the input occurred to the time instance at which output occurs should be "deterministic" or predictable. It just requires that the system should always respond within a known period of time.

Communication Elements Embedded devices and appliances can no longer remain as islands of information storage. They cannot remain isolated; rather they need to communicate with each other to perform any operation that is desired by the user. These communications could be done with some wireless networking protocols like Bluetooth, Wireless LAN etc. The communication element adds "intelligence" to simple embedded realtime system.

The other important parameters that define an embedded system are:

Cost Cost is the major driving factor behind many embedded systems. This requires the designer to be extremely conscious about memory, peripherals etc.. This factor plays an important role in high volume products.

Reliability Some products require 99.999% proper processing. But some doesn't like a bread piece in a toaster. Reliability may require the designer to opt for some level of redundancy.

Lifetime Products that have a longer lifetime must be built with robust and proven components.

Power Consumption This factor has become an important area of research in itself. With growing number of mobile instruments, power consumption has become a major concern. The design of mobile devices is such that the power consumption is reduced to the minimum. Some of the popular tactics used include shutting down those peripherals which are not immediately required. These tactics are highly dependent on software. The programmer for mobile devices is becoming increasingly aware of the power saving features in their programming platform.

Design of Embedded Systems

Design of embedded software, precise specification of the software that has to be built, includes:

1) Assumptions on the environment and on other components in the system;

2) Requirements on the implementation such as available hardware, memory usage, safety, error handling and use of energy;

3) Analysis of realtime behavior should be done to ensure that the deadlines are met on time. This requires proper allocation of processes to processor, scheduling of the tasks and assignment of priorities;

4) The selection of protocols for establishing communication between components, and the analysis to check the correctness and performance of the selected protocols;

5) The use of advanced methods for the designs should be ensured to achieve the required level of correctness and reliability.

Reasons for the Difficulty in Designing Embedded Systems

Several reasons that development of embedded systems is more difficult than development of any other software are as follows:

Complexity   The designing of embedded system is more complex than any other software.

Testing   There are more failure cases, therefore correctness is often more important.

Predictability   The performance of a system must be predictable in order for people to have confidence in it. If the performance is variable, then one is never sure that the system will meet its resource requirements on any given execution.

Specification   The specifications of an embedded system must be more detailed.

Domain Knowledge   The programmers must have more domain knowledge than usual.

Fault Tolerance   Embedded systems frequently require effective fault tolerance, so that when one or two things go wrong the system is able to at least partially recover. Failure of embedded systems often may have serious consequences (loss of lives, huge financial losses), so correctness and reliability are of vital importance.

WORDS AND TERMS

hydraulic   *adj.* 水压的
exponential   *adj.* 指数的
defrost   *v.* 除霜
deodorize   *v.* 除……臭
lapse   *n.* （时间等）流逝
drastic   *adj.* 激烈的
determinism   *n.* 决定论
deterministic   *adj.* 确定性的
redundancy   *n.* 冗余
tactic   *n.* 策略，战略
domain   *n.* 范围，领域

NOTES

[1] We cannot afford expansion slots as in PC for embedded system due to embedded-hardware design constraints.

由于嵌入式硬件设计的限制，我们不能像 PC 那样为嵌入式系统提供扩展槽。

[2] These systems are characterized as: A late answer is a wrong answer.

这些系统的特性可以描述为：迟到的响应就是错误的响应。

C 电子邮件

一、英文电子邮件格式

英文电子邮件的基本要素是：主题、称谓、正文、结尾用语及署名。

电子邮件最重要的部分是主题，假设我们都是很忙的人，在打开邮箱阅读邮件时，第一眼看到的就是邮件的主题。所以，主题应当做到言简意赅并突出邮件重要性。英文邮件的主题需要注意不超过 35 个字母，将位于句首的单词和专有名词首字母大写。比如：Some questions about C++。在比较正式的格式中，需要把每个单词的首字母都大写（介词、冠词和连词除外）。

称谓。如果是第一次给对方写信，那么称谓最好用"Dear + 全名"，这样会让人感觉比较正式。如果对方以非正式口吻来信，我们也可以类似非正式地回复。比如："Hello/Hi David"。在实际通信中可能遇到不知道对方姓名的情况，可以用"Dear + 对方头衔"，如"Dear President"，或者"Dear + Sir/Madam"形式。英国人习惯在称呼后加"，"，美国人习惯在称谓后加"："，有时也可以不加任何标点，视具体情况而定。

在书写正文时，把最重要的事情写在正文最前面或者邮件内容较长时写在第一段。为了让收件人阅读邮件时比较舒服，我们需要注意邮件正文结构的美感，邮件段落最好控制在两三段之内。如果一封电子邮件涉及多个信息点，则我们可以采用分条目的方法，如符号、小标题、编号来使得邮件想要表达的内容层次清晰。邮件内容应注意单词的拼写、大小写、标点、语法等。

结尾语在正文之后添加。注意一般结尾语中只有第一个单词首字母要大写，而剩余单词都小写，此处与称呼不同。

一般电子邮件：Sincerely, Sincerely yours, Best regards。

私人电子邮件：Regards, Best wishes, With best wishes, Wish best regards, Yours, Cheers, As ever, With love, Affectionately。

在正文最后需要署名，可以写全名，也可以只写名字。需要辨明性别时可以在姓名后面注明（Mr./Ms.）。对于中国人的姓名而言，为了区分姓和名，可以把姓的字母全部大写，例如 TONG David。如果写信人代表的是一个组织或部门，应在名字下一行写上自己的职位和所属部门。

邮件中常用词汇：

附件 attachment   I am attaching the report. 我把报告放在了附件中。

转发 forwarded   I've forwarded your e-mail to Alice. 我已经将你的邮件转发给 Alice 了。

保持联系 stay /keep in touch

请对方告知决定 Please let me know your decision as soon as possible.

写信给别人时一般常用 How are you been recently?

回复别人邮件时一般常用 I am happy to receive your e-mail.

二、实用邮件举例

1. 工作求职

工作求职类的邮件大体分为申请职位、介绍自己、期待回复、表达谢意几个部分。

Dear Sir/Madam,

  I am writing to apply for the position of … as posted in your website.

  Introduce yourself…

  Please find attached a copy of my resume for your review. 我的简历请见附件，作为参考。

  或者：Please refer to my attached resume for a summary of my skills and experience.

  I am looking forward to you. Thank you very much for your time and consideration.

Yours sincerely,
David Tong

2. 表达谢意

日常生活中经常会使用到表达谢意的邮件，在国外，参加完朋友邀请去的一次 party，或者接受了别人馈赠的礼物等情况下都需要写一封 thanks letter。而此类邮件也非常好写，只要能够表达自己对收件人的真挚的感激之情即可。

Dear Alice,

  Thanks so much for the lovely dinner last night. It was so thoughtful of you. I would like to invite both of you to my house when you are available.

  Best regards,

David

3. 咨询了解

咨询了解类邮件的目的是想要得到关于某人、某物的一些信息。邮件具体内容大致包含如下：

1）告知对方你是如何得知其信息的，比如报纸、网站等。

2）说明自己想要了解的东西，当咨询信息较多时，最好列举出来。

3）末尾用一句话总结这次咨询。

I am writing to request information about sth.

I would like to request a copy of sth.

最后表达希望时用语：I look forward to hearing from you soon. 期待尽快收到您的回复。

PART 4

Process Control

UNIT 1

A A Process Control System

The principal purpose of this part is to present you, the reader, with the need for automatic process control and to motivate you to study it. Automatic process control is concerned with maintaining process variables, temperatures, pressures, flows, compositions, and the like at some desired operating value. As we shall see in the ensuing pages, processes are dynamic in nature. Changes are always occurring, and if actions are not taken, the important process variables—those related to safety, product quality, and production rates—will not achieve design conditions.

In order to fix ideas, let us consider a heat exchanger in which a process stream is heated by condensing steam. The process is sketched in Fig. 4-1A-1.

The purpose of this unit is to heat the process fluid from some inlet temperature, $T_i(t)$, up to a certain desired outlet temperature, $T(t)$. As mentioned, the heating medium is condensing steam.

The energy gained by the process fluid is equal to the heat released by the steam, provided there are no heat losses to surroundings,[1] that is, the heat exchanger and piping are well insulated. In this case the heat released is the latent heat of condensation of the steam.

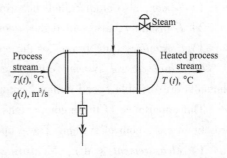

Fig. 4-1A-1 Heat exchanger

In this process there are many variables that can change, causing the outlet temperature to deviate from its desired value. If this happens, some action must be taken to correct for this deviation. That is, the objective is to control the outlet process temperature to maintain its desired value.

One way to accomplish this objective is by first measuring the temperature $T(t)$, then comparing it to its desired value, and, based on this comparison, deciding what to do to correct for any deviation. The flow of steam can be used to correct for the deviation. This is, if the temperature is above its desired value, then the steam valve can be throttled back to cut the steam flow (energy) to the heat exchanger. If the temperature is below its desired value, then the steam valve could be opened some more to increase the steam flow (energy) to the exchanger. All of these can be done manually by the operator, and since the procedure is fairly straightforward, it should present no problem. However, since in most process plants there are hundreds of variables that must be maintained at some desired value, this correction procedure would require a tremendous number of operators. Consequently, we would like to accomplish this control automatically. That is, we want to have instruments that control the variables without requiring intervention from the operator.[2] This is what we mean by *automatic process control*.

To accomplish this objective a *control system* must be designed and implemented. A possible control system and its basic components are shown in Fig. 4-1A-2. The first thing to do is to measure the outlet temperature of the process stream. A sensor (thermocouple, resistance temperature device, filled system thermometers, thermistors, etc) does this. This sensor is connected physically to a *transmitter*, which takes the output from the sensor and converts it to a signal strong enough to be transmitted to a *controller*. The controller then receives the signal, which is related to the temperature, and compares it with desired value.

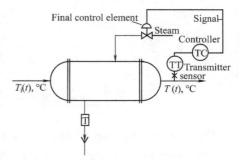

Fig. 4-1A-2 Heat exchanger control loop

Depending on this comparison, the controller decides what to do to maintain the temperature at its desired value. Base on this decision, the controller then sends another signal to *final control element*, which in turn manipulates the steam flow.

The preceding paragraph presents the four basic components of all control systems. They are

1) *Sensor*, also often called the primary element;

2) *Transmitter*, also called the secondary element;

3) *Controller*, the "brain" of the control system;

4) *Final control system*, often a control valve but not always. Other common final control elements are variable speed pumps, conveyors, and electric motors.

The importance of these components is that they perform the three basic operations that *must* be present in *every* control system. These operations are

1) *Measurement* (M): Measuring the variable to be controlled is usually done by the combination of sensor and transmitter.

2) *Decision* (D): Based on the measurement, the controller must then decide what to do to maintain the variable at its desired value.

3) *Action* (A): As a result of the controller's decision, the system must then take an action. This is usually accomplished by the final control element.

As mentioned, these three operations, M, D, and A, *must* be present in *every* control system. The decision-making operation in some system is rather simple, while in others it is more complex. The engineer designing a control system must be sure that the action taken affects the variable to be controlled, that is, that the action taken affects the measured value. Otherwise, the system is not controlling and will probably do more harm than good.

WORDS AND TERMS

ensuing   *adj.* 相继的
variable   *adj.* 变化的，可变的；*n.* 变量
latent heat   潜伏热
deviation   *n.* 偏差

throttle   *v.* 调节（阀门）
thermocouple   *n.* 热电偶
transmitter   *n.* 热敏电阻
conveyor   *n.* 传送带，传送机

NOTES

[1] ... provided there are no heat losses to surroundings...
provided 意为"只要"，是表示条件关系的连词。

[2] That is, we want to have instruments that control the variables without requiring intervention from the operator.
这就是说，我们需要一些不用操作人员介入就可控制这些变量的设备。

B   Fundamentals of Process Control

Important Terms of Automatic Process Control

At this time it is necessary to define some terms used in the field of automatic process control. The first term is *controlled variable*. This is the variable that must be maintained or controlled at some desired value. In the preceding example, the process outlet temperature, $T(t)$, is the controlled variable. The second term is *set point*, the desired value of controlled variable. The *manipulated variable* is the variable used to maintain the controlled variable at its set point. In the example, the flow of steam is the manipulated variable. Finally, any variable that can cause the controlled variable to deviate away from set point is defined as a *disturbance or upset*. In most process there are a number of different disturbances. As an example, in the heat exchanger shown in Fig. 4-1A-2, possible disturbances are inlet process temperature, $T_i(t)$, the process flow $q(t)$, the quality of the energy of the steam, ambient conditions, process fluid composition, fouling, and so on. What is important here is to understand that in the process industries, most often it is because of this disturbance that automatic process control is needed. If there were no disturbances, design-operating conditions would prevail and there would be no necessity of continuously "policing" the process.

The following additional terms are also important. *Open loop* refers to the condition in which the controller is disconnected from the process. That is, the controller is not making the decision of how to maintain the controlled variable at set point. Another instance in which open-loop control exists is when the action (A) taken by controller does not affect the measurement (M). This is indeed a major flaw in the control system design. *Close-loop control* refers to the condition in which the controller is connected to the process, comparing the set point to the controlled variable and determining corrective action.

With these terms defined, the objective of an automatic process control system can be stated as follows: the objective of an automatic process control system is to use the manipulated variable to maintain the controlled variable at its point in spite of disturbances.

Regulatory and Servo Control

In some processes the controlled variable deviates from a constant set point because of disturbance. Regulatory control refers to systems designed to compensate for these disturbances. In some other instances the most important disturbance is the set point itself. That is the set point may be changed as a function of time (typical of this are batch processes), and therefore the controlled variable must follow the set point. Servo control refers to control systems designed for this purpose.

Regulatory control is by far more common than servo control in the process industries. However, the basic approach to designing either of them is essentially the same. Thus, the many principles in automatic process control apply to both cases.[1]

Transmission Signals

Let us now say a few words about the signal used to provide communication between instruments of control system. There are three principal types of signals used in the process industry today. The *pneumatic signal*, or air pressure, ranges normally between 3 and 15 psig.[2] Less often, signals of 6 to 30 psig or 3 to 27 psig are used. The usual representation in piping and instrument diagrams (P&ID) for pneumatic signal is ─#─#─. The *electrical*, or *electronic*, *signal* ranges normally between 4 and 20 mA. Less often 10 to 50 mA, 1 to 5V or 0 to 10V are used. The usual representation in P&ID's for this signal is ---------. The third type of signal, which is becoming common, is the digital, or discrete, signal (zeros and ones). The use of process-control systems based on large-scale computers, minicomputers, or microprocessors is forcing increased use of this type of signal.

It is often necessary to change one type of signal into another type. This is done by a *transducer*. For example, there may be a need to change from an electrical signal, mA, to a pneumatic signal, psig. This is done by the use of a current (I) to pneumatic (P) transducer (I/P). This is shown graphically in Fig. 4-1B-1. The input signal may be 4 to 20 mA and the output 3 to 15 psig.[3] There are many other types of transducers: pneumatic to current (P/I), voltage-to-pneumatic (E/P), pneumatic-to-voltage (P/E), and so on.

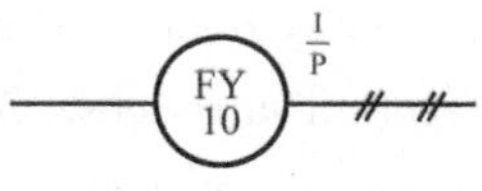

Fig. 4-1B-1　I/P transducer

Background Needed for Process Control

To be successful in practice of automatic process control, the engineer must first understand the principles of process engineering. Therefore, this unit assumes that the reader is familiar with the basic principles of process engineering of thermodynamics, fluid flow, heat transfer, separation process, reaction process, and the like.

For the study of process automatic control it is also important to understand how processes behave dynamically. Consequently, it is necessary to develop the set of equations that describe different processes. This is called modeling. To do this, the knowledge of the basic principles mentioned in the previous paragraph and of mathematics through differential equations is needed. In process control the Laplace transforms are used heavily. This greatly simplifies the solution of differential equations and the dynamic analysis of processes and their control systems.

Another important "tool" for the study and practice of process control is computer simulation. Many of the equations developed to describe process are nonlinear in nature and, consequently, the most exact way to solve them is by numerical methods; this means computer solution. The computer solution of process models is called *simulation*.

WORDS AND TERMS

manipulated variable　操纵量
disturbance　*n*. 扰动
upset　*n*. 干扰
fouling　*n*. 阻塞
regulatory control　调节控制
servo control　伺服控制

pneumatic　*adj*. 气动的
representation　*n*. 表示符号
transducer　*n*. 传感器
thermodynamics　*n*. 热力学
modeling　*n*. 建模
simulation　*n*. 仿真

NOTES

［1］Thus, the many principles in automatic process control apply to both cases.
因而，自动过程控制中的很多原理对两者都适用。
［2］psig: pounds per square inch gauge 表压（计算压强，剩余压强）[磅每平方英寸]
psia: pounds per square inch absolute 绝对压强 [磅每平方英寸]
　　　pounds per square inch of area 磅每平方英寸面积
psi: pounds per square inch 磅每平方英寸（$1\text{psi} = 0.070\text{kg/cm}^2$）
psid: pounds per square inch differential 压差 [磅每平方英寸]
［3］The input signal may be 4 to 20 mA and the output 3 to 15 psig.
输入信号在 4 到 20mA 之间，输出信号在 3 到 15 psig 之间。

C　通　知

通知表达信息应简明扼要，其内容包括所通知活动的时间、地点和内容，以及其他信息，如学术报告会的演讲者介绍、活动的组织者等。范例：

Lecture on Internet

By Prof. Bettey Powell
Rm 301 of Center Building
Sat. Oct. 5th
2:30 – 4:00
All welcome!

NOTICE

    As informed by Hebei University of Technology, Professor Rao Fangquan, academician of the Chinese Academy of Technology won't be able to come at appointed time owing to a urgent meeting. His report on control theory is put off to next Friday, May 15, at 2:00P.M. in the assembly hall of our college.

Department of Automation

UNIT 2

A Sensors and Transmitters

In Unit 1 we learned that the four basic components of control systems are the sensors, transmitter, controller, and final control elements. We also learned that these components perform the three basic operations of every control system: measurement (M), decision (D), and action (A). This section takes a brief look at sensors and transmitter, followed by a more detailed study, of controllers.

Sensors and transmitters perform the measurements (M) operation of control system. The sensor produces a phenomenon, mechanical, or the like related to the process variable that it measures. The transmitter in turn converts this phenomenon into a signal that can be transmitted. The signal, therefore, is related to the process variable.

There are three important terms related to the sensor/transmitter combination. The range of the instrument is given by the low and high values of the process variable that is measured. That is, consider a pressure sensor/transmitter that has been calibrated to measure a process pressure between the values of 20 psig and 50 psig. We say that the range of the sensor/transmitter combination is 20 – 50 psig. The *span* of the instrument is different between the high and low values of the range. For the mentioned pressure instrument the span is 30 psi. To summarize, we must specify a low and a high value to define the range of an instrument. That is, two numbers must be given. The span of the instrument is the difference between the two values. Finally, the low value of the range is referred to as the *zero* of the instrument. The value does not have to be zero in order to be called the zero of the instrument.[1] For the above example, the "zero" of the instrument is 20 psig.

There are some other common industrial sensors: pressure, flow, temperature, and level. Sometimes it is important for system analysis to obtain the parameters that describe the sensor/transmitter behavior. The gain term is fairly simple to obtain once the span is known. Consider an electronic pressure sensor/transmitter with a range of 0 – 200 psig. The gain is defined as the change in output, or responding variable, divided by the change in input, or forcing function, In this case the output is the electronic signal (4 – 20mA) and the input is the process pressure (0 – 200psig). Thus

$$K_T = \frac{20\text{mA} - 4\text{mA}}{200\text{psig} - 0\text{psig}} = \frac{16\text{mA}}{200\text{psig}} = 0.08 \frac{\text{mA}}{\text{psig}}$$

As another example consider a pneumatic temperature sensor/transmitter with a range of 100 – 300°F. The gain is

$$K_T = \frac{15\text{psig} - 3\text{psig}}{300°\text{F} - 100°\text{F}} = \frac{12\text{psig}}{200°\text{F}} = 0.06 \frac{\text{psig}}{°\text{F}}$$

That is, we can say that the gain of sensor/transmitter is ratio of the span of the output to the span of input.

The two cases presented show that the gain of the sensor/transmitter is constant over its complete operating range. For most sensor/transmitter this is the case; however, there are some instances, such as a differential pressure sensor used to measure flow, when this is not the case. A differential pressure sensor measures the differential pressure, h, across an orifice. This differential pressure is related to the square of the volumetric flow rate F. That is $F^2 \propto h$.

The equation that describes the output signal form an electronic differential pressure transmitter when used to measure volumetric flow with a range of $0 - F_{max}$ gpm[2] is

$$M_F = 4 + \frac{16}{(F_{max})^2} F^2$$

where M_F = output signal in mA, F = Volumetric flow.

From this equation the gain of the transmitter is obtained as follows:

$$K_T = \frac{d\overline{M}_F}{dF} = \frac{2 \times 16}{(F_{max})^2} \overline{F}$$

with a nominal gain

$$K'_T = \frac{16}{F_{max}}$$

This expression shows that the gain is not constant but rather a function of flow.[3] The greater the flow is, the greater the gain. Specifically,

$$At \left(\frac{\overline{F}}{F_{max}} \right) \quad 0 \quad 0.1 \quad 0.5 \quad 0.75 \quad 1.0$$

$$\left(\frac{K_T}{K'_T} \right) \quad 0 \quad 0.2 \quad 1.0 \quad 1.50 \quad 2.0$$

So the actual gain varies from zero to twice the nominal gain.

This fact results in a nonlinearly in flow control system. Nowadays most manufactures offer differential pressure transmitters with built-in square root extractors yielding a liner transmitter.

The dynamic response of most sensor/transmitters is much faster than the process. Consequently, their time constants and dead time can often be considered negligible and thus, their transfer function is given by a pure gain. However, when the dynamics must be considered, it is usual practice to represent the transfer function of the instrument by a first-order or second-order system:

$$G(s) = \frac{K_T}{\tau s + 1}$$

or

$$G(s) = \frac{K_T}{\tau^2 s + 2\tau\zeta s + 1}$$

WORDS AND TERMS

sensor   *n.* 传感器
span   *n.* 测量范围
calibrate   *v.* 校准
gain   *n.* 增益

orifice n. 孔,口
square root extractor 开方器

dynamics n. 动态特性

NOTES

[1] The value does not have to be zero in order to be called the zero of the instrument.
仪表的零点并不一定就是零。

[2] gpm = gallons per minute
每分钟加仑数。

[3] This expression shows that the gain is not constant but rather a function of flow.
此表达式说明增益不是常量,而是一个时间的函数。
not... but rather... 不是……而是……

B Final Control Elements and Controllers

Control Valves

Control valves are the most common final control elements. They are found in process plants manipulating flows to maintain controlled variables at their set points. In this section an introduction to the most important aspects of control valves as applied to process control is presented.

A control valve acts as a *variable restriction* in a process pipe. By changing its opening it changes the resistance to flow and, thus, the flow itself. Throttling flow is what control valves are all about. [1] This section presents the subject of control valve action (fail condition), control valve sizing, and their characteristics.

The first question the engineer must answer when choosing a control valve is: What do I want the valve to do when the energy supply to it fails? The question is concerned with the "fail position" of the valve. The main consideration in answering this question is, or should be, *safety*. If the engineer decides that for safety considerations the valve should close, he must then specify a "fail-closed" (FC) valve. The other possibility is that of a "fail-open" (FO) valve. When the energy supply fails, this valve will move to open its restriction to flow. The great majority of control valves are pneumatically operated and, consequently, the energy supply is air pressure. Fail-closed valves require energy to open it; therefore, they are also referred to as "air-to-open" (AO) valves. The fail-open valves that require energy to close are also referred to as "air-to-close" (AC) valves.

Let us look at an example to illustrate the choosing of the action of control valves. The example is the process shown in Fig. 4-2B-1. In this process the outlet

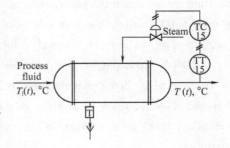

Fig. 4-2B-1 Heat exchanger control loop

temperature of a process fluid is controlled by manipulating the steam flow to heat exchanger. The question is: What do we want the steam valve to do when the air supply to it fails?

As explained above, we want the steam valve to move to the safest position. It seems that the safest condition may be the one that stops the steam flow; that is, we do not want to have steam flowing around in an unsafe operation. This means that a fail-closed valve should be specified. In making this decision we have not taken into consideration the effect of not heating the process fluid by closing the valve. In some cases this may not present any problems; however, in other cases it may have to be considered. As an example, consider the case when the steam is maintaining the temperature of a certain polymer. If the steam valve closes, the temperature will drop and the polymer may solidify in exchanger. In this case it might be decided that a fail-open valve provides the safest condition.

It is important to note that in this example only the safe condition around the heat exchanger has been taken into consideration. This may not necessarily be the safest overall operation; that is, the engineer needs to look at the complete plant rather than only one piece of equipment.[2] He must look at the effect on the heat exchanger and also on any other equipment, from which the steam and process fluid are coming or going. To repeat, he must take into consideration the *complete plant safety*.

Feedback Controllers

This section presents the most important types of industrial controllers. The physical significance of their parameters is stressed to aid in the understanding of how they work. The presentation holds true for both pneumatic electronic controller and most microprocessor-based controllers.

Briefly, the controller is the "brain" of the control loop. As mentioned in unit 1, the controller is the device that performs the decision (D) making in the control system. To do this, the controller:

1) Compares the process signal from the transmitter, the controller variable, with the set point.

2) Sends an appropriate signal to the control valve, or any other final control element, in order to maintain the controlled variable at its set point.

Consider the heat exchanger control loop shown in Fig. 4-2B-1. If the outlet temperature of the hot fluid moves above the point, the controller must close the steam valve. Since the valve is an air-to-open (AO) valve, the controller must reduce its output (air pressure or current) signal (see the arrows in the Fig. 4-2B-1). To make this decision the controller must be set to *reverse action*. Some manufactures refer to this action as *decrease*. That is, upon an *increase* in the input signal to the controller, there is a *decrease* in the output signal from the controller.

Consider now the level control loop show in Fig. 4-2B-2.

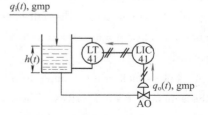

Fig. 4-2B-2 Liquid level control loop

If the liquid level moves above its set point, the controller must open the valve to bring the level back to set point. Since the valve is an air-to-open (AO) valve, the controller must increase its output signal (see the arrows in the figure). To make this decision the controller must be set to *direct action*. Some manufactures refer to this action as *increase*. That is, upon an *increase* in the input signal to the controller, there is an *increase* in the output signal from the controller.

In summary, to determine the action of a controller, the engineer must know:

1) The process requirements for control;
2) The action of the control valve or other final control element.

Both things *must* be taken into consideration. The reader can ask himself what the correct action of the level controller is if an air-to-close (AC) valve is used or if the level is controlled with the inlet flow instead of outlet flow. In the first case the control valve action changes while in the second case it is the process requirements for control that change.

The controller action is usually set by a switch on the side of pneumatics or electronic controllers, or by a configuration bit on most microprocessor-based controllers.

WORDS AND TERMS

throttle　　*v.* 调节（阀门），用（阀门）调节
fail-closed (FC)　　*adj.* 无信号则关的
fail-open (FO)　　*adj.* 无信号则开的
air-to-open (AO)　　*adj.* 气开的
air-to-close (AC)　　*adj.* 气关的
polymer　　*n.* 聚合物
hot exchanger　　热交换器

NOTES

[1] Throttling flow is what control valves are all about.
控制阀的作用就是调节流量。

[2] This may not necessarily be the safest overall operation; that is, the engineer needs to look at the complete plant rather than only one piece of equipment.
这并不一定是最安全的全局操作，也就是说，工程师应从整个工厂着眼，而不是仅注意某一台设备。

C　简　历

简历（Resume 或 Curriculum vitae）主要由个人简况（personal data）、学历（education）和经历（experience）三个部分组成，还可以根据不同需要增加一些项目，如求职目标（job objective）、外语技能（foreign language skills）、兴趣爱好（hobbies and interests）、奖励（honors and awards）等。清楚详细的简历是求职、求学者所必备的。请看下例：

RESUME

Name：　　　　Wang Jingru

Sex:	Male	;或 Female
Address:	Rm. 3-358, 12 Guangrong Road, Tianjin, 300130	
Telephone:	(O)26545801-189	;O 表示 office
	(H)23456789	;H 表示 home
Mobile phone	13800138000	
E-mail:	Wjr@heut.edu.cn	
Date of birth:	March 16, 1990	
Nationality:	Chinese	
Marital status:	Single	;或 Married (no/two children)
Health:	Good	

Education:
    9/2012-12/2014   Master of Automation
                               Hebei University of Technology
    9/2008-7/2012   Bachelor of Automation
                               Tianjin Institute of Science and Technology

Work experience:
    3/2015-present   Siemens Electrical Drives Ltd.   Tianjin
    7/2011-8/2011   Motorola Electronic Ltd.   Tianjin
    7/2010-8/2010   Lenovo Computer Ltd.   Beijing

Scholastic honors: Pan Chengxiao Scholarship in Hebei Univ. of Tech. 96–97
                                First-prize winner in Tianjin College Students Computer Contest 97

Foreign language: English, passed CET-6 and fluent in spoken English
                                Japanese, able to read

从上例可看出，英文简历的写作比较灵活，多使用名词词组表达。项目内容可根据需要做适当的调整，增加或去掉一些条目，总体格式较为自由。和中文简历不同的是，英文简历中的学历和工作经历可以按正序也可以按倒序排列，即越近的经历排得越靠前。

一般情况下，简历不宜手抄，以打印件为佳。

UNIT 3

A P Controllers and PI Controllers

The way feedback controllers make a decision in order to maintain the set point is by calculating the output on the basis of the difference between the controlled variable and set point. In this unit we will look at the most common types of controllers by looking at the equations that describe their operation.

Proportional Controller (P)

The proportional controller is the simplest type of controller, with the exception of the on-off controller, which we shall not discuss here. The equation that describes its operation is the following:

$$m(t) = \overline{m} + K_c[r(t) - c(t)] \qquad (4\text{-}3\text{A-}1)$$

or

$$m(t) = \overline{m} + K_c e(t) \qquad (4\text{-}3\text{A-}2)$$

where

$m(t)$ = output from the controller, psig or mA;

$r(t)$ = set point, psig or mA;

$c(t)$ = controlled variable, psig or mA. This is the signal from the transmitter;

$e(t)$ = error signal, psi or mA. This is the difference between the set point and the controlled variable;

K_c = controller gain, $\frac{\text{psi}}{\text{psi}}$ or $\frac{\text{mA}}{\text{mA}}$;

$\overline{m}$ = bias value, psig or mA. The significance of this value is the output from the controller when the error is zero. This value is usually set, during calibration of the controller, at mid-scale, 9 psig or 12 mA.

Since the input and output ranges are the same (3 – 15 psig or 4 – 20 mA), the input and output signals, as well as the set point, are sometimes also expressed in either fraction or percent of range.

It is interesting to note that Eq. (4-3A-1) is written for a reverse-acting controller. If the controller variable, $c(t)$, increases above the set point, $r(t)$, the error becomes negative and, the equation shows the output from the controller, $m(t)$, decreases. The usual way to show a direct-acting controller mathematically is by letting the controller gain, K_c, be negative. We must remember, however, that in industrial controllers there are no negative gains, only positive ones. The reverse/direct switch takes care of this. The negative K_c is used when doing a mathematical analysis of a control system that requires a direct-acting controller.

Eqs. (4-3A-1) and (4-3A-2) show that the output of the controller is proportional to the error between the set point and controlled variable. The proportionality is given by controller gain, K_c. This gain, or controller sensitivity, determines how much the output from the controller changes for a given change in error. This is shown graphically in Fig. 4-3A-1.

Proportional only controllers have the advantage of only one tuning parameter, K_c.[1] However, they suffer a major disadvantage, that of operating the controlled variable with an OFFEST, or "steady-state error". To show this offset graphically, consider the liquid level control loop show in Fig. 4-2B-2. Assume that the design operating conditions are $\bar{q}_i = \bar{q}_o = 150 g/m$, and $\bar{h} = 6 ft$. Let us also assume that in order for the outlet valve to pass 150 gpm, the air pressure over it must be 9 psig. If the inlet flow, q_i, increases, the response of the system with proportional only controller may look like Fig. 4-3A-2. The controller brings the controlled variable back to a steady value, but this value is not the required set point. The difference between the set point and the steady-state value of the controlled variable is the offset. Fig. 4-3A-2 shows two response curves corresponding to two different values of turning parameter K_c. This figure shows that the lager the value of K_c, the smaller the offset but the more oscillatory the response of the process becomes. For most processes, however, there is a maximum value of K_c beyond which the process will go unstable. The calculation of this maximum value of gain is what we shall call the ultimate gain, k_{cu}.

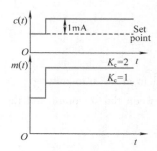

Fig. 4-3A-1 Effect of controller gain on output of controller

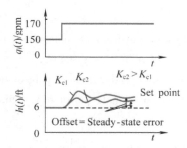

Fig. 4-3A-2 Response of liquid level system

To obtain the transfer function for proportional only controller, Eq. (4-3A-1) can be written as
$$m(t) - \bar{m} = K_c[e(t) - 0]$$
We define the following two deviation variables:
$$M(t) = m(t) - \bar{m} \qquad (4\text{-}3A\text{-}3)$$
$$E(t) = e(t) - 0 \qquad (4\text{-}3A\text{-}4)$$
Then
$$M(t) = K_c E(t)$$
Taking the Laplace transform gives the following transfer function:
$$\frac{M(s)}{E(s)} = K_c$$

To summarize briefly, proportional only controllers are the simplest controllers with the advantage of only one tuning parameter, K_c or PB. The disadvantage of those controllers is the operation with an offset in the controller variable. In some processes, such as a surge tank, this may not be of any major consequence. In cases in which the process can be controlled within a band from the set point, proportional only controllers are sufficient. However, in processes in which the control must be at set point, proportional only controllers will not provide satisfactory control.

Proportional-Integral Controller (PI)

Most processes can not be controlled with an offset; that is, they must be control at set point. In these instances an extra amount of intelligence must be added to the proportional only controller to remove the offset. The new intelligence, or new mode of control, is the integral, or reset, action; consequently, the controller becomes a proportional-integral controller (PI). The descriptive equation is as follows:

$$m(t) = \overline{m} + K_c[r(t) - c(t)] + \frac{K_c}{\tau_I}\int [r(t) - c(t)]dt \qquad (4\text{-}3A\text{-}5)$$

or

$$m(t) = \overline{m} + K_c e(t) + \frac{K_c}{\tau_I}\int e(t)dt \qquad (4\text{-}3A\text{-}6)$$

where τ_I = integral or reset time, minutes/repeat.

Therefore, the PI controller has two parameters, K_c and τ_I, that must be tuned to obtain satisfactory control.

To understand the physical significance of the reset time, τ_I, consider the hypothetical example shown in Fig. 4-3A-3. τ_I is the time that it takes the controller to repeat the proportional action, and consequently the units are minutes/repeat. The smaller the value of τ_I, the steeper the response curve, which means the faster the controller response becomes. Another way of explaining this is by looking at Eq. (4-3A-6). The small the value of τ_I, the large the term in front of the integral, K_c/τ_I, and, consequently, the more weight is given to integration, or reset, action.

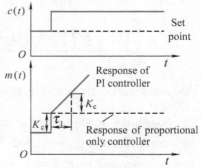

Fig. 4-3A-3 Response of proportional-integral (PI) controller to a step change in error

Also, from Eq. (4-3A-6) we note that as long as the error term is present, the controller will keep changing its output, thereby integrating the error, to remove the error.[2] Remember that integration also means summation.

To obtain the transfer function for the PI controller Eq. (4-3A-6) is written as follows:

$$m(t) - \overline{m} = K_c[e(t) - 0] + \frac{K_c}{\tau_I}\int [e(t) - 0]dt$$

Using the same definitions of deviation variables given by Eqs. (4-3A-3) and (4-3A-4), taking the Laplace transform, and rearranging yields

$$\frac{M(s)}{E(s)} = K_c(1 + \frac{s}{\tau_I})$$

To summarize, proportional-integral controllers have two parameters: the gain or proportional band, and the reset time or reset rate. The advantage of this controller is that the integral, or reset, action removes the offset.

WORDS AND TERMS

calibration *n.* 校准，检查
fraction *n.* 分数，小数
offset *n.* 静差
ultimate *adj.* 临界的

weight *n.* 权
reset time 复位时间
reset rate 复位速率

NOTES

[1] Proportional only controllers have the advantage of only one tuning parameter, K_c.
纯比例控制器的优点是只需整定一个参数 K_c。

[2] Also, from Eq. (4-3A-6) we note that as long as the error term is present, the controller will keep changing its output, thereby integrating the error, to remove the error.
从式（4-3A-6）中，我们注意到只要误差项存在，控制器就会不断改变其输出，从而通过对误差积分来消除误差。

B PID Controllers and Other Controllers

Proportional-Integral-Derivative Controller (PID)

Sometimes another mode of control is added to the PI controller. This new mode of control is the derivative action, also called the rate action or preact. Its purpose is to anticipate where the process is heading by looking at the time rate of change of the error, its derivative. The descriptive equation is the following:

$$m(t) = m + K_c e(t) + \frac{K_c}{\tau_I} \int e(t) \mathrm{d}t + K_c \tau_D \frac{\mathrm{d}e(t)}{\mathrm{d}t} \qquad (4\text{-}3\text{B-}1)$$

where τ_D = derivative, or rate, time in minutes.

Therefore, the PID controller has three parameters K_c or PB, τ_I or τ_I^R, and τ_D, that must be tuned to obtain satisfactory control. Notice that there is only one derivative tuning parameter, τ_D. It has the same units, minutes, for all manufactures.

As just mentioned, the derivative action gives the controller the capability to anticipate where the process is heading, that is, to "look ahead", by calculating the derivative of the error. The amount of "anticipation" is decided by the value of the tuning parameter, τ_D.

Let us consider the heat exchanger shown in Fig. 4-2B-1 and use it to clarify what is meant by "anticipating where the process is heading." Assume that the inlet process temperature decreases by some amount and the outlet temperature starts to decrease correspondingly as shown in Fig. 4-3B-1. At time t_a the amount of the error is positive and may be small. Consequently, the amount of control correction provided by the PI mode is small. However, the derivative of this error, the slope of the error curve, is large and positive, making the control correction provided by the derivative mode large. By looking at the derivative of the error, the controller knows that the controlled variable is heading away from the set point rather fast and, consequently, it uses this fact to help in controlling. At time t_b the error is still positive and larger than before. The amount of control correction provided by the proportional and integral modes is also larger than before and is still adding to the output of the controller to further open the steam valve. However, the derivative of the error at this time is negative, signifying that the error is decreasing; that is, the controlled variable has started to come down to set point. Again, using this fact, the derivative mode starts to subtract from the other two modes since it recognizes that the error is decreasing. By doing this it takes longer for the process to return to set point; however, the overshoot and oscillations around set point are reduced.

PID controllers are used in processes with long time constants. Typical examples are temperature and concentration loops. Processes with short time constants (small capacitance) are fast and susceptive to process noise. Typical of these processes are flow loops and loops controlling the pressure of liquid streams. Consider the recording of a flow shown in Fig. 4-3B-2. The application of the derivative mode will only result in the amplification of the noise because the derivative of the fast changing noise is a large value. Long time constant processes (large capacitance) are usually damped and, consequently, are less susceptible to noise. Be aware, however, that you may have a long time constant process, a temperature loop for example, with a noisy transmitter. In this case, before the PID controller is used the transmitter must be fixed.

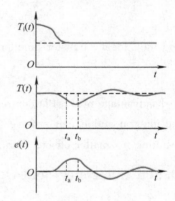

Fig. 4-3B-1 Heat exchanger control

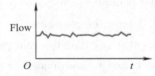

Fig. 4-3B-2 Recording of a flow loop

The transfer function of an "ideal" PID controller is obtain from Eq. (4-3B-1) by rearranging it as follows:

$$m(t) - \overline{m} = K_c[e(t) - 0] + \frac{K_c}{\tau_I}\int [e(t) - 0]dt + K_c\tau_D \frac{d[e(t) - 0]}{dt}$$

Using the same definitions of deviation variables given by Eqs. (4-3A-3) and (4-3A-4), taking the Laplace transform, and rearranging yields:

$$\frac{M(s)}{E(s)} = K_c(1 + \frac{1}{\tau_I s} + \tau_D s)$$

This transfer function is called "ideal" because in practice the implementation of the derivative calculation is impossible to obtain. The derivative is then approximated by the use of a lead/lag, resulting in the "actual" transfer function:

$$\frac{M(s)}{E(s)} = K_c\left(1 + \frac{1}{\tau_I}\right)\left(\frac{\tau_D + 1}{\alpha\tau_D + 1}\right) \quad (4\text{-}3\text{B-}2)$$

Typical value of α range between 0.05 and 0.1.

To summarize, PID controllers have three tuning parameters: the gain or proportional band, the reset time or reset rate, and the rate time. Rate, or derivative, time is always in minutes. PID controllers are recommended for long time constant loops which are free of noise.[1] The advantage of the derivative mode is that it provides the capability of "looking where the process is heading".

Proportional-Derivative Controller (PD)

This controller is used in processes where a proportional only controller can be used but some amount of "anticipation" is desired. The descriptive equation is

$$m(t) = \overline{m} + K_c e(t) + K_c \tau_D \frac{de(t)}{dt}$$

and the "ideal" transfer function is

$$\frac{M(s)}{E(s)} = K_c(1 + \tau_D s)$$

A disadvantage of the PD controller is the operation with an offset in the controlled variable. Only the integral action can remove the offset. However, a PD controller can stand higher gain, thus resulting in smaller offset, than a proportional only controller on the same loop.

Digital Controllers and Other Comments

As previously mentioned, Eq. (4-3B-2) is the transfer function of industrial analog controllers; however, the equation of digital controller is the discrete form of Eq. (4-3B-1). The methods to tune digital controllers are not very different from those to tune analog controllers.

A few other comments on controllers are in order before closing this section.[2] Eq. (4-3B-1) shows that any time the gain parameter K_c is changed, this affects the integral and derivative actions since τ_I and τ_D are divided or multiplied by it. This means that if you want to change only the gain action and not the amount of reset or anticipation, then the parameters τ_I and τ_D must also be changed to accommodate for the change in K_c. All analog controllers are of this type and are sometimes referred to as "interacting controller". Most microprocessor-based controllers are also of this type. However there are some that have avoided this problem by substituting the K_c/τ_I term by the single term K_I and the $K_c\tau_D$ term by K_D. This means that the three tuning parameter are K_c, K_I, and K_D.

The final comment is related to the derivative action. The typical way to change the controller set point is by introducing a change as shown in Fig. 4-3B-3a. When this happens, a step change in error is also introduced, as shown in Fig. 4-3B-3b. Since the controller takes the derivative of the error, this derivative produces a sudden change in the controller output, as shown in Fig. 4-3B-3c. This change in controller output is unnecessary and possibly detrimental to the process operation. It has been proposed that a way to get around this problem is by using the negative of the derivative of the controlled variable

$$-\frac{dc(t)}{dt}$$

instead of the derivative of the error. The two derivatives are equal when the set point remains constant, as can be shown by the following:

$$\frac{de(t)}{dt} = \frac{d[r(t) - c(t)]}{dt} = \frac{dr(t)}{dt} - \frac{dc(t)}{dt}$$

where $\frac{dr(t)}{dt} = 0$.

At the moment the set point change is introduced, this "new" derivative does not produce the sudden change. Immediately afterward, the behavior becomes the same as before. This option is offered by some analog and microprocessor-based controllers and is referred to as derivative-on-controlled variable.

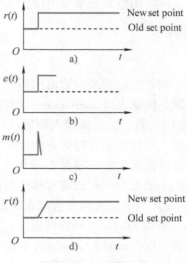

Fig. 4-3B-3 Effect of set-point change

Another possible way to avoid this derivative problem can be easily obtained by the use of digital controllers. This option changes the set point in ramp shape, even though the operator may have change the set point in s step fashion, as shown in Fig. 4-3B-3d. The slope of the ramp is predetermined by the operating personnel.[3]

This unit has presented the subject of process controllers. We have seen the purpose of the controllers: to make the decision of how to use the manipulated variable to maintain the controlled variable at set point. The different types of controllers were also studied, stressing the significance of the tuning parameters gain (K_c) or proportional band (PB), reset time (τ_I) or reset rate (τ_I^R), and rate time (τ_D).

WORDS AND TERMS

preact   v. 提前；n. 超前，提前修正量
susceptive   adj. 对……敏感的，易受……影响的
lead   n. 超前

lag   n. 滞后
discrete   adj. 离散的，分离的
get around   回避，躲开
detrimental   adj. 不利的

NOTES

[1] PID controllers are recommended for long time constant loops which are free of noise.
PID 控制器适用于不易受噪声干扰的具有长时间常数的环路系统。

[2] A few other comments on controllers are in order before closing this section.
在结束这节之前，有必要再对控制器做一些说明。
be in order 意为"完备的，适用的，适宜的，必要的"。

[3] The slope of the ramp is predetermined by the operating personnel.
斜坡函数的斜率要由操作人员预先设定。

C 面　　试

　　到外企应聘或申请出国留学签证，面试（personal interview）常常用英语进行。短短的面试是面试人（interviewer）与应试人（applicant）之间达到双向了解的过程，应试人尽力"推销"自己，而面试人则通过对应试人的外表、举止与谈吐了解对方的专业技能、性格、志向、人际关系处理等方面是否符合要求，同时也让应试人有机会了解情况。招聘单位往往通过面试来最后决定是否录用应聘人员。在与西方人交谈时，要注意东西方之间的文化差异，不能把中国式的客套移植到英语中去，以免造成误解。特别应该注意的是：

　　1）当面试人问及应试人所具备的能力时，要如实相告，不可谦虚。如面试人问：Can you speak English? 回答可以是：Yes, I can. I've learned English for 5 years. 但不要故作谦虚地说：I'm afraid I know very little English.

　　2）当应试人听到赞扬时，不可用自我否定的方法来表示谦虚。如面试人说：I can see you're quite experienced in this. 正确的回答是：Thank you. 不要说：No, I'm not good at that.

　　3）当提及报酬时，不要用中国"君子不言财"的观念去回避。如面试人问：Do you think that sum of money is OK for you? 应试人不要说：Oh, money is not important to me. I don't come here for money. 应该看到自身的价值，可以说：Good, I'll prove to be worth of it.

　　当应试人在回答问题时，要尽量与未来的工作联系起来，给对方一种印象，认为你的性格、专业知识、能力、抱负都是适合做该工作的，回答问题要坦率、诚实。面试前，应试人可以预测会发生什么情况或者将被问及的一些共同性的问题以及与特殊目的相关的特殊问题，把这些问题排列一下，做些面试前的练习。下面的情况可供应试人参考：

　　1）问候与介绍。应试人初次会见面试人时应该主动向对方致意，可以说"How do you do?"，不说"How are you?"。如不知面试人的姓名，可称呼"Sir""Madam"或"Ms."（已婚或未婚女士），以示尊敬。称呼与头衔应该与姓氏（last name）或姓名（full name）合用，不可以与名字（first name）合用。

　　2）个人情况。应试人的姓名（name）、出生年月和地点（date and place of birth）、地址（address）、婚姻（marital）、家庭（family background）、目前工作（present job）和健康状况（health）等。

　　3）文化程度。通常指应试人的学历（record of formal schooling）和学位（academic degree）。但是，面试人往往还会问及应试人的学习专业（field of study or major）、学习成绩（test score or academic record）、毕业论文（graduation thesis）、学科爱好（subject preference）、奖惩情况（awards and punishments）和社会活动（social activities）等。

　　4）工作经历。在许多用人单位，雇主对求职者的工作经历比对它的文化程度更为重视。应试人的工作经历涉及面很广，包括应试人曾供职的单位（place of work/office）、部门（department）及其从事过的职业（occupation），还包括应试人的职务（post）、职称（professional title）、专长（special skill）、成果（achievement）、著述（publication）和奖惩情况等。如果应试人是在校或刚毕业的学生，可以将自己的实习（fieldwork）或兼职打工

(part-time job) 的情况——叙述，以证明自己具有一定的工作经历。

5) 应试人的动机。所谓动机是指应试人求职、求学或申请签证的原因 (reasons for application or transfer)。此外，面试人还会就应试人所给予的理由提出进一步的问题，如应试人对招聘单位的了解程度，是否还同时向其他单位申请工作等。如果是申请出国签证，面试人还会向应试人问及在国外的经济来源，在国外停留多久以及是否打算回国等。

6) 个性与爱好。一个人的个性与爱好对他所从事的工作影响很大，而且，一个人的业余爱好还反映一个人的文化修养。正因为如此，面试人常常会问及应试人的业余爱好。

7) 漫谈是面试中的又一种方式。通过漫谈，面试人可以进一步了解应试人的知识面、相关能力和品德。漫谈的话题涉及一般常识 (general knowledge)、时事 (current affairs)、个人看法 (personal outlook) 和志向与抱负 (ambitions and aspirations) 等。

8) 在面试中，应试人往往处于被动地位，只是回答面试人的问题。实际上，面试人通常都会留一些时间让应试人提问，应试人应该抓住这个机会，主动地对自己急需了解的方面提些问题。放弃提问意味着面试中的一处"败笔"。但是，应试人提问时一定要注意礼貌，不能随意打断对方的谈话。提问前应该说"Excuse me."，"May I ask a question?"。

9) 面试结束后，面试人与应试人互相道别，并交待以后的联系方法。一般来说，面试结束时不会马上就有结果，但应试人仍然应该有礼貌地向面试人表示感谢。面试后，应写封短信给招聘单位或面试人，一方面表示对能有这次面试机会表示感谢，同时再一次表示自己希望能得到这份工作的愿望，但信要写得简要明了。

以下是一个申请签证的例子。冯涛去美国领事馆申请赴美留学的签证，领事约翰逊先生在与他面试。(J = 约翰逊，F = 冯涛)

J: Have a seat, please. May I know your name?

F: My name is Feng Tao.

J: What are you doing now?

F: I'm working in a design department in a machine tool factory.

J: Why do you want to go to the U. S.?

F: For two reasons. First, I want to study computer science. Because with the rapid development of our machine tool factory, CAD (Computer Aid Design) is beginning to be used in the factory. I hope to be a specialist in this area. The other reason is that America is the most advanced country in the computer science and computer application.

J: Who's your sponsor?

F: My uncle will finance all my expenses. He is a doctor in a children hospital in California. And I've also got the first year scholarship from the university.

J: How long will you stay in the States?

F: I want to get a master's degree, so I plan to be there for two years.

J: Do you think your English level can help you study in the American university?

F: Yes, I've received high score on the TOEFL and GRE. And now I'm attending an English conversation course to improve my spoken English.

J: Fine. We'll certainly consider your case. That's all. Good-bye.

F: Thank you. Good-bye.

UNIT 4

A Indicating Instruments

General Introduction

Generally an indicator consists of an assembly for producing and controlling ① the motion of a pointer with relation to a stationary scale, or ② the motion of a scale with relation to a fixed point or line of reference, or ③ the presentation of data in digital form (numeric or alphanumeric). Within this fundamental framework there are scores of variations.

Another classification would break indicating instruments down into three categories: ① mechanical types of devices as for example, the familiar dial-type pressure gage or the rise or fall of a column of mercury in a thermometer or barometer, ② optical types of designs wherein the moving pointer may be a weightless beam of radiation, as for example, a light-beam-type galvanometer or one of the many configurations of cathode-ray tube (CRT) and associated electronic display tube indicators, and ③ combinations of electromechanical, electrooptical, and even electrochemical principles.

For convenience, indicators also may be classified as ① analog indicators in which all or portion of a calibrated scale is in evidence so that the observer sees the indication in continuum so to speak-against all or part of the total range within which the indicator may swing, and ② digital indicators in which information is indicated a piece at a time. In addition to precision digital-type indicators, alarms and annunicators also operate on a pseudo-digital principle, in that they only indicate certain Go/No-Go types of situations as in the use of colored lights or horns and bells to indicate limits that have been exceeded. [1]

There are several forms of indication, reasonably uncommon except for very specific purposes, which illustrate the ingenuity that instrument engineers have applied to data display problems. The pyrometric cone and colored crayons and paints which indicate (historically) whether or not a certain temperature range has been exceeded serve a real need in certain high-temperature measurement situations, particularly where motion, as in tunnels and kilns, is involved. Manually operated, color-matching optical pyrometers also present an unusual and interesting indicating format.

Indicating instruments also can be classified in terms of speed which, unless associated with a rapid-recording means, must fall within the capabilities of human identification and resolution. [2] Although these characteristics vary considerably from one human being to the next, the standard of 24 frames per second as projected by motion picture equipment is indicative of the human limitation in time sensing individual events. In selecting the most appropriate (including economic justification) indicating mechanism for an instrument, the designer must consider the overall response of the measuring equipment as well as the time related importance of the measured data and the operator/instrument interface—and, in so doing, thus will not over-engineer or under-design the mode of indication. [3]

Moveable-pointer Indicators

Conventional fixed-scale movable-pointer indicators vary in appearance, form of scale, and the plane in which the indicating pointer moves. In all cases it is essential that the scale be graduated and mounted so as to minimize reading errors. To avoid parallax errors a precision indicating instrument, which usually is read at very close range, often employs a mirror mounted beneath the pointer and adjacent to the scale graduations. In another arrangement the graduated portion of the scale is raised to the level (nearly) of the pointer tip, thus enabling accurate readings from greater angles of observation. The index may be in the form of a thin edgewise strip, a hairline scratched on each side of a transparent member, or a sharp arrow-shaped tip. Illumination, external or internal, is extremely important. Indicators that are intended for distant reading present entirely different design criteria, but usually have fewer graduations and pointers with proportionately large tip areas.

Frequently, the scales are used with two indicating pointers. The first pointer can be used to indicate the value of a first variable. The second pointer may indicate the value of a second variable, or it may function as a manually set index or reference spotter. All the forms essentially are purely mechanical in nature, and most are widely used. Some typical movable-pointer indicators include straight scale, arc scale and segmental scale types, the scales of which are either vertically or horizontally mounted.

Moveable-scale Indicators

The two common forms of movable-scale fixed-pointer configurations are rotating-drum-type scale and rotating-dial-type scale. The rotating-dial-type scale indicator consists of a flat disk scale and a stationary index. In practice, the angle of rotation seldom exceeds 340(with a maximum calibrated scale length of about 30 inches. While in the precision indicator, the scale is finely divided from 300 to 600 divisions in a 24-inch scale length. The stationary index generally is in the form of a sharp tip or hairline. The rotating-drum-type scale indicator consists of a drum scale and a stationary index as well. The angle of rotation, scale length, and graduations are similar to those of the rotating-dial-type scale. Both rotating-drum and dial-type indicators, as a general rule, are servo-operated. Drum and dial-type indicators permit close observations for relatively precise readings. Another type of movable-scale indicator uses a two-color painted metal strip which is moved in a vertical plane and is often used in connection with liquid-level measurements, with the dividing line representing the top of the liquid.

There is a typical panel meter in which moving scale is optically projected. The operation is accomplished by mounting a scale of photographic film to the moving coil of a core magnet mechanism in place of the usual pointer. A lamp, lens system, and mirror project this moving scale onto a coated window. The optical system expands the scale ten times its original value and the portion to be read appears on both sides of a hairline on the coated window. This design automatically avoids problems of parallax in reading. Furthermore, although long-scale lengths are available in projection, a shortened deflection angle of the mechanism is permitted, allowing high sensitivity.

There are still some other indicators in use, such as multipoint indicators, multirange indicators, digital instruments, and so on.

WORDS AND TERMS

alphanumeric   *adj.* 字母数字混合的
galvanometer   *n.* 电流计，安培计
CRT   阴极射线管
electromechanical   *adj.* 机电的，电机的
electrooptical   *adj.* 电光的
electrochemical   *adj.* 电化学的
scale   *n.* 刻度
hairline   *n.* 游丝，细测量线

pyrometric   *adj.* 高温测量的
moveable-pointer indicator   动针式仪表
moveable-scale indicator   动圈式仪表
rotating-dial indicator   旋盘式仪表
rotating-drum indicator   旋鼓式仪表
multipoint indicator   多点式仪表
multirange indicator   多量程式仪表

NOTES

[1] In addition to precision digital-type indicators, alarms and annunciators also operate on a pseudo-digital principle, in that they only indicate certain Go/No-Go types of situations as in the use of colored lights or horns and bells to indicate limits that have been exceeded.

除了精确数字式仪表外，报警器和指示器同样根据近似于数字式的原理工作，因为它们仅指示某些运行或停止的状态，比如使用彩色指示灯、喇叭和电铃来指示某些超出界限的情况。

in that = because 意为"因为"。

[2] Indicating instruments also can be classified in terms of speed which, unless associated with a rapid-recording means, must fall within the capabilities of human identification and resolution.

指示性仪表也可以根据速度来分类，除了一些同快速记录有关的仪表以外，所有的仪表，其速度都必须在人的辨别力和分辨力以内。

[3] ..., thus will not over-engineer or under-design the mode of indication.

……因而显示的模式既不会过分复杂，也不会设计欠周。

B Control Panels

The effectiveness of a carefully engineered process control system can be reduced if inadequate attention is given to the design and construction of the control panel. The control panel, whether a single panel of simplified design for local mounting or part of an extensive control center, literally is the interface between the operator and the process.[1]

Control panel design embraces several objectives:

1) To simplify the operator/process interface so that:

a. The critical, most important process conditions are immediately available to the operator without conscious effort.

b. Less important information can be called upon by the operator in accordance with operating schedules in a routine fashion, but not in any way interfere with supervision of key variables. The panel should be designed to effectively sort out and separate information elements so that the operator's attention during any time of emergency or special situations will not be diverted to the information of lesser importance, or worse, of complete irrelevance.

2) To accommodate the special control situations which arise periodically, such as process start-up, shutdown, and major maintenance either of process or control equipment and instrumentation.

3) To provide information for cost accounting and other factors of managerial importance that may not be reflected immediately in the day-to-day adjustments of operating conditions.

4) To allow for expansion and modernization of the process with time, including later ties with computerized supervision or direct computer control, but considering carefully the cost of providing for such alterations versus the probability of their need.

5) To permit convenience of installation and case of maintenance of all panel-associated equipment.

6) To provide an aesthetically pleasing appearance (particularly in the case of central panels) because the control room essentially is the "window to the process" and usually is the center of visits by VIPs. Further, an attractive control center, well lighted and air-conditioned, contributes to the efficiency of the operators and protection of the equipment.

All the foregoing objectives and more are subject, of course, to trade-offs wherein cost must be weighed against projected gains.[2] Once made, these trade-off decisions result in such practical factors as:

1) Selection of instruments in terms of their display and controllability features as well as size and maintainability.

2) Functional arrangements of the instruments, including annunciators and all the other accessories. For example, pilot lights and selector switches must be located in accordance with the anthropomorphic principles developed in the study of human engineering to minimize mental and physical fatigue.

3) Lighting of the panel and control room and other environmental factors, including air conditioning.

4) Most effective means for bringing in and out all the connecting hardware—piping, cabling, and so on—involved in the routing of information and energy to and from the control panel.

Control panel design, with so many objectives to be achieved, including the economic objectives, thus involves scores of trade-off decision. In 1960's, panel design was largely determined by what was available in standard instrumentation. Usually, the instruments were contained in very large cases and not particularly designed for remotely controlling a complex process from a control panel. Earlier instruments essentially were not designed with their mounting on a panel as a factor of ultimate importance. The situation had reversed in the late 1970's, but not without having had to go through several steps in a panel design revolution, including full-graphic and semigraphic panel design.

Numerous panel design experiments have succeeded and are now commonplace; other failed and were dropped. A major step in the improved design of panels was the direct result of miniaturized electronic, electrical, mechanical instrument components and especially the computer technique which, in turn, enabled the design of practical, miniature instruments.

In the modern design of control panel, there are some rules according to the panel design philosophy: ① controllers are mounted at eye level with their manual adjustments within a few degrees of the shoulder arm arc; ② instruments are grouped into logical operational section with spaces between process entities; ③ the visual slope of the semigraphic allows an increase in height to permit placement of the alarm window close to eye level; ④ selector switches, push buttons, and pilot lights are mounted less conveniently, as they are utilized only during abnormal unit operation; ⑤ recorders, below their associated controller, are at a height where process trends may be readily noticed, and yet do not use valuable operating space; ⑥ although opened for the purpose of the photograph, the rear of the panel may be locked to deny access to unauthorized personnel.

As mentioned before, the anthropomorphic principles of human engineering are applied in the instrument panel design in order to obtain the effective and optimum operation. In addition, the grouping of process entities assists the operator in locating the desired instrument.

WORDS AND TERMS

control panel 控制盘
interface *n.* 界面
supervision *n.* 监视
irrelevance *n.* 不相干，不切题
aesthetically *adv.* 美术地，美学地

VIP = very important person
foregoing *adj.* 前面的，以上的
accessory *n.* 附件
semigraphic *adj.* 半图解的
unauthorized *adj.* 未授权的，未经批准的

NOTES

[1] The control panel, whether a single panel of simplified design for local mounting or part of an extensive control center, literally is the interface between the operator and the process.

不论是作为局部安装简化设计的单块屏，还是作为大范围控制中心的一部分，控制盘实际上就是操作人员与控制过程之间的界面。

[2] All the foregoing objectives and more are subject, of course, to trade-offs wherein cost must be weighed against projected gains.

所有前面提到的目标和其他问题当然要经过比较，其中，投资花费和预期收益必须经过权衡。

C 自动化专业信息检索

目前，各个高校和科研单位的内部网站都通过直接购买权限或访问镜像网站等方式，提供不同的网络科技文献数据库的链接，主要的科技文献网站有：

（1）CNKI 科技文献数据库 收录国内近 8000 种综合期刊与专业特色期刊的全文、优秀硕士和博士论文全文、国内重要会议论文和中国年鉴全文等科技文献资料，是目前世界上最大的连续动态更新的中国科技文献全文数据库。访问地址：http：//www.cnki.net/。

（2）万方期刊全文库 收录国内涉及理、工、农、医、哲学、人文、社会科学、经济管理与教科文艺等 8 大类 100 多个类目的近 6000 种各学科领域核心期刊，实现全文上网，论文引文关联检索和指标统计。访问地址：http：//bj.wanfangdata.com.cn/。

（3）中文科技期刊数据库——维普资讯 包含了 1989 年以来的 9000 余种中文科技期刊刊载的文献，每年以 250 万篇的速度递增，分为 8 个专辑：社会科学、自然科学、工程技术、农业科学、医药卫生、经济管理、教育科学和图书情报。访问地址：http：//www.cqvip.com/。

（4）超星数字图书网 全球最大的中文数字图书网，提供丰富的电子图书阅读，其中包括文学、经济、计算机等五十余大类。访问地址：http：//www.chaoxing.com/。

（5）北京图书馆（中国国家图书馆） 访问地址：http：//www.nlc.gov.cn/。

（6）中国科学院文献情报中心 访问地址：http：//www.las.ac.cn/。

（7）IEEE Explore Database 提供 IEEE 协会数据库中包括电气工程、计算机科学和电子学等领域的高水平学术论文全文。访问地址：http：//ieeexplore.ieee.org/Xp'ore/home.jsp。

（8）Ei Village 由 Ei 公司在互联网上提供的网络数据库，其核心 Engineering Index（工程索引）闻名于世。Ei Village 是以 Ei CompendexWeb 为核心数据库，将世界范围内的工程信息资源组织、筛选、集成在一起，向用户提供便捷式服务，是目前全球最全面的工程领域二次文献数据库，侧重提供应用科学和工程领域的文摘索引信息。访问地址：http：//www.engineeringvillage2.org/。

（9）ProQuest Digital Dissertations（PQDD） 世界著名的学位论文数据库，收录有欧美 1000 余所大学文、理、工、农、医等领域的博士、硕士学位论文，是学术研究中十分重要的信息资源。访问地址：http：//wwwlib.global.umi.com/dissertations。

（10）谷歌学术 该项索引包括了世界上绝大部分出版的学术期刊，可广泛搜索学术文献。可以从一个位置搜索众多学科和资料来源：来自学术著作出版商、专业性社团、预印本、各大学及其他学术组织的经同行评论的文章、论文、图书、摘要和文章。谷歌学术搜索可帮助您在整个学术领域中确定相关性最强的研究。谷歌学术搜索的功能有：①从一个位置方便地搜索各种资源；②查找报告、摘要及引用内容；③通过您的图书馆或在 Web 上查找完整的论文；④了解任何科研领域的重要论文。访问地址：http：//scholar.google.com。

PART 5

Control Based on Network and Information

UNIT 1

A Automation Networking Application Areas

Networking is used in all areas of automation. In factory automation, process automation and building automation networks perform diverse tasks. Likewise, there are distinct differences between tasks performed for applications in different industry sectors that all have unique characteristics and consequently varying requirements. The way devices are connected, configured, and exchange data also differ.

There is no one-size-fits-all for industrial networks; rather, buses are optimized for different characteristics. For example, factory automation and process automation are often used in harsh and hazardous environments where people, nature, and expensive machinery are at stake or where a production interruption is costly. These requirements contrast significantly with building automation, for example, where keeping costs low is a main driving force. [1]

Factory Automation

Factories with assembly-line manufacturing, as in the automotive, bottling, and machinery industries, are predominantly controlled using discrete logic and sensors that sense whether or not, for example, a process machine has a box standing in front of it. The network types ideal for simple discrete I/O focus on low overhead and small data packets, but they are unsuitable for larger messages like configuration download and the like. [2] Examples of this network type are Seriplex®, Interbus-S, and AS-I (AS-Interface), which are sometimes called *sensor buses* or *bit level buses*. Other more advanced protocols oriented toward discrete logic include DeviceNet™, ControlNet™, and PROFIBUS (DP and FMS application profiles). These buses are sometimes referred to as *device buses* or *byte-level buses*. Factory automation involves fast-moving machinery and therefore requires quicker response than slower processes. Traditionally, these tasks have been handled by PLCs.

Process Automation

Process plants in industry segments like refining, pulp & paper, power, and chemicals are dominated by continuous regulatory control. Measurement is analog (here meaning scalar values transmitted digitally), and actuation is modulating. Of course, process industries also use some discrete control and the predominantly discrete manufacturing industries use some modulating. Fieldbus on/off valves are already available in the market, as are small remotely mounted I/O modules for discrete sensors. In the past, a DCS or single-loop controller did this.

Process-related networks include FOUNDATION Fieldbus, PROFIBUS (PA application profile), and HART. All these buses as a category are now typically referred to as fieldbus (without the capital F), though some would argue that one or the other does not belong. These three protocols were specifically designed for bus-powered field instruments with predefined parameters and commands for asset management information like identification, diagnostics, materials of construction, and functions for calibration and commissioning. In terms of size, the networks used in industrial automation are considered to constitute local area networks (LAN) spanning areas no greater than a kilometer or two in diameter and typically confined to a single building or a group of buildings. Networks that extend only a few meters are insufficient, and networks that span cities or even the globe are overkill.

Field and Host Tier Networks

Even within control systems for the process sector there is a need for different network characteristics at each tier of the control system hierarchy. At the field end there are instruments such as transmitters and valve positioners that have their specific needs, and at the host level there are workstations, linking devices, and controllers that have other needs (Fig. 5-1A-1). When fieldbus began to evolve, the process industry put a large number of requirements on the field-level network that were not met by other types of networks. Many new design considerations needed to be taken into account. On the upper tier, data from all the field-level networks have to be marshaled onto a single host-level network that also serves any tasks the plant may have that seem related to factory automation.

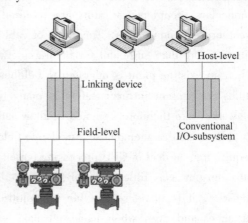

Fig. 5-1A-1 Two-tiered automation network architecture

Field Level

At the field level, the dominant protocols for process instruments are HART, FOUNDATION Fieldbus H1, and PROFIBUS PA. HART is significantly different from the other two in that it is a so-called smart protocol, that is a combination of digital communication simultaneously superimposed on a conventional 4-20 mA signal. As such, the HART protocol has been an ideal intermediate solution in the transition from analog. HART is compatible with existing analog recorders, controllers, and indicators while at the same time it makes possible remote configuration and diagnostics using digital communication. The HART protocol does allow several devices to be multidropped on a single pair of wires, but this is a capability infrequently explored because of the low update speed, typically half a second per device. For a vast majority of installations HART devices are connected point to point, that is, one pair of wires for each device and a handheld connected temporarily from time to time for configuration and maintenance. Both FOUNDATION Fieldbus H1 and PROFIBUS PA are completely digital and even use identical wiring, following the IEC 61158-2 standard. However, beyond that there are major differences between these two protocols, and depending on the desired system architecture one may be more suitable than the other.

At the field level, instruments appear in large quantities, often in the hundreds or thousands. The wire runs are very long, as the network cable must run from the control room all the way into the field, up towers, and then branching out to devices scattered throughout the site. Because there is a limit to the number of devices that can be multidropped on each network, even a medium-sized plant may have many network cables running into the field, although substantially fewer than if point-to-point wiring was used. The field-level networks were therefore designed to enable very long wire runs and to allow field devices to take their power from the network. Only a single pair of wires carries both the device's power and the digital communications signal. This eliminates the need for a separate power cable, thus keeping the wiring simple and inexpensive.

As another measure to keep costs down, designers chose a moderate field-level network speed so normal instrument-grade cable could be used instead of special data cable. No special connectors, couplers, or hubs are required either, which makes it possible to use rugged and weatherproof connections. The grade of cable used for conventional instrument connections on most sites is more than sufficient for fieldbus networking. As a result, it is possible to reuse that cable when an existing plant is migrated to fieldbus. In hazardous process environments where flammable fluids are present intrinsic safety is many times the preferred protection method. The field-level networks were therefore designed to allow safety barriers to be installed on the bus.

Because designers chose a moderate field-level network speed the devices connected to it do not require a great deal of CPU processing power to handle the communication quickly. As a result, they also consume very little power. Because the low power consumption results in low voltage drop along the wire, it is therefore possible to multidrop several devices on the network even for long wire distances and even when using intrinsic safety barriers. Another great advantage of field-level networks is that they provide a lot of freedom when it comes to network topology since wires can be run quite freely. Finally, these fieldbus networks were also designed to operate in the often rather harsh, electrically noisy environment found on site.

Host Level

At the host level, the Ethernet network standard is already the dominant wiring technology (Fig. 5-1A-2). There are many protocols built on Ethernet wiring, including FOUNDATION Fieldbus HSE, PROFInet, Modbus/TCP, and the like. Sites employing fieldbus instrumentation and asset management software can expect to encounter a steep rise in bandwidth requirements and must therefore have a high-speed network at the host level.

The field-level networks have made it possible to retrieve so much more data from the field instruments that an information explosion has resulted, one that old proprietary control level networks are unable to cope with. Ethernet provides the throughput required to transfer the large amount of data used for traditional plant operation and historical trending; for new capabilities for remote diagnostics, maintenance, and configuration; and for the quick response necessary for factory automation task. Ethernet was chosen for these applications because its high speed enables it to carry all this information. Moreover, Ethernet is already a standard and consequently is well understood and widely used. A large variety of equipment and solutions for Ethernet is available. In many applications, one of the key requirements for the host level protocol is availability. The network must be fault tolerant—up and running even in the presence of a fault. This is extremely critical at the host level since the entire site is operated and supervised over this network. Downtime can be very disruptive and cause heavy losses; a complete breakdown of the network would be extremely serious. Though Ethernet originated in the office environment, rugged industrial-grade (as opposed to commercial-grade) accessories and wiring schemes can be used. The host-level network was designed so redundancy may be used, making the network fault tolerant. Industrial-grade networks that use several layers of redundancy and industrial-hardened components can handle many simultaneous faults.

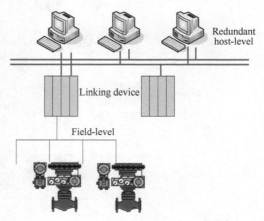

Fig. 5-1A-2 Host-level network redundancy for availability

Physical remoteness is less important for the host-level network because it is typically confined within the control room, and the distance Ethernet provides is therefore sufficient. An advantage of an established standard like Ethernet is that several media options are available. On copper wire Ethernet is unsuitable for the field because it does not run long distances. It is therefore limited to use within the control room (i. e., a "hostbus" rather than a fieldbus). However, optical fiber Ethernet can run very long distances, as can radio signals, making Ethernet suitable for remote applications.

The host-level network ties together all the subsystems the process automation system might have. In addition to the basic control function, a plant often has package units for auxiliary functions such as boilers or compressors that are bought ready-made. They have their own controls that need to be integrated with the rest of the system (Fig. 5-1A-3). For example, a refinery may have a safety shutdown system, a paper mill may have a web scanner, and a chemical plant may have an advanced control system. Subsystems based on a standard protocol on Ethernet can simply be plugged into the rest of the system.

The host-level network tier makes large systems possible by linking together field-level networks from different areas around the site. Intra-area control and supervision becomes possible. The host-level protocol is also the link to business systems, either directly or via historians and other plant

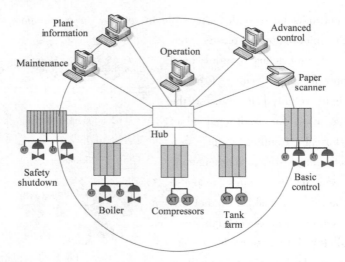

Fig. 5-1A-3 A standard host-level network ties distributed subsystems together

information software.

It is important to remember that the Ethernet standard is not a complete protocol. Essentially, Ethernet only specifies different options for cables and how devices on the network access the bus. Ethernet does not specify data formats or the semantics of the data. Even when used with other technologies like TCP/IP and UDP the protocol is incomplete. Several control system manufacturers have been using Ethernet for many years, but each one has implemented it with data formats and functionality different from the others. Even with TCP/IP, most of the Ethernet networks used in control systems on the market today are in fact proprietary since other devices cannot access and interpret the information even though connected on the same wire and existing without conflict. As a result, take great care when buying products and systems for Ethernet; they are often not as they appear to be. TCP, UDP, and IP are discussed in chapter 4 of Fieldbuses for Process Control.

It is a good idea to look for complete open protocols based on Ethernet so devices and subsystems from different sources can talk to each other, even peer to peer.

Homogeneous Network Architecture

Because of their almost opposite requirements, different network features are required at the field and host levels. Because the field-level network is slow it is unsuitable for the host level, and because the host level has too limited a distance it is unlikely it will be seen in the field. The field-level network takes the place of the traditional protocols for smart instruments and I/O subsystems, and the host-level network takes the place of the control network and business network. The host-level network in the control system uses the same networking technology as the business network so they can be integrated seamlessly. A simple router between the networks safeguards performance by keeping pure business communication traffic separate from pure control communication traffic.

For easy and tight system integration it is important to select a homogeneous network architecture in which the protocols at the higher and lower tiers are essentially the same but just traveling on different media. This will ensure transparency and a minimum of problems with communication mapping and interoperability. Fortunately, there are protocols available in such "suites". Good combinations would be FOUNDATION Fieldbus H1 and HSE or PROFIBUS PA and PROFInet. If a proprietary protocol is used at the host level or somewhere in the link between the instruments and the operator important functionality and interoperability may be lost. This may force engineers to perform time-consuming mapping of parameters between protocols.

The use of the same technology throughout the system greatly simplifies the initial engineering and deployment of the system as well as its ongoing operation and management. Engineers can readily work with different parts of the system without retraining.

WORDS AND TERMS

building automation	楼宇自动化	marshal *v.*	整顿，配置
harsh *adj.*	苛刻的	superimposed *adj.*	有层次的
modulate *v.*	调制，调节	moderate *adj.*	缓和的
calibration *n.*	校准，标度	fault tolerant	容错
commission *n.*	试车，试运转	transparency *n.*	透明
hierarchy *n.*	层次，级别	ongoing *adj.*	进行中的，不间断的

NOTES

[1] For example, factory automation and process automation are often used in harsh and hazardous environments where people, nature, and expensive machinery are at stake or where a production interruption is costly. These requirements contrast significantly with building automation, for example, where keeping costs low is a main driving force.

举例而言，工厂自动化与过程自动化通常应用在恶劣和危险的场合。而在这些场合，人员、环境及昂贵的机器设备处于危险之中，或者生产的中断会造成巨大损失。这些需求同追逐低成本的楼宇自动化相比，有明显的差异。

[2] The network types ideal for simple discrete I/O focus on low overhead and small data packets, but they are unsuitable for larger messages like configuration download and the like.

对简单的离散 I/O 较为理想的网络类型注重低开销及小的数据包，但它们不适用于组态下载等较大的报文。

B Evolution of Control System Architecture

Field signaling and system architecture developed in very close-knit fashion. Every improvement in signal transmission has subsequently led to an increased level of system decentralization and better access to field information. In the pneumatic era the controller was typically situated in the field and there operated locally. There was therefore no system to speak of.[1] With the analog current loop it became easier to bring a signal from the transmitters in the field to a central controller in the control room and then from there back out to the valves again. In the completely centralized direct digital control (DDC) architecture the complete control strategy was executed in a computer. Because all the functions were concentrated into a computer the entire system with all of its loops would fail if there were even a single fault. For this reason, it was not uncommon to have local pneumatic controllers existing in the field on standby, ready to be put in operation once the DDC failed. Clearly, the centralized architecture had some serious availability issues, which led in the early 1970s to the introduction of more decentralized programmable logic controller (PLC) and distributed control system (DCS) architecture.

DCS and PLC Architecture

The DCS and PLC emerged with the advent of digital communication, but these architectures were also designed based on 4-20 mA for field transmitters and valve positioners. However, the DCS was a great improvement over the DDC in that the controls were now distributed over several smaller controllers that shared the tasks, each one handling perhaps thirty control loops. This had the immediate benefit that a single fault would only affect part of the plant, not all of it as with the DDC. In other words, a higher level of distribution increased the availability of the system. A secondary benefit was that the configuration could be better structured where separate plant units were also kept separate in configuration and controllers. The DCS and PLC architectures are characterized by conventional I/O (input/output) subsystems or "nests" in which racks of I/O modules are networked to their respective centralized controller via an I/O-subsystem network. Field instruments were predominantly conventional analog devices. The controllers are networked with each other and to the workstations via a control-level network. There may also be a plant-level network at the very top that links the workstations to the business environment. The DCS evolved over many years, and such capabilities as communications interfaces for smart instruments that used the manufacturer's proprietary protocol became an option. This allowed some degree of configuration and check. Not all of the smart instrument protocols allowed simultaneous 4-20 mA and communication. For this reason, many were unable to use the communication feature. However, most DCS models did not provide HART interface because all the system manufacturers had their own competing proprietary protocols. Thus, plants were inclined to buy the field instruments from the system supplier rather than from third parties. A DCS can often have, in all, as many as four different tiers of networking, each with a different technology: device, I/O subsystem, controllers, and plant-wide integration to

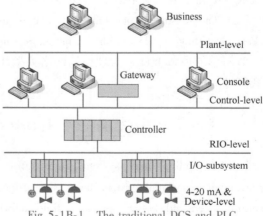

Fig. 5-1B-1 The traditional DCS and PLC architecture has multiple network levels

business applications (Fig. 5-1B-1). All these levels of hardware and networking result in a rather complex and costly system. When introduced, the DCS was christened "distributed" because it was less centralized than the DDC architecture. By today's standards, however, the DCS is considered centralized.[2] This architecture is relatively vulnerable because just one failure may have widespread consequences. Because of this vulnerability, redundancy of controllers, I/O-subsystem networking, I/O modules, and the like is a must to avoid a total loss of control. Of course, redundancy at every level means complexity and high price.

FCS Architecture

The FOUNDATION Fieldbus specification is uniquely different from other networking technologies in that it is not only a communications protocol but also a programming language for building control strategies. One of the possibilities that a standard programming language and powerful communications features enable is the ability to perform control that is distributed into the field devices rather than a central controller.[3] For example, it is common for the valve positioner to act as a controller for the loop it is part of. It executes the PID function block but only for its own loop, not for other loops. This new architecture based on field device capability is called Field Control System (FCS) and is an alternative to DCS (Fig. 5-1B-2) in that the architecture is not controller-centric. It does not treat every field device as a peripheral. Because of its decentralized nature the FCS architecture has advantages like high availability, greater scalability, and lower cost. The FCS architecture has evolved from the concept of the DCS carrying the original concept further, and the result is a system that is more distributed and therefore less vulnerable to faults. In the FCS architecture the instruments on the field-level networks are connected to the workstations via a linking device to the host-level network. Thus, there are only two network tiers in a FCS. Typically, the field instruments perform the regulatory control that in the process industries accounts for the bulk of the automation tasks. The linking device or a central controller may perform discrete logic and sequence controls. When control is performed in the field devices the number of central controllers that is required is drastically reduced and in some cases eliminated altogether. This dramatically cuts the cost of the system. In other words, wire savings are not the only hardware savings that can be achieved by using bus technology. Since the central controllers have the computation-intensive regulatory controls offloaded they are freed up to execute other controls with higher performance, thus improving controls. Because in the FCS no one controller handles multiple loops the problem of a single fault affecting a large part of the plant is largely eliminated. However, even in an FCS a centralized controller can often be found handling discrete I/O and controls since these functions are still seldom networked. Whenever a plant uses centralized controllers, it should employ redundancy if availability is a necessity. It may at first be hard to comprehend how small field device controllers could replace a "unit controller" to control a large plant. The secret behind this concept is that each device handles only one loop. By networking hundreds or thousands of devices together the combined power of the microprocessors exceeds that found in earlier systems. The control task is broken up into its components and distributed among the field devices working in parallel, with each device responsible for its loop. Since these devices work simultaneously, a true multitasking system is achieved, something that cannot be realized using only a single processor. The net result is therefore very good performance, and the more devices that are added the more powerful the system becomes. This increased power has made it possible to eliminate the need to scale analog values. For centralized systems, this scaling had not always been possible because it loaded the processor too much. Floating-point format is now used throughout the control strategy.

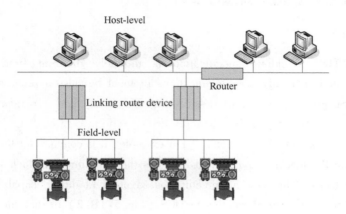

Fig. 5-1B-2   FCS architecture with control in the field devices

Host Versus System

Because a 4-20 mA signal carries only a single piece of information and only in one direction, operators had no way of determining what was going on within analog field devices. It was impossible to perform configuration, diagnostics, and other checks from the system console. In the cases where smart instruments had been adopted a handheld terminal was usually used to extract any additional information. Conventional and even smart devices were not integrated within the control system. The operator's view extended down to the controllers and possibly to the I/O subsystem, but no further. Because the field instruments were isolated entities, they were treated as separate from the control system rather than part of it.

In an FCS the field instruments are an integral part of the system as a whole. All that remains of what used to be called the system is the workstations and linking devices. The workstations that connect directly to the host-level network are simply referred to as the host (Fig. 5-1B-3).

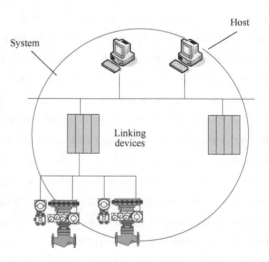

Fig. 5-1B-3   Field devices, and host, are integral parts of the system

WORDS AND TERMS

close-knit　　*adj.*　紧密的
pneumatic　　*adj.*　气动的
predominantly　　*adv.*　卓越地，突出地
incline to　　倾向于
christen　　*v.*　为……命名

vulnerability　　*n.*　弱点
scalability　　*n.*　可测量性
redundancy　　*n.*　冗余
handheld terminal　　手持终端

NOTES

［1］In the pneumatic era the controller was typically situated in the field and there operated locally. There was therefore no system to speak of.

在气动时代，控制器一般位于现场并就地进行操作，因此毫无系统可言。

［2］When introduced, the DCS was christened "distributed" because it was less centralized than the DDC architecture. By today's standards, however, the DCS is considered centralized.

DCS 之所以一开始被冠名为"分散的"，是因为它没有 DDC 体系结构那么集中。但是，如按今天的标准来衡量，DCS 应该被认为是集中的。

［3］One of the possibilities that a standard programming language and powerful communications features enable is the ability to perform control that is distributed into the field devices rather than a central controller.

标准的编程语言同强有力的通信功能相结合的结果之一是将控制分散到现场设备中，而不集中在一个控制器上。

C　国内自动化专业主要期刊

国内期刊刊名	出版地	主办单位
中国电机工程学报	北京	中国电机工程学会
电工电能新技术	北京	中国科学院电工研究所
电工技术学报	北京	中国电工技术学会
电力电子技术	西安	西安电力电子技术研究所
电气传动	天津	天津电气传动设计研究所，中国自动化学会
电气传动自动化	天水	天水电气传动研究所
电气时代	北京	机械工业信息研究院
电世界	上海	上海电气（集团）总公司、上海市电机工程学会
电气自动化	上海	上海市自动化学会，上海电气自动化设计研究所有限公司
电子测量与仪器学报	北京	中国电子学会
工业控制计算机	南京	江苏省计算研究所有限责任公司
工业仪表与自动化装置	西安	陕西鼓风机（集团）有限公司
机器人	沈阳	中国科学院沈阳自动化研究所

机器人技术与应用	北京	国家高技术计划智能机器人专家组，兵器工业部第210研究所
计算技术与自动化	长沙	中国自动化学会等
控制工程	沈阳	东北大学
控制理论与应用	广州	华南理工大学，中国科学院系统科学研究所
控制与决策	沈阳	东北大学
模式识别与人工智能	合肥	中国自动化学会，国家智能计算机研究开发中心
系统仿真学报	北京	中国航天科工集团第二研究院706所，中国系统仿真学会
系统工程学报	天津	中国系统工程学会
信息与控制	沈阳	中国科学院沈阳自动化研究所
冶金自动化	北京	冶金部自动化研究院
仪器仪表学报	北京	中国仪器仪表学会
自动化博览	北京	中国自动化学会，北京中煤电气有限公司
自动化学报	北京	中国自动化学会
自动化仪表	上海	上海工业自动化仪表研究所，中国仪器仪表学会
计算机学报	北京	中国计算机学会，中国科学院计算技术研究所

UNIT 2

A Fundamental Issues in Networked Control Systems

Feedback control systems wherein the control loops are closed through a real-time network are called networked control systems (NCSs). The defining feature of an NCS is that information (reference input, plant output, control input, etc.) is exchanged using a network among control system components (sensors, controller, actuators, etc.). Fig. 5-2A-1 illustrates a typical setup and the information flows of an NCS. The primary advantages of an NCS are reduced system wiring, ease of system diagnosis and maintenance, and increased system agility.

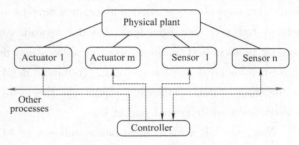

Fig. 5-2A-1 A typical NCS setup and information flows

The insertion of the communication network in the feedback control loop makes the analysis and design of an NCS complex. Conventional control theories with many ideal assumptions, such as synchronized control and nondelayed sensing and actuation, must be reevaluated before they can be applied to NCSs. Specifically, the following issues need to be addressed. The first issue is the network-induced delay (sensor-to-controller delay and controller-to-actuator delay) that occurs while exchanging data among devices connected to the shared medium. This delay, either constant (up to jitter) or time varying, can degrade the performance of control systems designed without considering the delay and can even destabilize the system. Next, the network can be viewed as a web of unreliable transmission paths. Some packets not only suffer transmission delay but, even worse, can be lost during transmission. Thus, how much packet dropouts affect the performance of an NCS is an issue that must be considered. Another issue is that plant outputs may be transmitted using multiple network packets (so-called multiple-packet transmission), due to the bandwidth and packet size constraints of the network. Because of the arbitration of the network medium with other nodes on the network, chances are that all/part/none of the packets could arrive by the time of control calculation.

The implementation of distributed control can be traced back at least to the early 1970s when Honeywell's Distributed Control System (DCS) was introduced. Control modules in a DCS are loosely connected because most of the real-time control tasks (sensing, calculation, and actuation) are carried out within individual modules. Only on/off signals, monitoring information, alarm information, and the like are transmitted on the serial network. Today, with help from ASIC chip design and significant price drops in silicon, sensors and actuators can be equipped with a network interface and thus can become independent nodes on a real-time control network. Hence, in NCSs, real-time sensing and control data are transmitted on the network, and network nodes need to work closely together to perform control tasks.

Current candidate networks for NCS implementations are DeviceNet, Ethernet, and FireWire, to name a few. Each network has its own protocols that are designed for a specific range of applications. Also, the behavior of an NCS largely depends on the performance parameters of the underlying network, which include transmission rate, medium access protocol, packet length, and so on.

There are two main approaches for accommodating all of these issues in NCS design. One way is to design the control system without regard to the packet delay and loss but design a communication protocol that minimizes the likelihood of these events. For example, various congestion control and avoidance algorithms have been proposed to gain better performance when the network traffic is above the limit that the network can handle. The other approach is to treat the network protocol and traffic as given conditions and design control strategies that explicitly take the above-mentioned issues into account. To handle delay, one might formulate control strategies based on the study of delay-differential equations. Here, we discuss analysis and design strategies for both network-induced delay and packet loss.

Now, we will analyze some basic problems in NCSs, including network-induced delay, single-packet or multiple-packet transmission of plant inputs and outputs, and dropping of network packets.

Network-induced Delay The network-induced delay in NCSs occurs when sensors, actuators, and controllers exchange data across the network. This delay can degrade the performance of control systems designed without considering it and can even destabilize the system.

Depending on the medium access control (MAC) protocol of the control network, network-induced delay can be constant, time varying, or even random. MAC protocols generally fall into two categories: random access and scheduling. Carrier sense multiple access (CSMA) is most often used in random access networks, whereas token passing (TP) and time division multiple access (TDMA) are commonly employed in scheduling networks.

Control networks using CSMA protocols include DeviceNet and Ethernet. Fig. 5-2A-2 illustrates various possible situations for this type of network. The figure depicts two nodes continually transmitting messages (with respect to a fixed time line). A node on a CSMA network monitors the network before each transmission. When the network is idle, it begins transmission immediately, as shown in Case 1 of Fig. 5-2A-2. Otherwise it waits until the network is not busy. When two or more nodes try to transmit simultaneously, a collision occurs.

The way to resolve the collision is protocol dependent. DeviceNet, which is a controller area network (CAN), uses CSMA with a bitwise arbitration (CSMA/BA) protocol. Since CAN messages are prioritized, the message with the highest priority is transmitted without interruption when a collision occurs, and transmission of the lower priority message is terminated and will be retried when the network is idle, as shown in Case 2 of Fig. 5-2A-2. Ethernet employs a CSMA with collision detection (CSMA/CD) protocol. When there is a collision, all of the affected nodes will back off, wait a random time (usually decided by the *binary exponential backoff* algorithm), and retransmit, as shown in Case 3 of Fig. 5-2A-2. [1] Packets on these types of networks are affected by random delays, and the worst-case transmission time of packets is unbounded. Therefore, CSMA networks are generally considered nondeterministic. However, if network messages are prioritized, higher priority messages have a better chance of timely transmission.

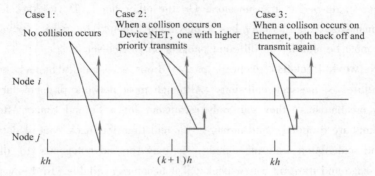

Fig. 5-2A-2 Timing diagram for two nodes on a random access

The TP protocol appears in token bus (IEEE Standard 802.4), token ring (IEEE Standard 802.5), and the fiber distributed data interface (FDDI) MAC architectures; TDMA is used in FireWire. A timing diagram for this type of network is shown in Fig. 5-2A-3. These protocols eliminate the contention for the shared network medium by allowing each node on the network to transmit according to a predetermined schedule. In a token bus, the token is passed around a logical ring, whereas in a token ring, it is passed around a physical ring. In scheduling networks, it is possible to arrange for periodic transmission of messages. For example, FireWire has a transmission cycle (125μs) divided into small time slots, where each *isochronous transaction* is guaranteed a time slot to transmit in every cycle. Packet transmission delays on scheduling networks occur while waiting for the token or time slot. They can be made both bounded and constant by transmitting packets periodically.

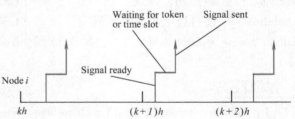

Fig. 5-2A-3 Timing diagram for an arbitrary node on a scheduling network

Single-Packet Versus Multiple-Packet Transmission Single-packet transmission means that sensor or actuator data are lumped together into one network packet and transmitted at the same time, whereas in multiple-packet transmission, sensor or actuator data are transmitted in separate network packets, and they may not arrive at the controller and plant simultaneously. One reason for multiple-packet transmission is that packet-switched networks can only carry limited information in a single packet due to packet size constraints. Thus, large amounts of data must be broken into multiple packets to be transmitted. The other reason is that sensors and actuators in an NCS are often distributed over a large physical area, and it is impossible to put the data into one network packet. Conventional sampled-data systems assume that plant outputs and control inputs are delivered at the same time, which may not be true for NCSs with multiple-packet transmissions. Due to network access delays, the controller may not be able to receive all of the plant output updates at the time of the control calculation. Different networks are suitable for different types of transmissions. Ethernet, originally designed for transmitting information such as data files, can hold a maximum of 1500 bytes of data in a single packet. Hence, it is more efficient to lump the sensor data into one packet and transmit it together-single-packet transmission. On the other hand, DeviceNet, featuring frequent transmission of small-size control data, has a maximum 8-byte data field in each packet; thus, sensor data often must be shuttled in different packets on DeviceNet.

Dropping Network Packets Network packet drops occasionally happen on an NCS when there are node failures or message collisions. Although most network protocols are equipped with transmission-retry mechanisms, they can only retransmit for a limited time. After this time has expired, the packets are dropped. Furthermore, for real-time feedback control data such as sensor measurements and calculated control signals, it may be advantageous to discard the old, untransmitted message and transmit a new packet if it becomes available. In this way, the controller always receives fresh data for control calculation. Normally, feedback-controlled plants can tolerate a certain amount of data loss, but it is valuable to determine whether the system is stable when only transmitting the packets at a certain rate and to compute acceptable lower bounds on the packet transmission rate.

WORDS AND TERMS

agility n. 灵活，便捷
jitter n. 抖（颤）动，颠簸
destabilize v. 使打破平衡，使不稳定
medium access control (MAC) 媒质访问控制
CSMA/BA 载波侦听多路访问/位仲裁
CSMA/CD 载波侦听多路访问/冲突检测
isochronous adj. 同步的，等时的

NOTES

[1] Ethernet employs a CSMA with collision detection (CSMA/CD) protocol. When there is a collision, all of the affected nodes will back off, wait a random time (usually decided by the binary exponential backoff algorithm), and retransmit, as shown in Case 3 of Fig. 5-1A-2.

以太网采用CSMA/CD协议，当产生冲突时，所有冲突节点都将避让等待一段时间（所等待的时长是由二进制指数避让算法决定的），然后重新发送，如图5-2A-2中所示的第三种情况。

B Stability of NCSs with Network-Induced Delay

Modeling NCSs with Network-Induced Delay

The NCS model with network-induced delay is shown in Fig. 5-2B-1. The model consists of a continuous plant

$$\dot{x}(t) = Ax(t) + Bu(t)$$
$$y(t) = Cx(t)$$

and a discrete controller

$$u(kh) = -Kx(kh), \quad k = 0,1,2,\cdots$$

Here, $x \in R^n$, $u \in R^m$, $y \in R^p$, and A, B, C, K are of compatible dimensions.

There are two sources of delays from the network: sensor-to-controller τ_{sc} and controller-to-actuator τ_{ca}. Any controller computational delay can be absorbed into either τ_{sc} or τ_{ca} without loss of generality. For fixed control law (time-invariant controllers), the sensor-to-controller delay and controller-to-actuator delay can be lumped together as $\tau = \tau_{sc} + \tau_{ca}$ for analysis purposes.

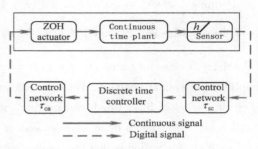

Fig. 5-2B-1 NCS model with network-induced delay

We consider the setup with ① *clock-driven* sensors that sample the plant outputs periodically at sampling instants; ② an *event-driven* controller, which can be implemented by an external event interrupt mechanism and which calculates the control signal as soon as the sensor data arrives; and ③ event-driven actuators, which means the plant inputs are changed as soon as the data become available. The timing of signals of the setup with $\tau_k < h$ is shown in Fig. 5-2B-2.

Delay Less than One Sampling Period First consider the case where the

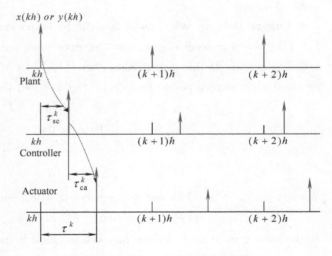

Fig. 5-2B-2 Networked-induced delay

delay of each sample, τ_k, is less than one sampling period, h. (Here the subscript represents the sampling instant.) This constraint means that at most two control samples, $u((k-1)h)$ and $u(kh)$, need be applied during the kth sampling period. The system equations can be written as

$$\dot{x}(t) = Ax(t) + Bu(t), \quad t \in [kh + \tau_k, (k+1)h + \tau_{k+1}],$$
$$y(t) = Cx(t),$$
$$u(t^+) = -Kx(t - \tau_k), \quad t \in \{kh + \tau_k, k = 0,1,2,\cdots\}$$

where $u(t^+)$ is piecewise continuous and only changes value at $kh + \tau_k$. Sampling the system with period h we obtain

$$x((k+1)h) = \Phi x(kh) + \Gamma_0(\tau_k)u(kh) + \Gamma_1(\tau_k)u((k-1)h)$$
$$y(kh) = Cx(kh)$$

where $\Phi = e^{Ah}$, $\Gamma_0(\tau_k) = \int_0^{h-\tau_k} e^{As} B \, ds$, $\Gamma_1(\tau_k) = \int_{h-\tau_k}^{h} e^{As} B \, ds$.

Defining $z(kh) = [x^T(kh), u^T((k-1)h)]^T$ as the augmented state vector, the augmented closed-loop system is $z((k+1)h) = \widetilde{\Phi}(k) z(kh)$ where $\widetilde{\Phi}_1(k) = \begin{bmatrix} \Phi + \Gamma_0(\tau_k) K & \Gamma_1(\tau_k) \\ -K & 0 \end{bmatrix}$.

If the delay is constant (i.e., $\tau_k = \tau$ for $k = 0, 1, 2, \cdots$), the system is still time invariant, which simplifies the system analysis. Thus we can envision static scheduling network protocols, such as token ring or token bus, which can provide constant delay. Even in this simplified setup, the next question is, "How much delay can the system tolerate?"

Another observation is that the sensor-controller delay can be compensated by an estimator if the messages sent out by sensors are time stamped. Traditional one-step prediction estimation can compensate delays less than one sampling period, since the estimate of $x(kh)$ only depends on the value of $y((k-1)h)$. We will revisit this problem in the section on compensation for network-induced delay.

Longer Delays When the delays can be longer than one sampling period (say, $0 < \tau_k < lh$, $l > 1$), one may receive zero, one, or more than one (up to l) control sample(s) in a single sampling period. In the special case where $(l-1)h < \tau_k < lh$ for all k, one control sample is received every sample period for $k > l$. In this case, the analysis resulted in

$$\widetilde{\Phi}(k) = \begin{bmatrix} \Phi & \Gamma_1(\tau_k') & \Gamma_0(\tau_k') & \cdots & 0 \\ 0 & 0 & I & \cdots & 0 \\ \vdots & \vdots & \vdots & \ddots & \vdots \\ -K & 0 & 0 & \cdots & 0 \end{bmatrix}$$

Where $\tau_k' = \tau_k - (l-1)h$ and the augmented state vector is

$$z(kh) = [x^T(kh), u^T((k-1)h), \cdots, u^T((k-1)h)]^T$$

In the more general case, tedious bookkeeping must be performed, as even the block structure of the matrix $\widetilde{\Phi}$ is time varying, since it depends on the schedule of the receipt of the control samples.

Stability Regions

Conventionally, a faster sampling rate is desirable in sampled-data systems so the discrete-time control design and performance can approximate that of the continuous system. But in NCSs, a faster sampling rate can increase the network load, which in turn results in longer delay of the signals. Thus finding a sampling rate that can both tolerate the network-induced delay and achieve desired system performance is important in NCS design.

Plotting the stability region of an NCS with respect to the sampling rate, h, and network delay, τ, is helpful to see the relationship between these two parameters. Note that here we are considering constant delay, which can be achieved by using an appropriate network protocol.

Integrator Case The relationship between h and τ can be derived analytically for simple scalar systems.

Example 1: Consider the integrator example

$$\dot{x}(t) = u(t), t \in [kh + \tau, (k+1)h + \tau], \tau < h$$
$$u(t^+) = -Kx(t - \tau), t \in \{kh + \tau, k = 0, 1, 2, \cdots\}, K > 0$$

Defining $z(kh)$ as before

$$\widetilde{\boldsymbol{\Phi}}_1 = \begin{bmatrix} 1 - hK + \tau K & \tau \\ -K & 0 \end{bmatrix}$$

For this 2×2 case, we can use the stability triangle to explicitly calculate the relation between τ and h. [1] For a stable NCS, the delay τ must satisfy

$$\max\left\{\frac{1}{2}h - \frac{1}{K}, 0\right\} < \tau < \min\left\{\frac{1}{K}, h\right\} \quad \text{or} \quad \max\left\{\frac{1}{2} - \frac{1}{Kh}, 0\right\} < \frac{\tau}{h} < \min\left\{\frac{1}{Kh}, 1\right\}$$

The analytically determined stability region for $0 \leq \tau < h$ is shown in Fig. 5-2B-3. We can see from the stability region that when the sampling period h is small, the system can tolerate a delay up to one full sampling period. As h becomes larger, the upper bound on τ/h becomes smaller. Note that for $K > 2/h$, even the system with no delay is unstable.

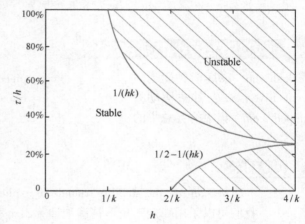

Fig. 5-2B-3 Stability region of a controlled integrator

General Scalar System It may be analytically infeasible to derive the exact stability region for general systems; however, stability regions for such systems can still be determined by simulation. [2] The stability region is plotted by incrementally increasing the delay, τ, and testing the closed-loop system matrix, as formulated before. If the closed-loop system matrix is stable, a point is marked in that location of the stability region.

Example 2: For a general scalar system

$$\dot{x}(t) = ax(t) + u(t), t \in [kh + \tau, (k+1)h + \tau], \tau < h$$
$$u(t^+) = -Kx(t - \tau), t \in \{kh + \tau, k = 0, 1, 2, \cdots\}, K > 0$$

$$\widetilde{\boldsymbol{\Phi}}_1 = \begin{bmatrix} e^{ah} - \frac{K}{a}(e^{a(h-\tau)} - 1) & \frac{1}{a}e^{ah}(1 - e^{-a\tau}) \\ -K & 0 \end{bmatrix}$$

The stability region can be determined by simulation. A special scalar case with $a = 1$ and $K = 2$ is shown in Fig. 5-2B-4. For this simulation, we considered delays between 0 and $4h$. We can see that when $0 \leq \tau < h$, the region has a shape similar to the integrator case. The shape of the stability region is also affected by the feedback controller (in this case, the scalar feedback gain).

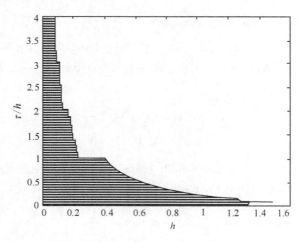

Fig. 5-2B-4 Stability region integrator

WORDS AND TERMS

clock-driven *adj.* 时钟驱动的
event-driven *adj.* 事件驱动的
lumped *adj.* 集中的；总集的

piecewise continuous 分段连续
infeasible *adj.* 不可行的

NOTES

[1] ... we can use the stability triangle to explicitly calculate the relation between τ and h.
……我们可以采用稳定三角形精确地计算出 τ 和 h 之间的关系。

[2] It may be analytically infeasible to derive the exact stability region for general systems; however, stability regions for such systems can still be determined by simulation.
对于一般的系统不太可能推导出精确的稳定区域，然而却可以通过仿真的方法来确定。

C 国外自动化专业主要期刊

国别	中译刊名	外文刊名	刊号
美国	IEEE 电力输送汇刊	IEEE Transactions on Power Delivery	730B0001TPD
美国	日本电工	Electrical Engineering in Japan	732B0068
美国	电气世界	Electrical World	721B0003
美国	控制与系统	Cybernetics and Systems	737B0055
美国	自动化	Automation	737B0015
美国	电机与动力系统	Electric Machines Power Systems	732B0007
美国	控制论	Cybernetics	737B0005
美国	IEEE 工业电子学汇刊	IEEE Transactions on Industrial Electronics	730B0001
美国	IEEE 能量转换汇刊	IEEE Transactions on Energy Conversion	730B0001-TEC

国别	中译刊名	外文刊名	刊号
美国	控制工程	Control Engineering	737B0013
美国	IEEE 工业应用汇刊	IEEE Transactions on Industry Applications	730B0001
美国	IEEE 控制系统杂志	IEEE Control Systems Magazine	730B0001
美国	IEEE 自动控制汇刊	IEEE Transactions on Automatic Control	730B0001
美国	仪表与控制系统	I & CS (Instruments & Control Systems)	798B0003
美国	IEEE 功率电子学汇刊	IEEE Transactions on Power electronics	730B0001TPE
美国	IEEE 动力系统汇刊	IEEE Transactions on Power Systems	730B0001-TPWRS
美国	IEEE 仪表与测量汇刊	IEEE Transactions on Instrumentation and Measurement	730B0001-TIM
美国	IEEE 机器人与自动化杂志	IEEE Journal of Robotics and Automation	730B0001-JRA
美国	IEEE 机器人与自动化汇刊	IEEE Transactions on Robotics and Automation	730B0001
美国	IEEE 系统、人与控制论汇刊	IEEE Transactions on Systems, Man and Cybernetics	730B0001TSMC
美国	IEEE 模式分析与机器信息汇刊	IEEE Transaction on Pattern, Analysis and Machine Intelligence	737B0021
英国	现代动力系统	Modern Power Systems	720C0069
英国	电气工程师协会评论	IEE Review	736C0058
英国	电气评论	Electrical Review	730C0002
英国	自动化	Automation	737C0006
英国	自动化	Automatica	737C0003
英国	控制与仪表应用	Control & Instrumentation	737C0002
英国	国际控制杂志	International Journal of Control	737C0001
英国	模式识别	Pattern Recognition	738C0065
英国	国际电力与能源系统杂志	International Journal of Electrical Power Energy Systems	730C0083
英国	英国电气工程师学会志 B 辑：电力应用	IEE Proceedings, B: Electric Power Application	730C0004-B
英国	电气工程师协会会报 D 辑：控制理论与应用	OEEL Institution of Electrical Engineers Proceedings, D: Control theory and Applications	730C0004
德国	电工学杂志	Eledtrotechnische Zeitschrift	730E0005
德国	电工技术文献	ETZ Arehiv	721E0004
德国	能量与自动化	Energy & Automation	ISSN0931-6221
德国	电能	Elektrie	721A0001
德国	电技术	Elektrodechnik	730E0053
德国	计量－控制－调节	MSR (Messen, Steuern, Regeln)	737A0001
德国	自动化技术	Automatisierungs Technik	738E0005

国别	中译刊名	外文刊名	刊号
德国	电子学	Elektronik（Fachzelitschrift fur Entwickle und Industrielle Anwender）	736E0001
德国	德国工程师协会志	VDI-A（Zeitschrift fur Maschinenbau und Metallbear-beilung）	710E0002
法国	电学杂志	Revue Generale de l'Electricite	730F0004
瑞士	电技术	Elektrotechnik	730LD057
瑞士	ABB 评论	ABB Review	730LD002
荷兰	人工智能	Artificial Intelligence	738LB003
意大利	自动化与计分表装置	Automazione e Strumentazione	737MC052
日本	计测与控制	Journal of the Society of Instrument & Control Engineer	737D0055
日本	三菱电机技报	Mitsubishi Denki Giho	732D0005
日本	电气评论	Electrical Review	721D0052
日本	欧姆电气杂志	OHM /电气杂志	732D0006
日本	电气学会杂志	The Journal of the Institute of Electrical Engineers of Japan	730D0087
日本	电气学会论文志，B	The Transaction of the Institute of Electrical Engineers of Japan	730D0087-B
日本	仪表应用与控制工程	Instrumentation and Control Engineering	ISSN：0955-9531
日本	机器人	Robot	737D0060
日本	系统控制情报	Systems, Control and Information	737D0053
日本	人工智能学会志	Journal of Japanese Society for Artificial Intelligence	737D0074

附：国内外自动化学术团体

1. 美国电气与电子工程师学会（The Institute of Electrical and Electronics Engineers，简称 IEEE）

IEEE 是电气与电子工程方面目前世界上最大的学术团体。它与 100 多个国家建立了人事往来和学术交流活动，并已逐渐发展成一个国际性的机构，现有国内外会员 20 余万人。IEEE 活动主要是举办大量国际性或全国性会议并出版期刊和会议录，举办各种展览会，组织会员访问科研单位和工厂企业的实验室、进行职业调查并出版相应的调查报告等。IEEE 按专业活动划分为：计算机，控制系统，电介质与电气绝缘，电子器件，工业应用，工业电子学，信息理论，测试设备与测量，大功率电子学（委员会），机器人与自动学（委员会），系统、人与控制论等 10 个部和 36 个技术协会和委员会。

2. 英国工程技术学会（IET, The Institution of Engineering and Technology）

IET 是工程技术领域全球领先的专业学术学会。目前在全球 127 个国家拥有 15 多万名会员。IET 的前身 IEE（电机工程师学会）创建于 1871 年，是电子电气领域的国际知名专业学术团体。电机工程师学会（IEE）和企业工程师学会（IIE）于 2006 年 3 月合并组成工程技术学会（IET）。IET 拥有的全球工程技术文献索引（INSPEC，Information Service in Physics, Electro-Technology, Computer and Control）是占世界主导地位的英文工程出版物索引，资讯涵盖全球范围内 1000 万篇科技论文，专业技术杂志以及其他多种语言的出版物，内容涉及电子、电气、制造、生物、物理、电信、资讯技术等多个工程技术领域。

3. 国内主要相关专业性学会

中国电子学会　　　中国计算机学会　　　中国仪器仪表学会
中国电机工程学会　中国自动化学会
中国系统工程学会　中国电工技术学会

UNIT 3

A Fundamentals of the Database System

Grasping the Concept of a Database

A *database* is a collection of information—preferably related information and preferably organized. [1] A database consists of the physical files you set up on a computer when installing the database software. On the other hand, a database model is more of a concept than a physical object and is used to create the tables in your database. This section examines the database, not the database model.

By definition, a database is a structured object. It can be a pile of papers, but most likely in the modern world it exists on a computer system. That structured object consists of *data* and *metadata*, with metadata being the structured part. Data in a database is the actual stored descriptive information, such as all the names and addresses of your customers. Metadata describes the structure applied by the database to the customer data. In other words, the metadata is the customer table definition. The customer table definition contains the fields for the names and addresses, the lengths of each of those fields, and datatypes. (A *datatype* restricts values in fields, such as allowing only a date, or a number). Metadata applies structure and organization to raw data.

Fig. 5-3A-1 shows a general overview of a database. A database is often represented graphically by a cylindrical disk, as shown on the left of the diagram. The database contains both metadata and raw data. The database itself is stored and executed on a database server computer.

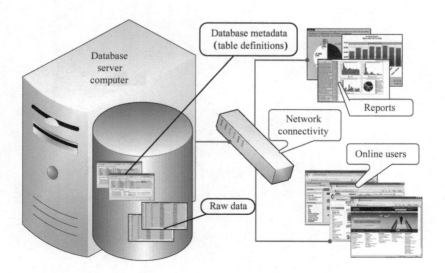

Fig. 5-3A-1 General overview of a database

In Fig. 5-3A-1, the database server computer is connected across a network to end-users running reports, and online browser users browsing your Web site (among many other application types).

Understanding a Database Model

There are numerous, precise explanations as to what exactly a *database model* or *data model* is. A database model can be loosely used to describe an organized and ordered set of information stored on a computer. This ordered set of data is often structured using a data modeling solution in such a way as to make the retrieval of and changes to that data more efficient. Depending on the type of applications using the database, the database structure can be modified to allow for efficient changes to that data. It is appropriate to discover how different database modeling techniques have developed over the past 50 years to accommodate efficiency, in terms of both data retrieval and data changes. [2] Before examining database modeling and its evolution, a brief look at applications is important.

What Is an Application?

In computer jargon, an *application* is a piece of software that runs on a computer and performs a task. That task can be interactive and use a graphical user interface (GUI), and can execute reports requiring the click of a button and subsequent retrieval from a printer. Or it can be completely transparent to end-users. *Transparency* in computer jargon means that end-users see just the pretty boxes on their screens and not the inner workings of the database, such as the tables. From the perspective of database modeling, different application types can somewhat (if not completely) determine the requirements for the design of a database model.

An *online transaction processing* (OLTP) database is usually a specialized, highly *concurrent* (shareable) architecture requiring rapid access to very small amounts of data. OLTP applications are often well served by rigidly structured OLTP transactional database models. A *transactional database model* is designed to process lots of small pieces of information for lots of different people, all at the same time.

On the other side of the coin, a *data warehouse* application that requires frequent updates and extensive reporting must have large amounts of properly sorted data, low concurrency, and relatively low response times. A data warehouse database modeling solution is often best served by implementing a denormalized duplication of an OLTP source database. [3]

Fig. 5-3A-2 shows the same image as in Fig. 5-3A-1, except that in Fig. 5-3A-2, the reporting and online browser applications are made more prominent. The most important point to remember is that database modeling requirements are generally determined by application needs. It's all about the applications. End-users use your applications. If you have no end-users, you have no business.

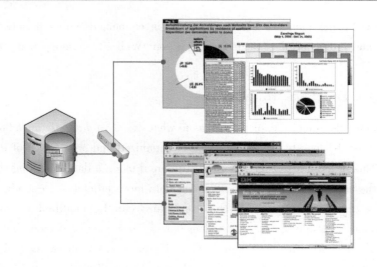

Fig. 5-3A-2　Graphic image of an application

WORDS AND TERMS

metadata　*n.*　元数据
field　*n.*　字段
datatype　*n.*　数据类型
cylindrical　*adj.*　圆柱（形）的
retrieval　*n.*　取回
application　*n.*　应用（程序）
jargon　*n.*　行话

interactive　*adj.*　交互式的
graphical user interface（GUI）图形用户界面
online transaction processing（OLTP）联机事务处理
concurrent　*adj.*　并发的
data warehouse　数据仓库
prominent　*adj.*　卓越的，突出的

NOTES

［1］A *database* is a collection of information—preferably related information and preferably organized.

数据库是一个相互关联且组织良好的信息的集合。

［2］It is appropriate to discover how different database modeling techniques have developed over the past 50 years to accommodate efficiency, in terms of both data retrieval and data changes.

在过去的 50 年中，各种数据库建模技术（尤其是在数据检索和数据交换方面）的发展大相径庭，但为了满足提高效率的要求，这些做法都是恰如其分的。

［3］A data warehouse database modeling solution is often best served by implementing a denormalized duplication of an OLTP source database.

数据仓库的数据库建模方案通过实施一个 OLTP 源数据库的非规范化的副本来实现。

B Virtual Manufacturing—A Growing Trend in Automation

Industrial automation is developing quickly, but product life cycles are getting shorter. Virtual manufacturing may relieve this strain on manufacturers. [1]

Business

Everything we consume or buy has, at some point, come in contact with some sort of automated process. Whether it is in the factory or at the transportation stage, our mobile phones, yoghurts, clothes and newspapers have all been whisked from one place to another by powerful, intelligent and, often, networked conveyor systems.

Industrial automation originates from the industrial revolution, when assembly lines were developed to manufacture products. Since then, industrial automation has progressively taken menial jobs away from humans and mechanised them. It has grown into a complex field in which computers, sensors, networks and Ethernet have progressively entered to control and monitor production. [2]

However, says Björn Langbeck, at the IVF Industrial Research and Development Corporation, a research institute in Stockholm, "it doesn't have to be that complicated. A car is complicated under the hood, but it is easy to drive. IT systems in the factory are complicated, but they ought to be easy for the user."

According to the Siemens Web site, in the near future it will be possible to control production at a factory by sitting in front of a Web browser. "The drivers in industrial automation are microelectronics and software." says Klaus Wucherer, member of the corporate executive committee at Siemens AG, in an article in IEE Computing & Control Engineering magazine in July 2003.

Digital Dream

"The dream of every factory designer and automation specialist is a digital factory," says Wucherer. "Soon, designers at an auto manufacturer will watch the virtual vehicle emerge from the virtual factory. In fact, auto manufacturers will approve a new model only after seeing the digital vehicle pass through the digital factory—and after they have exhausted all product design and production options."

But it is the pace at which companies are churning out new products with new functionalities and tastes that has put a strain on existing factory automation. As life cycles get shorter, cost-conscious industries are slowly rethinking manufacturing routines.

In 2002, Daimler Chrysler launched a pilot project at two of its automobile factories based on this digital factory concept. The idea was to simulate at an early stage all the steps and processes that go into making a car. By doing this, the company was able to foresee problems before they happened on the factory floor, decrease time to market by 40 percent and achieve a 70 percent gain in efficiency, according to the German car magazine *Auto Magazin*.

"More products are going to market, (and they are going) much faster than they ever have before," says Anders Kinnander, an expert in production technology at the Chalmers Institute of Technology in Gothenburg, Sweden. "Shorter life cycles also mean that production lines have to be more flexible than ever. And this has big consequences for companies that need industrial automation for their production."

Modular Automation

Another clear trend in the field is a move towards more modular automation technologies that are easy to reconfigure, in order to satisfy particular customer needs. Approximately 70 percent of the cost of a product comes from the cost of its manufacture, and flexibility in automation can substantially cut costs.

According to Fredrik Jönsson, CEO of FlexLink, a supplier of industrial automation solutions, modular automation and virtual production (the digital factory) offer the most significant cost savings for manufacturers.

Together with IBM and others, FlexLink launched its own digital factory concept in late 2002. By combining all the aspects of manufacturing—computer-assisted design, production data management, manufacturing execution systems, supply-chain management and enterprise resource planning—into one system, FlexLink and its partners have developed one of the most flexible automation systems in the world. FlexLink calls it the PLM factory, short for "product and process life-cycle management."

"How do you optimise future production of a prototype, quickly move to a full-scale production and maximise profits during the phase out of old products simultaneously in the same production line?... You need to manage product, production and order data effectively, and (you need to) use modular, flexible and reusable equipment and software. The key is to evaluate all options in the simulated production environment before you start in real life," says Jönsson.

"Instead of building up production capacity dimensioned for an expected maximum volume, the idea with the PLM factory is to progressively build up capacity according to what the market needs," he says. "If these (needs) change, the modules can easily be used for another product. In other words, we are building flexibility into the automation process, and in this fashion drastically diminishing the capital expenditures in the factory."

Too Much of a Good Thing?

Developments in industrial automation follow advances in IT, and, as such, are inevitable. But there is a latent worry in the business that factory procedures are getting over-engineered. In fact, there have been cases in recent history in which manufacturing plants have gone over to robotic assembly, only to realize that manual labor was much more efficient.[3]

In fact, Chalmers' Kinnander tells of a large truck manufacturer that decided to partially automate an axle factory in the late 1980s because of a shortage of workers. Unfortunately, axle assembly was a difficult task for the robots. The product was not designed for automatic assembly, and after seven years, the factory returned to manual assembly.

WORDS AND TERMS

yoghurt　*n.*　乳酸酪
whisk　*v.*　飞奔
menial　*adj.*　卑微的，仆人的
churn out　艰苦地做出
prototype　*n.*　原型

drastically　*adv.*　激烈地，彻底地
diminish　*v.*　（使）减少，（使）变小
CEO = Chief Executive Officer　执行总裁，首席执行官

NOTES

[1] Industrial automation is developing quickly, but product life cycles are getting shorter. Virtual manufacturing may relieve this strain on manufacturers.

尽管工业自动化飞速发展，但产品的生命周期变得越来越短。虚拟制造可以减轻这种对于制造者的压力。

[2] It has grown into a complex field in which computers, sensors, networks and Ethernet have progressively entered to control and monitor production.

随着越来越多地在控制和监控产品中采用计算机、传感器、网络和以太网技术，工业自动化已经成为一个复杂的领域。

[3] In fact, there have been cases in recent history in which manufacturing plants have gone over to robotic assembly, only to realize that manual labor was much more efficient.

事实上，最近的历史上已有些案例，那些已经转向使用自动机械装配的制造工厂未曾料到人力劳动是更高效的。

C　自动化专业的科技前沿

国家自然科学基金是国家创新体系的重要组成部分，其战略定位是"支持基础研究，坚持自由探索，发挥导向作用"。国家自然科学基金委员会每年向社会发布《项目指南》，引导广大科研人员申请项目，积极鼓励科研人员开展具有重要科学意义的、瞄准国际科学发展前沿的研究，以及开展针对我国国民经济和社会可持续发展中关键科学问题的创新性研究，重点支持传统领域中创新性强的基础研究和关键技术研究，鼓励支持从社会经济发展和国防安全中提炼出来的基础科学问题的研究，并鼓励支持交叉学科领域方面的研究。

自动化专业的相关项目由国家自然科学基金委员会信息科学三处负责管理，主要资助的面上项目分为七大领域，包括：控制理论及应用，导航、制导与先进传感器，系统科学与系统工程，模式识别及应用，人工智能与知识科学，机器人学，认知科学及智能信息处理。在"十一五"期间鼓励的研究课题主要涉及：面向节能、降耗与降低污染的生产过程一体化控制与管理，网络化系统及网络系统控制，复杂系统的建模、分析与安全控制，系统生物学中的控制问题，新型传感器与多源信息融合，模式识别的新理论与新方法，人工智能的新理论与新方法，先进机器人系统及其关键技术，认知过程及智能信息处理。

国家自然科学基金重点项目主要支持科技工作者结合国家需求，把握世界科学前沿，针对我国已有较好基础和积累的重要研究领域或新学科生长点开展深入、系统的创新性研究工作。近几年来，纳米科学、生命科学、认知科学和复杂性科学等在世界范围内得到了迅速的发展。其中，微-纳操作与量子系统控制、生物信息学与系统生物学、认知过程及智能信息处理、复杂系统建模、分析与控制等不仅为自动化科学与技术提出了新的机遇和挑战，而且正在成为自动化科学与技术新的研究领域。此外，国内生产领域中高消耗、高排放、高污染和低效益的现状以及社会发展对绿色经济和循环经济的需求，迫切需要自动化科学与技术方面的专家从中提炼科学问题，进行创新性研究和关键技术的突破。针对以上问题，国家自然科学基金重点项目主要资助的项目包括：面向节能降耗的工业过程控制，运动目标动态模型的在线辨识与估计，纳米尺度物体的检测、控制和操作的基础理论与方法，工程与产品现代设计理论与方法，水下移动传感器网络的关键技术，机器学习中的若干重要问题研究，高效可伸缩视频编解码的基础理论和方法研究。

国家自然科学基金委员会网站：http：//www.nsfc.gov.cn/。

UNIT 4

A Overview of Artificial Intelligence

The overall research goal of artificial intelligence is to create technology that allows computers and machines to function in an intelligent manner. The general problem of simulating (or creating) intelligence has been broken down into sub-problems. These consist of particular traits or capabilities that researchers expect an intelligent system to display. The traits described below have received the most attention.

Reasoning, Problem Solving

Early researchers developed algorithms that imitated step-by-step reasoning that humans use when they solve puzzles or make logical deductions. By the late 1980s and 1990s, AI research had developed methods for dealing with uncertain or incomplete information, employing concepts from probability and economics.

For difficult problems, algorithms can require enormous computational resources—most experience a "combinatorial explosion": the amount of memory or computer time required becomes astronomical for problems of a certain size.[1] The search for more efficient problem-solving algorithms is a high priority.

Human beings ordinarily use fast, intuitive judgments rather than step-by-step deduction that early AI research was able to model. AI has progressed using "sub-symbolic" problem solving: embodied agent approaches emphasize the importance of sensorimotor skills to higher reasoning; neural net research attempts to simulate the structures inside the brain that give rise to this skill; statistical approaches to AI mimic the human ability to guess.

Knowledge Representation

Knowledge representation and knowledge engineering are central to AI research. Many of the problems machines are expected to solve will require extensive knowledge about the world. Among the things that AI needs to represent are: objects, properties, categories and relations between objects; situations, events, states and time; causes and effects; knowledge about knowledge (what we know about what other people know); and many other, less well researched domains. A representation of "what exists" is an ontology: the set of objects, relations, concepts, and properties formally described so that software agents can interpret them. The semantics of these are captured as description logic concepts, roles, and individuals, and typically implemented as classes, properties, and individuals in the Web Ontology Language. The most general ontologies are called upper ontologies, which attempt to provide a foundation for all other knowledge by acting as mediators between domain ontologies that cover specific knowledge about a particular knowledge domain (field of interest or area of concern).[2] Such formal knowledge representations are suitable for content-based indexing and retrieval, scene interpretation, clinical decision support, knowledge discovery via automated reasoning (inferring new statements based on explicitly stated knowledge), etc. Video events are often represented as SWRL rules, which can be used, among others, to automatically generate subtitles for constrained videos. Among the most difficult problems in knowledge representation are listed below.

1) Default reasoning and the qualification problem

Many of the things people know take the form of "working assumptions". For example, if a bird comes up in conversation, people typically picture an animal that is fist sized, sings, and flies. None of these things are true about all birds. John McCarthy identified this problem in 1969 as the qualification problem: for any commonsense rule that AI researchers care to represent, there tend to be a huge number of exceptions. Almost nothing is simply true or false in the way that abstract logic requires. AI research has explored a number of solutions to this problem.

2) The breadth of commonsense knowledge

The number of atomic facts that the average person knows is very large. Research projects that attempt to build a complete knowledge base of commonsense knowledge (e. g., Cyc) requires enormous amounts of laborious ontological engineering—they must be built, by hand, one complicated concept at a time. A major goal is to have the computer understand enough concepts to be able to learn by reading from sources like the Internet, and thus be able to add to its own ontology.

3) The subsymbolic form of some commonsense knowledge

Much of what people know is not represented as "facts" or "statements" that they could express verbally. For example, a chess master will avoid a particular chess position because it "feels too exposed" or an art critic can take one look at a statue and realize that it is a fake. These are non-conscious and sub-symbolic intuitions or tendencies in the human brain. Knowledge like this informs, supports and provides a context for symbolic, conscious knowledge. As with the related problem of sub-symbolic reasoning, it is hoped that situated AI, computational intelligence, or statistical AI will provide ways to represent this kind of knowledge.

Planning

Intelligent agents must be able to set goals and achieve them. They need a way to visualize the future—a representation of the state of the world and be able to make predictions about how their actions will change it—and be able to make choices that maximize the utility (or "value") of available choices.

In classical planning problems, the agent can assume that it is the only system acting in the world, allowing the agent to be certain of the consequences of its actions. However, if the agent is not the only actor, then it requires that the agent can reason under uncertainty. This calls for an agent that can not only assess its environment and make predictions, but also evaluate its predictions and adapt based on its assessment.

Multi-agent planning uses the cooperation and competition of many agents to achieve a given goal. Emergent behavior such as this is used by evolutionary algorithms and swarm intelligence.

Learning

Machine learning, a fundamental concept of AI research since the field's inception, is the study of computer algorithms that improve automatically through experience.

Unsupervised learning is the ability to find patterns in a stream of input. Supervised learning includes both classification and numerical regression. Classification is used to determine what category something belongs in, after seeing a number of examples of things from several categories. Regression is the attempt to produce a function that describes the relationship between inputs and outputs and predicts how the outputs should change as the inputs change. In reinforcement learning the agent is rewarded for good responses and punished for bad ones. The agent uses this sequence of rewards and punishments to form a strategy for operating in its problem space. These three types of learning can be analyzed in terms of decision theory, using concepts like utility. The mathematical analysis of machine learning algorithms and their performance is a branch of theoretical computer science known as computational learning theory.

Within developmental robotics, developmental learning approaches are elaborated upon to allow robots to accumulate repertoires of novel skills through autonomous self-exploration, social interaction with human teachers, and the use of guidance mechanisms (active learning, maturation, motor synergies, etc.).

Natural Language Processing

Natural language processing gives machines the ability to read and understand human language. A sufficiently powerful natural language processing system would enable natural language user interfaces and the acquisition of knowledge directly from human-written sources, such as newswire texts. Some straightforward applications of natural language processing include information retrieval, text mining, question answering and machine translation.

A common method of processing and extracting meaning from natural language is through semantic indexing. Although these indexes require a large volume of user input, it is expected that increases in processor speeds and decreases in data storage costs will result in greater efficiency.

Perception

Machine perception is the ability to use input from sensors (such as cameras, microphones, tactile sensors, sonar and others) to deduce aspects of the world. Computer vision is the ability to analyze visual input. A few selected subproblems are speech recognition, facial recognition and object recognition.

Motion and Manipulation

The field of robotics is closely related to AI. Intelligence is required for robots to handle tasks such as object manipulation and navigation, with sub-problems such as localization, mapping, and motion planning. These systems require that an agent is able to: Be spatially cognizant of its surroundings, learn from and build a map of its environment, figure out how to get from one point in space to another, and execute that movement (which often involves compliant motion, a process where movement requires maintaining physical contact with an object).

Social Intelligence

Affective computing is the study and development of systems that can recognize, interpret, process, and simulate human affects. It is an interdisciplinary field spanning computer sciences, psychology, and cognitive science. While the origins of the field may be traced as far back as the early philosophical inquiries into emotion, the more modern branch of computer science originated with Rosalind Picard's 1995 paper on "affective computing". A motivation for the research is the ability to simulate empathy, where the machine would be able to interpret human emotions and adapts its behavior to give an appropriate response to those emotions.

Emotion and social skills are important to an intelligent agent for two reasons. First, being able to predict the actions of others by understanding their motives and emotional states allow an agent to make better decisions. Concepts such as game theory, decision theory, necessitate that an agent be able to detect and model human emotions. Second, in an effort to facilitate human-computer interaction, an intelligent machine may want to display emotions (even if it does not experience those emotions itself) to appear more sensitive to the emotional dynamics of human interaction.

Creativity

A sub-field of AI addresses creativity both theoretically (the philosophical psychological perspective) and practically (the specific implementation of systems that generate novel and useful outputs).

General Intelligence

Many researchers think that their work will eventually be incorporated into a machine with artificial general intelligence, combining all the skills mentioned above and even exceeding human ability in most or all these areas. A few believe that anthropomorphic features like artificial consciousness or an artificial brain may be required for such a project.

Many of the problems above also require that general intelligence be solved. For example, even specific straightforward tasks, likemachine translation, require that a machine read and write in both languages (NLP), follow the author's argument (reason), know what is being talked about (knowledge), and faithfully reproduce the author's original intent (social intelligence). A problem like machine translation is considered "AI-complete", but all of these problems need to be solved simultaneously in order to reach human-level machine performance.

WORDS AND TERMS

artificial intelligence   人工智能
ontology   *n.* 本体论，实体论

semantic   *adj.* 语义的，语义学的
clinical   *adj.* 临床的

machine learning　机器学习
reinforcement learning　强化学习
interdisciplinary　*adj*. 跨学科的
synergy　*n*. 协同，增效
repertoire　*n*. 计算机指令系统，本领
empathy　*n*. 神入，同感，共鸣

philosophical　*adj*. 哲学上的，哲学的
psychological　*adj*. 心理的，精神上的
anthropomorphic　*adj*. 拟人的，赋予人性的
default reasoning　缺省推理
verbally　*adv*. 言语上，口头上

NOTES

[1] For difficult problems, algorithms can require enormous computational resources—most experience a "combinatorial explosion": the amount of memory or computer time required becomes astronomical for problems of a certain size.

a certain size：具有一定规模，一定规模。

对于困难的问题，算法需要大量的计算资源——大多数算法经历了"组合式爆炸"：对于具有一定规模的问题，所需的内存或计算时间会非常庞大。

[2] The most general ontologies are called upper ontologies, which attempt to provide a foundation for all other knowledge by acting as mediators between domain ontologies that cover specific knowledge about a particular knowledge domain (field of interest or area of concern).

最广义的本体叫作上本体，上本体为所有涵盖特殊知识的特殊知识域（感兴趣或关注的域）的域本体提供基础平台，上本体的作用是作为域本体互相沟通的桥梁。

B　Applications of Artificial Intelligence in Machine Learning

The performance of machine learning methods is heavily dependent on the choice of data representation (or features) on which they are applied. For that reason, much of the actual effort in deploying machine learning algorithms goes into the design of preprocessing pipelines and data transformations that result in a representation of the data that can support effective machine learning. Such feature engineering is important but labor intensive and highlights the weakness of current learning algorithms: Their inability to extract and organize the discriminative information from the data. Feature engineering is a way to take advantage of human ingenuity and prior knowledge to compensate for that weakness. To expand the scope and ease of applicability of machine learning, it would be highly desirable to make learning algorithms less dependent on feature engineering so that novel applications could be constructed faster, and more importantly, to make progress toward artificial intelligence (AI). An AI must fundamentally understand the world around us, and we argue that this can only be achieved if it can learn to identify and disentangle the underlying explanatory factors hidden in the observed milieu of low-level sensory data.[1] This text is about representation learning, i.e., learning representations of the data that make it easier to extract useful information when building classifiers or other predictors. In the case of probabilistic models, a good representation is often one that captures the posterior distribution of the underlying explanatory factors for the observed input. A good representation is also one that is useful as input to a supervised predictor. Among the various ways of learning representations, this text focuses on deep learning methods: those that are formed by the composition of multiple nonlinear transformations with the goal of yielding more abstract—and ultimately more useful—representations. Here, we survey this rapidly developing area with special emphasis on some progress. We consider some of the fundamental questions that have been driving research in this area. Specifically, what makes one representation better than another? Given an example, how should we compute its representation, i. e., perform feature extraction? Also, what are appropriate objectives for learning good representations?

Why Should We Care about Learning Representations?

Representation learning has become a field in itself in the machine learning community, with regular workshops at the leading conferences such as NIPS and ICML, and a new conference dedicated to it, ICLR, sometimes under the header of Deep Learning or Feature Learning. Although depth is an important part of the story, many other priors are interesting and can be conveniently captured when the problem is cast as one of learning a representation, as discussed in the next section. The rapid increase in scientific activity on representation learning has been accompanied and nourished by a remarkable string of empirical successes both in academia and in industry. Below, we briefly highlight some of these high points.

Speech Recognition and Signal Processing

Speech was one of the early applications of neural networks, in particular convolutional (or time-delay) neural networks. The recent revival of interest in neural networks, deep learning, and representation learning has had a strong impact in the area of speech recognition, with breakthrough results obtained by several academics as well as researchers at industrial labs bringing these algorithms to a larger scale and into products. For example, Microsoft released a new version of their Microsoft Audio Video Indexing Service speech system based on deep learning. These authors managed to reduce the word error rate on four major benchmarks by about 30 percent (e. g., from 27. 4 to 18. 5 percent on RT03S) compared to state-of-the-art models based on Gaussian mixtures for the acoustic modeling and trained on the same amount of data (309 hours of speech).[2] The relative improvement in error rate obtained by Dahl et al. on a smaller large-vocabulary speech recognition benchmark (Bing mobile business search dataset, with 40 hours of speech) is between 16 and 23 percent.

Representation-learning algorithms have also been applied to music, substantially beating the state-of-the-art in polyphonic transcription, with relative error improvement between 5 and 30 percent on a standard benchmark of four datasets. Deep learning also helped to win MIREX (music information retrieval) competitions, for example, in 2011 on audio tagging.

Object Recognition

The beginnings of deep learning in 2006 focused on the MNIST digit image classification problem breaking the supremacy of SVMs (1.4 percent error) on this dataset. The latest records are still held by deep networks: Ciresan et al. claim the title of state-of-the-art for the unconstrained version of the task (e.g., using a convolutional architecture), with 0.27 percent error, and Rifai et al., is state-of-the-art for the knowledge-free version of MNIST, with 0.81 percent error.

In the last few years, deep learning has moved from digits to object recognition in natural images, and the latest breakthrough has been achieved on the ImageNet dataset, bringing down the state-of-the-art error rate from 26.1 to 15.3 percent.

Natural Language Processing (NLP)

Besides speech recognition, there are many other NLP applications of representation learning. Distributed representations for symbolic data were introduced by Hinton, and first developed in the context of statistical language modeling by Bengio et al. in the so-called neural net language models. They are all based on learning a distributed representation for each word, called a word embedding. Adding a convolutional architecture, Collobert et al. developed the SENNA system that shares representations across the tasks of language modeling, part-of-speech tagging, chunking, named entity recognition, semantic role labeling, and syntactic parsing. SENNA approaches or surpasses the state-of-the-art on these tasks, but is simpler and much faster than traditional predictors. Learning word embeddings can be combined with learning image representations in a way that allows associating text and images. This approach has been used successfully to build Google's image search, exploiting huge quantities of data to map images and queries in the same space, and it has recently been extended to deeper multimodal representations.

The neural net language model was also improved by adding recurrence to the hidden layers, allowing it to beat the state-of-the-art (smoothed n-gram models) not only in terms of perplexity (exponential of the average negative log likelihood of predicting the right next word, going down from 140 to 102), but also in terms of word error rate in speech recognition (since the language model is an important component of a speech recognition system), decreasing it from 17.2 percent (KN5 baseline) or 16.9 percent (discriminative language model) to 14.4 percent on the Wall Street Journal benchmark task.[3] Similar models have been applied in statistical machine translation improving perplexity and BLEU scores. Recursive autoencoders (which generalize recurrent networks) have also been used to beat the state-of-the-art in full sentence paraphrase detection, almost doubling the F1 score for paraphrase detection. Representation learning can also be used to perform word sense disambiguation, bringing up the accuracy from 67.8 to 70.2 percent on the subset of Senseval-3 where the system could be applied (with subject-verb-object sentences). Finally, it has also been successfully used to surpass the state-of-the-art in sentiment analysis.

Multitask and Transfer Learning, Domain Adaptation

Transfer learning is the ability of a learning algorithm to exploit commonalities between different learning tasks to share statistical strength and transfer knowledge across tasks. As further discussion, we could hypothesize that representation learning algorithms have an advantage for such tasks because they learn representations that capture underlying factors, a subset of which may be relevant for each particular task, as illustrated in Fig. 5-4B-1. This hypothesis seems confirmed by a number of empirical results showing the strengths of representation learning algorithms in transfer learning scenarios.

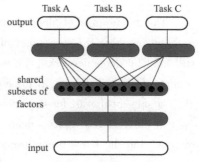

Fig. 5-4B-1 Illustration of representation-learning discovering explanatory factors

WORDS AND TERMS

learning representation　学习式表达
discriminative　*adj.* 区别的，有识别力的
disentangle　*v.* 理顺，解开纠结
probabilistic　*adj.* 概率性的，概率的
nourish　*v.* 营养，滋养
empirical　*adj.* 完全根据经验的，实证的
polyphonic　*adj.* 多音的，有韵律变化的
transcription　*n.* 翻译，抄本，录音

supremacy　*n.* 至高无上，最高地位
queries　*n.* 问号；*v.* 查询
entity recognition　实体识别
semantic role labeling　语义角色标注
syntactic parsing　句法分析
multimodal　*adj.* 多峰的，多模式的
domain adaptation　领域适应

NOTES

[1] An AI must fundamentally understand the world around us, and we argue that this can only be achieved if it can learn to identify and disentangle the underlying explanatory factors hidden in the observed milieu of low-level sensory data.

人工智能必须从根本上理解我们所处的世界，我们认为：只有当人工智能学会识别并区分隐藏在可观测的低端传感器数据环境中内在的可解释因素时，才能达到这一目标。

[2] These authors managed to reduce the word error rate on four major benchmarks by about 30 percent (e.g., from 27.4 to 18.5 percent on RT03S) compared to state-of-the-art models based on Gaussian mixtures for the acoustic modeling and trained on the same amount of data (309 hours of speech).

和基于高斯混合声学建模、代表当时技术水平的模型相比，在训练相同数据（309 小时语音）的情况下，这些作者已在四个主要的基准测试任务上将单词误差率减少约 30%（例如，在 RT03S 上从 27.4 到 18.5%）。

[3] The neural net language model was also improved by adding recurrence to the hidden layers, allowing it to beat the state of the art (smoothed n-gram models) not only in terms of

perplexity (exponential of the average negative log likelihood of predicting the right next word, going down from 140 to 102), but also in terms of word error rate in speech recognition (since the language model is an important component of a speech recognition system), decreasing it from 17.2 percent (KN5 baseline) or 16.9 percent (discriminative language model) to 14.4 percent on the Wall Street Journal benchmark task.

通过增加对隐藏层的递推来改善神经网络语言模型,不仅降低了困惑度(预测下一个单词的平均负对数似然度的指数,从 140 降到 102),还使语音识别单词误差率(由于语言模型是一个语音识别系统的重要组成部分)从 17.2%(KN5 基线)或 16.9%(鉴别式语言模型)降到《华尔街日报》基准测试任务的 14.4%,超越了现有的技术水平(平滑 n 元模型)。

C 自动化专业的学术会议

一、国内学术会议

1. 中国控制与决策学术年会

由《控制与决策》杂志编委会联合中国航空学会自动控制专业委员会、中国自动化学会应用专业委员会、中国系统仿真学会仿真方法与建模专业委员会、中国人工智能学会智能控制与智能管理专业委员会等学术组织主办,于 1989 年发起和主办的全国性大型学术会议,每年召开一次。

2. 中国控制会议

中国科学院数学与系统科学研究院的中国自动化学会控制理论专业委员会主办的国际性学术会议,每年举办一次。会议为海内外控制领域的专家、学者、研究生及工程设计人员提供了一个及时交流科研成果的机会和平台。会议以中文和英文为工作语言,采用大会报告、专题研讨会、会前专题讲座、分组报告和张贴论文等形式进行学术交流。

3. 中国智能自动化会议

由中国自动化学会智能自动化专业委员会主办,中国自动化学会机器人竞赛工作委员会、中国人工智能学会智能控制及智能管理专业委员会协办。

4. 中国过程控制会议

由中国自动化学会过程控制专业委员会主办的每年一次的全国性学术会议,会议为海内外过程控制领域的专家、学者、研究生和工程技术人员提供一个交流、研讨的平台。

5. 中国自动化学会青年学术年会(YAC)

由中国自动化学会青年工作委员会组织召开的全国性学术会议,自 1985 年起,每年举办一次,其宗旨是为自动化领域的青年学者、研究生以及工程技术人员提供一个学术交流的机会,推动我国自动化理论、技术与应用的发展。

二、国际学术会议

1. World Congress on Intelligent Control and Automation

全球智能控制与自动化大会,简称 WCICA,每两年在中国召开一次,其前身是全球华人智能控制与自动化大会。该会议是自动化专业国际会议中唯一同时接收中、英文稿件的。

2. International Federation of Automatic Control World Congress

国际自动控制联合会世界大会,简称 IFAC World Congress,每三年召开一次。

3. IEEE International Conference on Systems, Man and Cybernetics

国际系统、人与控制大会，简称 SMC，是控制领域中的权威会议，每年召开一次。

4. International Conference on Intelligent Robots and Systems

国际智能机器人和系统会议，简称 IROS，每年召开一次。

5. IEEE Conference on Decision and Control

IEEE 控制与决策大会，简称 IEEE CDC，是控制与决策领域最高水平的国际会议。

6. International Conference on Computer Vision and Pattern Recognition

国际计算机视觉与模式识别大会，简称 CVPR，是计算机视觉领域最高水平的年会之一。

7. IEEE International Conference on Intelligent Transport Systems

IEEE 国际智能交通系统大会，简称 IEEE ITSC，是最重要的智能交通方面国际年会。

8. IEEE Intelligent Vehicles Symposium

IEEE 智能车辆研讨会，简称 IEEE IVC，是智能交通方面最高水平的国际会议之一。

9. Asian Test Symposium

亚洲测试研讨会，简称 ATS，每年召开一次。

10. World Congress on Industrial Process Tomography

世界工业过程大会，简称 WCIPT，每两年召开一次。

11. American Control Conference

美国控制大会，简称 ACC，由 IEEE Control System Society 主办，每年召开一次。

12. IEEE International Conference on Robotics and Automation

IEEE 国际机器人学与自动化大会，简称 ICRA，每年一次。

13. IEEE Industry Applications Society Annual Meeting

IEEE 工业应用学会年会，简称 IAS，是过程控制、电气、电子、拖动在工业中应用的最高水平会议。

14. International Conference on Pattern Recognition

国际模式识别大会，简称 ICPR，是模式识别领域最高水平的国际学术年会。

15. International Conference on Acoustics, Speech, and Signal Processing

国际语音声音信号处理大会，简称 ICASSP，是语音声音信号处理研究领域中最高水平的国际会议。

16. International Conference on Computer Vision

国际计算机视觉大会，简称 ICCV，是计算机视觉最高水平的国际年会。

17. International Conference on Control, Automation, Robotics and Vision

国际控制、自动化、机器人学与视觉会议，简称 ICARCV，由新加坡南洋理工大学组织，得到 IEEE 协会多个分会支持，是东亚和东南亚地区高水平的控制科学领域学术交流年会。

18. Asian Control Conference

亚洲控制大会，简称 ASCC，是亚洲重要的控制科学领域学术会议。

19. International Conference on Machine Learning and Cybernetics

国际机器学习与控制论大会，简称 ICMLC，是机器学习、数据挖掘及控制应用方面的重要学术年会。

其他国际会议信息可以参考小木虫网站，网站地址：http://emuch.net/。

PART 6

Synthetic Applications of Automatic Technology

UNIT 1

A Scanning the Issue and Beyond: Toward ITS Knowledge Automation

Starting with the first issue of the bimonthly version of *IEEE Transactions On Intelligent Transportation Systems*, I will begin each issue by scanning and summarizing each article in a format that is suitable for presentation at *Weibo* (Micro blogs in Chinese), Twitter, and Facebook. Please check @ IEEE-TITS (http://www.weibo.com/u/3967923931) for *Weibo*, https://www.facebook.com/IEEEITS for Facebook, and@ IEEEITS(https://twitter.com/IEEEITS) for Twitter. In addition, I will go beyond the papers published here and give my thought on issues that I consider interesting or important for current or future research and development in the area of intelligent transportation.

Scanning the Issue

Symmetrical SURF and Its Applications to Vehicle Detection and Vehicle Make and Model Recognition
J. W. Hsieh, L. C. Chen, and D. Y. Chen

A new symmetrical SURF descriptor is presented to enhance SURF's power to detect all possible symmetrical matching pairs through a mirroring transformation. To deal with multiplicity and ambiguity, a grid division scheme is also proposed to separate a vehicle into several grids in which different weak classifiers are trained and then integrated to build a strong ensemble classifier.[1] Due to the rich representation power of the grid-based method and the high accuracy of vehicle detection, the ensemble classifier can accurately recognize each vehicle.

Sensor Fusion-Based Vacant Parking Slot Detection and Tracking
J. K. Suhr and H. G. Jung

A vacant parking slot detection and tracking system is proposed and expected to help drivers select available parking slots. The system fuses the sensors of an around view monitor system and automatic parking system. The experimental results show that the proposed method can recognize the positions and occupancies of various types of parking slot markings and stably track them in real time.

Portable Roadside Sensors for Vehicle Counting, Classification, and Speed Measurement
T. Saber and R. Rajesh

A portable roadside sensor system for measuring traffic flow rate, vehicle speeds, and vehicle classification is developed. An algorithm based on a magnetic field model is proposed to make the system robust. In addition, an algorithm to automatically correct for any small misalignment of the sensors is applied. The accuracy and benefits of the developed sensor system is discussed.

Analytical Hierarchy Process Using Fuzzy Inference Technique for Real-Time Route Guidance System
C. Li, S. Anavatti, and T. Ray

An optimum route search function in the in-vehicle routing guidance system is discussed. An analytical hierarchy process using Fuzzy inference technique based on the real-time traffic information is proposed to realize the dynamic route guidance. The proposed method can simplify the definition of decision strategy and represent the multiple criteria explicitly. A simulation system is developed based on the proposed method.

A Wireless Accelerometer-Based Automatic Vehicle Classification Prototype System
W. Ma, D. Xing, A. McKee, R. Bajwa, C. Flores, B. Fuller, and P. Varaiya

The problem of automatic vehicle classification (AVC) systems is addressed. A prototype axle count and spacing AVC system based on wireless accelerometers and magnetometers is introduced. The detected parameters and the installation of the system are presented. Through an experiment under various traffic conditions, the prototype AVC system is proved to be reliable in classifying vehicles with an accuracy of 99% even under congested traffic.

Stochastic Characterization of Information Propagation Process in Vehicular Ad-Hoc Networks
Z. Zhang, G. Mao, and B. Anderson

The information propagation process in vehicular ad-hoc networks on highways is addressed. It is assumed that vehicles in the network are categorized into different traffic streams with regard to their types and lanes, whose speed distributions are the same within the category and different from those of other categories. An analytical formula for the information propagation speed is obtained by analyzing the information propagation process. Using the formula, the impact of some parameters is studied. Simulations are conducted to validate the accuracy of the analytical results.

Reducing the Error Accumulation in Car-Following Models Calibrated with Vehicle Trajectory Data

J. Jin, D. Yang, and B. Ran

Considering the error accumulation problem in the calibration of car-following models using trajectory data, this paper proposes an error dynamic model. The stability conditions for the derived error dynamic model are different from the model stability conditions. The traditional and the proposed error measures through the calibration of representative car-following models are compared.

Automated Detection of Driver Fatigue Based on Entropy and Complexity Measures

C. Zhang, H. Wang, and R. Fu

Since some traffic accidents are caused by driver's fatigue, this paper proposes a real-time method to detect and identify driving fatigue based on various entropy and complexity measures from some records. It is shown that the proposed approach is effective and it is valuable for the application of avoiding some traffic accidents caused by driver's fatigue.

Two-Dimensional Sensor System for Automotive Crash Prediction

S. Taghvaeeyan and R. Rajamani

The use of magnetoresistive and sonar sensors for imminent collision detection in cars is investigated. An adaptive estimator is proposed and both sonar and magnetoresistive sensors are used to estimate the parameters to determine cars' position and orientation. Experimental results show this approach's effective for a range of relative motions at different oblique angles.

Utilizing Microscopic Traffic and Weather Data to Analyze Real-Time Crash Patterns in the Context of Active Traffic Management

R. Yu, M. A. Abdel-Aty, M. M. Ahmed, and X. Wang

The effects of microscopic traffic, weather, and roadway geometric factors on the occurrence of specific crash types for a freeway are studied. The authors propose to expand the purpose of the existing ITS system and suggest Active Traffic Management strategies by identifying the real-time crash patterns. Numerical simulation results show that single-vehicle crashes are more probable to occur in snow seasons, at moderate slopes, three-lane segments, and under free-flow conditions.

Modeling and Forecasting the Urban Volume Using Stochastic Differential Equations

R. Tahmasbi and S. M. Hashemi

To deal with the problem of short-term prediction of traffic flow, a methodology is developed in this paper using the stochastic differential equation. The Hull-White model is used to consider the time dependency of short term traffic volume. It may simulate traffic conditions easily and detecting incidents precisely. It is illustrated that a better fit to the traffic volume is obtained using the proposed method compared to the previous artworks.

Observer-Based Robust Control of Vehicle Dynamics for Rollover Mitigation in Critical Situations
H. Dahmani, O. Pagès, A. El Hajjaji, and N. Daraoui

  A fuzzy control method of vehicle dynamics to improve stability and minimize the rollover risk is proposed. The authors take into account several aspects to obtain a robust controller, where the nonlinearities of the lateral forces is represented using a Takagi-Sugeno (TS), changes in road friction is considered by introducing parameter uncertainties and road bank angle is set as an unknown input. The linear matrix inequalities constraints are solved to obtain the observer and controller gains.

Cooperative Adaptive Cruise Control in Real Traffic Situations
V. Milanés, S. E. Shladover, J. Spring, C. Nowakowski, H. Kawazoe, and M. Nakamura

  This paper presents the design, development, implementation, and testing of a cooperative adaptive cruise control system. This system has been implemented on four production Infiniti M56s vehicles to validate the performance of the controller and the improvements.

Coding or Not: Optimal Mobile Data Offloading in Opportunistic Vehicular Networks
Y. Li, D. Jin, Z. Wang, L. Zeng, and S. Chen

  The coding-based mobile data offloading problem is formulated as a users' interest satisfaction maximization problem with multiple linear constraints of limited storage. The problem is solved by an efficient scheme which provides a solution to decide when the coding should be used and how to allocate the network resource.[2] The effectiveness of the algorithm is demonstrated extensive simulations using two real vehicular traces.

Estimating Dynamic Queue Distribution in a Signalized Network Through a Probability Generating Model
Y. Lu and X. Yang

  A stochastic queue model using the probability generating function, which considers the strong interdependence relations between adjacent intersections, is proposed. Various traffic flow phenomena are formulated as stochastic events and obtain their distributions by iteratively computing through a stochastic network loading procedure. The effectiveness of the proposed approach is demonstrated by the theoretical derivation and numerical investigations.

The Process of Information Propagation along a Traffic Stream through Intervehicle Communication
W. Wang, S. S. Liao, X. Li, and J. S. Ren

  A model is proposed to calculate the average speed of transmission of inter-vehicle communication (IVC) messages in general traffic stream on highways in the early stage of deploying distributed traffic information systems (DTIS). The relationship between average IVC message speed and traffic parameters can be explained with this model. The correctness of the model is verified by simulation results, and the theoretical analysis is given.

Toward Real-Time Pedestrian Detection Based on a Deformable Template Model
M. Pedersoli, J. Gonzàlez, X. Hu, and X. Roca

The problem of pedestrian detection in driving assistance systems which has a trade-off between accuracy and real-time is investigated. A pedestrian detection system using a hierarchical multi resolution part-based model is proposed and implemented on GPU. The proposed system can achieve the state-of-the-art pedestrian detection accuracy and show a speed-up of more than one order of magnitude, which is suitable for pedestrian detection with respect to both precision and real time.

An Event-Triggered Receding-Horizon Scheme for Planning Rail Operations in Maritime Terminals
C. Caballini, C. Pasquale, S. Sacone, and S. Siri

The problem of planning rail port operations is investigated where unexpected events or disturbances often affect seaport terminals. Based on a queue-based discrete-time model, the paper defines a mixed integer linear mathematical programming problem and proposes an event-triggered receding-horizon optimization approach. The test of the proposed approach based on data of a real terminal is given and discussed.[3]

Understanding Bicycle Dynamics and Cyclist Behavior from Naturalistic Field Data
(November 2012)
M. Dozza and A. Fernandez

Considering the significant role but the absence of models of bicycles in intelligent transportation systems, this paper presents a platform, based on which the bicycle dynamics and bicyclist behavior can be researched. In this platform, field data is collected continuously from sensors and can be employed to derive, develop, and test intelligent transportation systems including bicycles.

Robust Control for Urban Road Traffic Networks
T. Tettamanti, T. Luspay, B. Kulcsár, T. Péni, and I. Varga

This paper proposes a robust real-time signal split algorithm to minimize the overall weighted queue lengths within an urban network area. The traffic control problem is formulated in a centralized rolling-horizon way, and the green time combination is obtained with an efficient constrained minimax optimization. The proposed algorithm is tested by using real-world traffic data and microscopic traffic simulation and compared with well-tuned fixed-time signal timing.

A Survey of Traffic Control with Vehicular Communications
L. Li, D. Wen, and D. Yao

The problem of deploying vehicle-to-vehicle communications and/or vehicle-to-infrastructure communications to coordinate vehicles and traffic signals in real time is addressed. A perspective of its research frontiers is given, early stage key technologies are identified, and the possible improvements are discussed. Furthermore, the prominence to scheduling based intersection control approaches is also given here. Moreover, this paper discusses two cultures including using rich information or concise information.

On Optimality Criteria for Reverse Charging of Electric Vehicles
S. Stüdli, W. Griggs, E. Crisostomi, and R. Shorten

The issue of the controllable loads and storage systems of electric vehicles, which can be utilized to mitigate the load on the grid during peak times by offering power, is investigated. The problem of returning electrical load to the grid as an optimization aiming at returning the desired energy in a fashion minimizing the cost on the environment is formulated. It is shown that this optimization is highly complex and the cost of vehicle to grid in some circumstances can be prohibitive.

ITS Knowledge Automation

While writing the summaries, I had really hoped that I could have an automatic summarization system that can do the job for me quickly and accurately, either by extraction or abstraction or both, and send the result to social media and interested readers immediately and automatically. It also reminded me of the recent report "Disruptive Technologies: Advances That Will Transform Life, Business, and the Global Economy" by McKinsey Global Institute (MGI), where "automation of knowledge work" or "intelligent software systems that can perform knowledge work tasks involving unstructured commands and subtle judgments" was listed as the second most economically disruptive technology with $5.2 trillion to $6.7 trillion in potential economic impact annually by 2025 and an estimated task performance that would be equal to the output of 110 million to 140 million full-time equivalents. Actually, my editorial for the last issue of Acta Automatica Sinica in 2013 was entitled "The Destiny: Toward Knowledge Automation". For IEEE-TITS and ITS R&D in general, my thoughts and goals are more specific: ITS knowledge automation.

WORDS AND TERMS

bimonthly　*adj.* 两月一次的
stochastic　*adj.* 随机的
hierarchy　*n.* 层次，等级制度
magnetoresistive　*adj.* 磁阻的
iteratively　*adv.* 迭代地，反复地
event-triggered　事件触发
prohibitive　*adj.* 禁止的

symmetrical　*adj.* 对称的
algorithm　*n.* 算法，预算法则
information propagation　信息传播
methodology　*n.* 方法学，方法论
optimization　*n.* 最佳化
microscopic　*adj.* 微观的
disruptive technology　破坏性技术

NOTES

[1] To deal with multiplicity and ambiguity, a grid division scheme is also proposed to separate a vehicle into several grids in which different weak classifiers are trained and then integrated to build a strong ensemble classifier.

为了应对多重性和模糊性,同时还提出一套网格划分方案将车辆划分为几种类型,其中对不同的薄弱环节进行培训,然后综合打造一个强大的集成分类。

[2] The problem is solved by an efficient scheme which provides a solution to decide when the coding should be used and how to allocate the network resource.

该问题通过一种有效的方案得到了解决,该方案提供了一种应该在什么时候进行编码和如何进行网络资源分配的方法。

[3] The test of the proposed approach based on data of a real terminal is given and discussed.

基于一个真实的终端数据,该方法的测试已经被给出并进行了讨论。

B Automation or Interaction: What's Best for Big Data?

Introduction

In the late 1800's telephone exchanges were manually operated and could only process a few callers a minute. As the volume of calls grew, a single operator could not handle the demand and manual exchanges gave way to automated ones. Today, operators still connect some calls, usually when the caller needs additional information (or money), but the vast majority can be handled by automated systems. History is littered with examples of systems that have become automated as technology improves.

This panel questions whether we, the visualization community, are on the right track by concentrating our research and development on interactive visualization tools and systems. After all, research programs like the Department of Energy's Accelerated Strategic Computing Initiative (ASCI) run computer simulations that produce terabytes of data every day. This raises the following questions:

1) Is it feasible to analyze terabyte data sets using interactive techniques?

2) Has visualization reached a level of maturity where most of the tasks can be automated?

3) Will automatic feature detection tools be able to find all the interesting phenomena?

Our distinguished panelists will provide answers to these questions. They have been asked to "choose sides" to stimulate the discussion and to provoke controversy. Steve Bryson and Robert van Liere make strong cases for interactive visualization tools, while Robert Haimes and David Banks will tell us why automation is required for big data. Sam Uselton brings balance to the debate by suggesting that both automatic and interactive techniques will play important roles in understanding big data sets.

Position Statements

David Banks
"Automation Suffices for 80% of Visualization"

Interactive visualization would be essential to those scientists who pursue unfettered exploration of unfamiliar data, the scientists who discover new phenomena in their simulation that they never suspected were there, the scientists who like to try new tools that other people have created for their use. As many of us have experienced first-hand, these scientists exist in the realm of science fiction and PBS specials, not in real life.

There are two primary applications of computer graphics in scientific computing: debugging and presentation.

Tom Crockett (ICASE) champions the paradigm of visualization as a 3D print statement to let you quickly hunt down an offending segment of code. An interactive debugger is great for finding errors, but most people only use one as a last resort. The automatically-generated compiler messages catch the large fraction of simple bugs, and print statements reveal most of the others. In the same way, automatic visualization tools are well suited for debugging scientific codes. With datasets reaching the terabyte scale, a scientist could spend hours exploring iso-surfaces, volume density mappings, or particle paths latent in a dataset. Interaction is the method of last resort.

Others consider visualization to be primarily a post-processing step to create a slick demo or a colorful poster or an animation for the Web. If the scientist's intent is to display certain features, then the visualization tool should be designed to locate them in the data. Research in visualization therefore includes characterizing discipline-specific features (tumors, blood vessels, vortices, shock surfaces, oil deposits) via robust algorithms. Setting up the right viewpoint and lighting and layout is important in preparing images for public presentation, but this requires interaction for the art department rather than for the scientist.

Steve Bryson
"Show me that" vs. "What's there?"

It is a well-worn adage that the question you ask in large part determines the answer you will get.[1] That is a fine thing if you are asking the right questions. But insight and discovery are driven as much by open-ended, curious exploration as by having your specific questions answered. This is not a place for the obvious ensuing philosophical discussion, but when your questions are very narrow exploration becomes much more difficult to perform. Automatic feature detection requires the framing of very specific questions: "show me the vortices" or "show me where a specific condition is satisfied". As an example, consider a room full of stuff. Automatic feature detection is like saying "show me the red boxes in the room". You'll find out where the red boxes are, but you'll miss knowledge of other objects in the room. Many more such specific questions are required for automatic feature detection techniques to give me a sense of all the objects in the room. On the other hand I could simply say "show me what's in the room".

Of course scientific visualization is not so simple, otherwise we would not have these annual conferences. Even physical simulation data can be very abstract, and there simply is no canonical or obvious way to "show me the data". Thus we have to ask questions of the data, in our business in the form of graphical representations of that data. Yes, the specifics of the graphical representation will determine the type of information we obtain on that data. But automatic feature detection asks very specific questions so that the results can be computed algorithmically and simply represented. Thus automated detection of features in a data set will always detect the features you ask for, and, nothing else. If this is what you want then you are done.

But if you want to have a broad understanding of the data set, in particular if you want to understand why certain features are in the data set it is rarely sufficient to just display features. A sense of the data "around" the features is critical to understanding their context and often their cause. Knowing the vortices in a flow has use, but understanding why those vortices are where they are requires knowledge of the flow around them. Put another way, much scientific investigation is of phenomena that, while subject to local laws, are determined by global considerations. Fluid flow is a very common example: the existence of a vortex is due to the shape of the object that the flow is moving around. A somewhat global sense of the flow is required to understand the subsequent vortices.

Getting a global sense of data is difficult, particularly when the data is in three or more dimensional space and may have several interacting components. As is well known, scenes can quickly get very cluttered when many aspects of data are presented at the same time. Interactive techniques, where you have a graphical representation that you may move about in a data set in near real time, allows you to rapidly sample different regions of the data in different ways. To continue the flow example, observing interactive streamlines around a vortex can give great insight into the cause of the vortex. Interactivity is required to allow a sense of exploring the data. The more intuitive the interaction interface, the better the exploration will be. A rapid exploration capability allows you to get a general sense of the data, which provides a context for any features that may be detected in your exploration.

In some circumstances, you don't even know what specific questions to ask: science advances when new questions are thought up in response to new ways of seeing things. In these cases interactive exploration is a valuable tool to allow you to ask old questions in new ways. Streamlines of a vector field, which originally represented the paths of particles in a (steady) flow, can be used to study the behavior of, for example, the gradient of a scalar such as pressure. Exploration of a gradient field via streamlines can provide new insights such as the maxima and minima structure and so on. (OK, a weak example, but it's hard to think of nontrivial examples of fundamentally new questions!) While the same game can be played with feature detection techniques, e. g. by looking for the vortices in a gradient field, it is not quickly apparent what such features represent.

So we are presented with a spectrum: At one end automated feature detection techniques provide specific answers to narrow questions. If the question is exactly appropriate to your problem the automated feature detection may be all you need. At the other extreme, you may be exploring data in a simulation in which you have little understanding and don't know the interesting questions.

In this case a suite of interactive visualization techniques will allow you to get a sense of the data and perhaps prompt interesting questions and understanding. In between, as in the example of flow around a complex object, feature detection techniques can give you a good starting point for detailed interactive exploration.

I'm reminded of the situation in robotics, where the initial hope was that robots could be completely autonomous. This turned out to be somewhat beyond our reach in general situations, but high-level human control of robots has been very successful. This mix of automated and interactive activity is, I feel, very informative for our field. While one may argue that if we were just a little smarter we could automate everything, I feel that it is precisely at the frontiers of our understanding that scientific visualization has the greatest leverage, and it is here that we know the least about what questions to ask. This will always be the case.

Robert Haimes

Beyond stone knives and bearskins

Programs like the Accelerated Strategic Computing Initiative (ASCI) represent a tremendous growth of large-scale computing applied to the analysis of scientific problems.[2] Most of the proposed ASCI simulations create output data sets containing billions of words of information (distributed on a 3D mesh) for the results of a single steady-state run. Clearly, transient simulations of the same spatial fidelity stress any available computer resources. The sheer size of this data results in an exceedingly difficult and time consuming analysis process. The task of interrogation and interpretation of this information is required so that the knowledge contained within the simulation can be extracted.

Traditional interactive visualization probes the data in order to locate and identify physical phenomena. In order to find important flow features, users must interactively explore their data using one or more of the visualization tools (iso-surfaces, geometric cuts, streamline, and etc.). Scientists and engineers that use them on a regular basis have reported the following drawbacks:

1) Exploration Time: Interactive exploration of large-scale 3D data sets is laborious and consumes hours or days of the scientists/engineers time.

2) Field Coverage: Interactive visualization techniques produce output based on local sample points in the grid or solution data. Important features may be missed if the user does not exhaustively search the data set.

3) Non-specific: Interactive techniques usually reveal the behavior in the neighborhood of a feature rather than displaying the feature itself.

4) Visual Clutter: After generating only a small number of visualization objects the display becomes cluttered and makes visual interpretation difficult.

It is clear that these tools do not directly answer the questions of the investigator. An expert is required to infer the underlying field topology from the imagery supplied. Getting a more specific answer is required. Direct, automated feature extraction has the following advantages over these exploratory visualization tools:

1) Deterministic Algorithms: If there are no 'parameters' that the user need adjust, then no

intervention is required.

2) Fully Automated: The analysis can be done off-line (without a visualization subsystem). It can be used by other components in the analysis suite (i.e., directly by a solver to adapt the mesh to better resolve the feature).

3) Local Analysis: These schemes, where possible, perform only local operations. Therefore, the computations for each cell are independent of any other cell and may be performed in parallel. This is clearly advantageous in distributed memory compute arenas.

4) Data Reduction: The output geometry is several orders of magnitude smaller than the input data set. This is an important characteristic for the size of a resultant output. High fidelity spatial and temporal results of the feature extraction can be stored on disk. This is usually not possible for the entire transient simulation.

5) Quantitative Information: Precise: locations for the extracted features are provided. Also, classification and measures of strength can be reported.

A simple analogy can be drawn to any complex code. A large-scale program (that runs for more than a couple of seconds) may perform billions of integer and float-point calculations. It is not necessary to examine each operation to know that the program is running properly. There is usually some metric that the user of the program can use to determine the results. Even large, long running scientific simulations report integrated values to the user as some measure of goodness. Unfortunately, most of these measures are based on numerics and not physics. The physics can be examined by automatically extracting the features of interest and then answering the question: Is this what I expected?

Only when something happens outside our expectations (our analogous large-scale program produces unanticipated results) do we need to more closely examine the operations. Interactive visualization is only the debugger of our 3D scientific simulation codes.

Robert van Liere

"Sorry, but I'm not really sure what I'm looking at. "

The importance of data visualization is clearly recognized in scientific computing. Display of simulation results and interactive steering of computation require interactive exploration environments in which a user can see relationships and test hypotheses.

To support this claim I will discuss two examples. Both examples are motivated by the lack of knowledge of what is contained in the data. The first example is from flow visualization: the exploration of a very large turbulent data set from a direct numerical simulation. The second example is from cell biology: the exploration of cell components acquired from a confocal microscope.

The need for exploration environments will increase as models become more complex, simulation solutions become more detailed or acquisition devices become more powerful.

Sam Uselton

Best bets for big data

"Automation or Interaction, what's best for big data?" is the wrong question! The data doesn't know or care! Seriously, the question should be rephrased to focus on what is best for the USERS of big data. And that requires understanding what the users are trying to accomplish, and why large amounts of data are involved.

The first thing to notice is that there are many users with a wide variety of reasons for interest in large amounts of data. A single user's interest, even in a particular data set, may also vary drastically over time. I like to characterize one dimension of the variation in users purposes as ranging between "scientific" and "engineering" purposes. Engineering purposes are characterized by specific goals that result in precise answers to be extracted from the data. "Where in this lease should I drill to get the most oil?" "What angle of attack results in the largest lift to drag ratio for this aircraft?" Scientific purposes are characterized by vague, qualitative goals or extremely broad and general goals, which result in a desire to browse through the data looking for something unusual or different? "How does turbulence develop in originally laminar fluid flow?" "How did the universe evolve to produce galaxies and stars?" Remember that this is a continuum, not a binary classification. It is clear that automated answer finding is easier for questions at the engineering end of this spectrum than at the scientific end.

We are now producing and collecting data of many different kinds at a rate that precludes thorough inter-active exploration as means for discovering the answers at the scientific end of the spectrum. Automatic methods suffer from typical computer "blind spots"— finding what they are directed to find, not everything the user might find interesting. Many people now favor using collections of tools that allow both kinds of activities. It is important to have such tools that "play well together". And that is not enough; we also need tools that are some new hybrid, using automatic methods to find less specific "things" in data sets, and suggesting places and dimensions in which interactive exploration is likely to be interesting.

Biographies

David Kenwright is a senior research scientist with MRJ Technology Solutions and works in the Data Analysis Group at NASA Ames Research Center. His current research interests include flow feature detection, vector field topology, and biomimetics. He received his BE degree with first class honors in 1988 and his Ph. D. in mechanical engineering in 1994 from the University of Auckland, New Zealand.

Steve Bryson is a research scientist in the Numerical Aerodynamic Simulation Systems Division at NASA Ames Research Center and currently leads the Data Analysis group. He does research in the application of virtual reality techniques for scientific visualization, of which the virtual wind-tunnel is the main focus. He is the general co-chair of IEEE Visualization'99.

Robert Haimes is a principal research engineer in the Department of Aeronautics and Astronautics at the Massachusetts Institute of Technology. He is the author of a number of scientific visualization software toolkits in use worldwide, including Visual3 and pV3. His professional interests include computational fluid dynamics, turbomachinery, numerical algorithms, parallel and distributed programming, and scientific visualization.

Robert van Liere is head of a small interactive visualization and virtual reality research group at the Center for Mathematics and Computer Science, CWI, in Amsterdam. The group's research activities focus on computational steering, high-performance visualization, and virtual reality. Robert has been at the center for 12 years. Before that Robert worked at TNO, at Dutch applied research organization.

Sam Uselton is a computer scientist in the Center for Applied Scientific Computing (CASC), and leads the research efforts in data exploration. He received his B. A. in Mathematics and Economics in 1973 from the University of Texas at Austin. He earned his M. S. in 1976 and his Ph. D. in 1981, both from the University of Texas at Dallas. His current research interests include interactive methods of exploring very large scientific data sets, methods for evaluating visualizations and visualization systems, data fusion, comparative analysis methods, feature specification and detection, pattern recognition, innovative user interfaces, direct volume rendering, parallel rendering and realistic image synthesis.

WORDS AND TERMS

interactive   *adj.* 交互的
geometry   *n.* 几何学
visualization   *n.* 可视化
drastically   *adv.* 彻底地
exploratory   *adj.* 探究的

blind spot   *n.* 盲点
well-worn   *adj.* 老生常谈的
parallel   *adj.* 并行的
robotics   *n.* 机器人学
browse   *v.* 浏览

NOTES

[1] It is a well-worn adage that the question you ask in large part determines the answer you will get.

这是一个老生常谈的格言：你问的问题在很大程度上决定了你会得到的答案。

It is... that 引导的是同位语从句。

[2] Programs like the Accelerated Strategic Computing Initiative (ASCI) represent a tremendous growth of large-scale computing applied to the analysis of scientific problems.

加速战略计算计划（ASCI）等项目，代表着应用于科学问题分析中的大规模的计算有着巨大的增长。

applied to... 过去分词做后置定语。

C   说明书常用术语

operational/operating instructions 操作说明书
major components and functions 主要部件及功能
assembles and controls 各部件及其操作机构
group designation 总类名称
operating flow chart 操作流程图

user's manual 用户手册
feature 特点
construction 构造
electric system 电气系统
coolant system 冷却系统

fine adjustment 微调
coarse adjustment 粗调
direction for use 使用方法
wear-life 抗磨损寿命
high voltage cautions 小心高电压
transportation 搬运，运输
instruction for erection 安装规程
power requirements 电源条件
service condition 工作条件
system diagram 系统示意图
wiring/circuit diagram 线路图
operating voltage 工作电压
factory services 工厂检修服务
specific wearability 磨损率
precautions/cautions 注意事项
measuring range 量程
data book 数据表
Don't cast 勿掷
standard accessories 标准附件
accessories supplied 备用附件
safety factor 安全系数
tested error free 经检验无质量问题
ground/GND terminal 接地端子
earth lead 地线
inflammable 易燃物，防火
Keep dry 保持干燥
Keep upright 勿倒置
To be protected from cold/heat 避免遇冷/热
Use rollers（出现在外包装箱上时）移动时使用滚子
warranty 保证书

test run 试运转
first commissioning 试车
maintenance 维护，维修
dimensions 尺寸
measurement 尺码
lubrication 润滑
inspection 检验
location 安装位置
fix screw 固定螺钉
specifications 规格
rated load 额定负载
rate capacity 额定容量
nominal speed 额定转速
nominal horsepower 额定马力
Gross/Gr. Wt. 毛重
Net Wt. 净重
fragile 易碎
cutting capacity 加工范围
humidity 湿度
oiling period 加油间隔期
work cycle 工作周期
recyclable 可回收利用的
Handle with care 小心装卸
Heave here 从此提起
haul 起吊，此处起吊
Keep in cool place 置于阴凉处
Not to be tipped 勿倾倒
stuffing 填充料
cleaning 清洗
guarantee 保证书，保修书

UNIT 2

A Smart Grid Standards for Home and Building Automation

I. Introduction

The deployment of smart home and building automation systems are getting popular with the advancement in Information and Communication Technology (ICT) applications. Smart systems offer convenience in both private residence and commercial buildings. These systems increase the comfort through remote control of heating, ventilation, air condition, lighting, and allow user to manage appliances without physical presence. A smart home and building automation network consists of devices that monitor and control technical systems in a home or/and building automatically. The smart home and building automation system aims at improving control, monitoring and administration of these systems using two way communication either through wireless or wired technologies. In addition, through smart grid a system enables a user to control the energy usage according to the price and demand. In doing so, these systems contribute toward energy saving. This is certainly one of the challenging global targets for researchers to tackle recently(Fig. 6-2A-1).

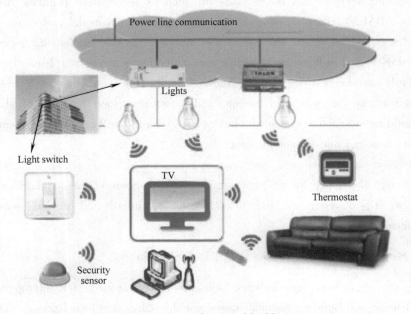

Fig. 6-2A-1 An overview of smart home and building automation system:
communication can be performed either through power line or wireless

With growing popularity of smart home and building automation systems, various organizations offer standards for interoperable products, enabling smart home and building automation systems that can control appliances, lighting, energy management, and security environment.[1] It also helps in the network expandability in order to connect with different networks. All these standards have been developed in parallel by different organizations. It is therefore necessary to arrange those standards in such a way that it is easier for potential readers to easily understand and select a particular standard according to their requirements without going into the depth of each standard, which often spans from hundreds to thousands of pages.

To the best of our knowledge, this paper is the first comprehensive survey conducted to compare and evaluate different types of smart grid standards related to home and building automation in terms of functions field. We introduce the main standards proposed by different organizations for home and building automation. In addition, we evaluate and compare the scope of interoperability, benefits and drawbacks of these standards.

The rest of the paper is organized as follows. Section II provides details of related works. Section III explains in details about various standards divided into different categories. Section IV concludes the paper.

II. Related Works

Some research works focused on home and building automation systems with reference to some chosen standards. The researchers provided a survey on building automation systems, directing attention to their communication systems in terms of BACnet, LonWorks and KNX. After a general overview on building services and the benefits provided by the current Building Automation and Control Systems (BACS), they introduced a three-level functional model and showed how control networks can be embedded inside the automation systems. Various challenging aspects related to wireless technologies used in home and building automation applications have been discussed and the relevant standards have been surveyed. The National Institute of Standards and Technology (NIST) framework and roadmap for smart grid interoperability standards have been specified. The authors outlined the building automation systems including BACnet and LonWorks that integrate the services like ventilation, heating, and air-conditioning.

There are a variety of smart grid standards for home and building automation, which are either open or proprietary, developed by different organizations.[2] Some of the standards are still at the developing stage. It is therefore necessary to collect these standards and provide comparison among the related standards.

III. Smart Home and Building Automation Standards

Over the years, various organizations have proposed standards for interoperable products enabling smart home and building automation systems that can control appliances, lighting, energy management and security environment. In this section, we explain each standard based on the function fields as shown in Fig. 6-2A-2. Following is the brief explanation of each standard.

1. Zigbee Home Automation Public Application Profile

ZigBee Home Automation (ZHA) Public Application profile is a Zigbee profile for home

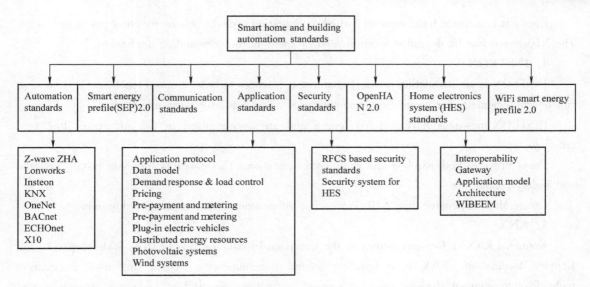

Fig. 6-2A-2 An overview of different categories of smart grid standards in home and building automation

automation applications. With the introduction of ZHA profile, home automation can move from currently limited implementations at homes to the higher volume products in the conventional market. The ZHA profile supports a variety of devices for the home including lighting, heating and cooling, and even window blind control. It provides interoperability from different vendors that allow a greater range of control and integration of different devices in the home. It mainly deals with sporadic real time control of devices.

Pros: ZHA is robust, secure and reliable solution in noisy Radio Frequency environments.

Cons: ZHA is still expensive and complicated to install. It may not be worth the cost for systems having few devices.

2. LonWorks

LonWorks is a distributed control system developed by the American company Echelon, which meets the peer-to-peer and/or master-slave communication needs of BACS networks. LonWorks is the leading market solution in the USA, whereas KNX has yet to make an impact.

Pros: Provides effective device level protocol, also the controllers adhering to LonMark profiles are cost effective.

Cons: Proprietary standard, which has little acceptance in certain systems such as fire, security, and Power Line Communication (PLC).

3. X10

X10 is one of the first home automation standards, developed in 1975. It uses power line wiring for signaling and control. X10 became the de facto standard for home automation over the years and it can still be seen today.

Pros: X10 is time proven technology which has been around more than 30 years. It is inexpensive, simple to install, with hundreds of compatible products. It does not require any new wiring.

Cons: It is a single band network, which is operating over the inherently noisy power line only. The X10 signal can be degraded or attenuated by many of the surrounding appliances.

4. INSTEON

INSTEON is an automation protocol enabling appliances to be networked together. This concept can be implemented over power line communication as well as over radio interface.

INSTEON messages have a fixed length and are synchronized to the AC power line zero crossings.

Pros: Provides both wireless and wired communication. The system is simple to install, reliable and widely compatible.

Cons: More expensive than X10 and having interference issues in the wired systems.

5. KNX

Konnex (KNX), formerly known as the European Installation Bus (EIB), is developed by the Konnex Association. KNX is a building control communication system that uses information technology to connect devices such as sensors, actuators, controllers, operating terminals and monitors.

Pros: KNX is a large association of corporation/professionals in the automation field.

Cons: Mostly limited to Europe and currently having radio quality issues.

6. OneNet

OneNet is an open standard (royalty-fee) and open source solution for home and building automation, which is based on the proprietary physical interface. It defines physical, network and message protocol in order to provide low-power, low-delay, low-cost and medium range wireless solution for devices and applications.

Pros: Royalty fee standard, low-power, low-delay, low-cost, and interoperable with X10 and INSTEON.

Cons: Not available for PLC.

7. BACnet

BACnet is a building automation and control networking protocol developed by the American Society of Heating, Re-frigeration and Air-Conditioning Engineers (ASHRAE). BACnet has been designed specifically as the data communication protocol of building automation and control systems for applications such as heating, ventilating, and air-conditioning control, lighting control, access control, and fire detection systems. The purpose is to define data communication service and protocols for computer equipments used for monitoring and control of Heating Ventilation Air Conditioning and Refrigeration (HVAC&R) and other building systems to define object-oriented representation of information communicated between those devices.

Pros: Available from most BAS and HVAC equipment vendors, cost effective integration solution, provides insulation from vendor system migrations and upgrades and allows for flexibility in implementation.

Cons: Configuration tools are proprietary and no device application profiles, system, and device control algorithms are unique to vendors. In addition, the database structure is not standardized and methods of implementing functionality vary from vendor to vendor.

IV. Conclusions

Smart home and building automation systems have gained popularity in recent time. The advancement in these systems became possible due to the development of various comprehensive smart grid standards. In this paper, we have introduced a wide range of smart grid standards, proposed for home and building automations systems. These standards span from communication layers, which include physical, data link and network layers, applications, security, to automation and deployment of the system. It is expected that more advanced standards will be introduced mainly IPV6 based, in the market in the near future due to some open loops in the existing standards. In order to better utilize these standards, both the regulatory and standardization bodies and the utility companies must first recognize the existing gaps in order to rectify them in the upcoming standards.[3]

WORDS AND TERMS

interoperable adj. 彼此协作的
deployment n. 部署，调度
private residence 私人住宅，私人居所
ventilation n. 通风设备，空气流通
appliances n. 电器用具，家用电器
expandability n. 可扩展性，延伸性
in parallel 并联的，并行的，平行的
comprehensive adj. 综合性的

embed v. 使嵌入，使载入
implementation n. 实现，履行，安装启用
sporadic adj. 零星的，零散的
degrade v. 贬低，使……降解
attenuate v. 使减弱，使纤细
application protocol 应用协议
photovoltaic adj. 光电的，光伏的
regulatory adj. 管理的，控制的

NOTES

[1] With growing popularity of smart home and building automation systems, various organizations offer standards for interoperable products, enabling smart home and building automation systems that can control appliances, lighting, energy management, and security environment.

随着智能家居和楼宇自动化系统的日益普及，各种组织提供可互操作的产品标准，使智能家居和楼宇自动化系统可以控制家电、照明、能源管理和环境的安全。

[2] There are a variety of smart grid standards for home and building automation, which are either open or proprietary, developed by different organizations.

家庭和楼宇自动化有各种各样的智能电网标准，这些标准是由不同组织开发的，它们或是公开的或是专有的。

[3] In order to better utilize these standards, both the regulatory and standardization bodies and the utility companies must first recognize the existing gaps in order to rectify them in the upcoming standards.

为了更好地利用这些标准，监督和标准化机构以及电力公司必须首先认识到存在的差距，以便在新的标准中对其加以纠正。

B Cloud Computing for Industrial Automation Systems—A Comprehensive Overview

Introduction

Today, automation systems are facing fast growing market demands where agility and flexibility in production plants is needed. In addition, over the next years, the fourth industrial revolution will grow based on "intelligent" production. The main focus for Industry 4.0 are smart objects, autonomous products and decision making processes using new technologies from Information Technology (IT) domain.

Cloud computing, as a new trend from the IT area, might become an enabler for these future automation systems, because it has recently influenced many areas such as office and enterprise systems.[1] Cloud computing has rapidly emerged as an accepted computing paradigm in many enterprises worldwide due to its flexibility and many other advantages. This poses two main questions: Is there any potential to adapt this technology for usage in industrial automation and why should manufacturing companies care about using it in their automation systems? This work aims at analyzing these questions and providing detailed answers in order to identify the existing gaps where this new technology can provide solutions for industrial automation systems.

The main goal of this paper is to provide an overview of cloud computing in automation by discussing current research activities. Therefore, cloud computing will be first evaluated as a solution to implement a global automation architecture by analyzing existing works in this context. These works will be categorized with respect to the different levels of the traditional automation pyramid hierarchy. After this, the current state and future work will be determined as a part of this work. Based on already proposed architectures for future automation systems, these architectures are extended accordingly, because the existing proposals are limited to very abstract models for the future.

Related Work

Basically, existing works related to cloud computing in automation could be categorized based on their focus on each individual automation level. Most of these works aim at migrating functions and services from the common hierarchical automation architecture to a flat architecture.

1. Enterprise Management and Manufacturing Execution Level

Among the works related to the higher levels of the automation especially enterprise management level, Xu et al. discussed some of the essential features for cloud computing with regard to apply them in manufacturing management systems. "Cloud Manufacturing" is proposed as a business model to transform traditional manufacturing business model and create intelligent factory networks that encourage effective collaborations. Similarly, Tao et al. combined existing advanced manufacturing models with cloud computing technology to achieve a new computing and service-oriented manufacturing model, named cloud manufacturing (CMfg).

Pe'rez et al. proposed a new manufacturing paradigm called "Cloud agile manufacturing". The main aim for this work is to offer industrial automation functions as a service to enable the users in higher levels to access available functionalities of the automation system with minimum complexity.

Gilart-Iglesias in proposed a service model for delivering industrial machinery as a service to incorporate them easily during production process in order to facilitate self-management and proactive management of the business logic for which it is responsible.

2. Process Control Level

For the process control level, several research projects especially related to Service Oriented Architectures (SOA) have been done already. Related to these works, Delsing et al. proposed an approach to migrate from legacy industrial systems to the next generation of SOA-based automation systems. Gerach et al. proposed a private cloud model to host and deliver SIMATIC PCS7 as a generic Distributed Control System (DCS).

The IMC-AESOP project is another interesting work which aims to develop SOA-based approach for next generation of SCADA/DCS systems targeting process control applications. Similarly, Combs et al. analyzed migration of the SCADA systems to cloud computing. Consequently, SCADA providers and users can reduce costs and achieve more scalability. Furthermore, Web-oriented Automation System (WOAS) project aims to research a new architecture for automation systems based on web and cloud technologies. Staggs et al. proposed a system including a computing cloud having at least one data storage unit and one processing unit as a simple demonstrator for an industrial automation application. Beside the works in this level, still concrete implementation and evaluation for the proposals are missing within most of the works.

3. Control and Field Level

Related to the lower levels of automation, i.e. control and field levels, numbers of existing works are mainly limited due to the tough requirements of these levels. These levels basically consist of physical devices and functions. These functions can be offered as services from the cloud. Therefore, the main goal for the works related to these levels is to provide interactions between the cyberspace (cloud) and the physical world (real field devices).

A cloud solution has been introduced in which shows an application to connect sensors and actuators inside wind turbines to a cloud infrastructure.

There are several solutions related to the sensor cloud which are described in for general purposes and IT applications. Most of these works focused on connecting devices with M2M standards and interfaces to the cloud and therefore local cloud solutions, e.g. private clouds, were not addressed.

A first use case for engineering applications as a service has been investigated in. The main focus was to deliver engineering applications from the cloud to the user without having installation and maintenance efforts.

4. Summary

After analyzing the state of the art, a lack of projects and research works in the area of cloud computing for automation was noticed, especially lower levels of automation are not sufficiently addressed. The main reason for this could be the leakage of information about this buzzword in automation industry which is known as a conservative domain against new technologies.

Since cloud computing is one of the potential solutions for integration between different automation levels, the necessity for research to provide integration between available applications must be considered. The summary of existing works is shown in Table 6-2B-1. As viewed, some of the requirements, e. g. real-time and security are not sufficiently addressed and remain as open research questions for the future.

Proposed Architectures for Industrial Automation

During recent years, industrial automation witnessed the new demands and trends in different areas. First solutions for agile and smart manufacturing and other new trends have appeared. Field devices become more intelligent by embedding new functionalities inside IO devices or sensors and actuators. These trends might require the definition of new automation architectures differing from the current hierarchical automation pyramid. Among these newly proposed architectures, Vogel-Heuser et. al. introduced a new architecture as global information architecture for industrial automation. It consists of two main layers presented with two cones which are placed between business and technical processes. The lower layer represents the migration of field and control levels in traditional pyramid including devices and functionalities, whereas the upper layer represents the process control and management levels on top of the control level in the automation pyramid. An information model is inserted between these two layers to standardize the information exchange in a structured way for achieving vertical and horizontal integration between devices and entities during the whole engineering life-cycle.[2]

At the end of this proposal, the authors described the new model as a short motivation and introduction for a huge change in automation. They promised a big improvement in engineering life-cycle and more flexible operations. Based on this, cloud computing technology has been applied in as a solution to implement this newly proposed architecture for automation. The architecture is shown in Fig. 6-2B-1 and was implemented using an example application. Special automation functions and services can be offered directly as SaaS from the IT-cloud as well as Automation-Cloud(AT-Cloud). Alternatively, a PaaS is used as an automation platform to deliver specific needs for integration, e. g. process logs and plug and play parameters. The authors conclude that this solution resulted in an improved information flow.

As shown in Fig. 6-2B-1, AT-Cloud is offered to provide functions and services in lower levels and the IT-Cloud hosts applications and services in upper levels of the automation. Integrations between these two infrastructures are provided by an Information Model. Considering basic cloud computing features and architecture, the Information Model could be replaced with a service bus and can be shifted to the cloud as displayed in Fig. 6-2B-2. As a result of this migration, along with the definition of standard services, it is possible to replace IT and Automation clouds with a single cloud proposal illustrated in Fig. 6-2B-2. The integration between the cloud and real devices will be enabled by encapsulating services and functions inside delivery standards. In our proposed architecture this is introduced as Everything-as-a-Service (XaaS) for automation.

Table 6-2B-1 Categorization of related work and related projects

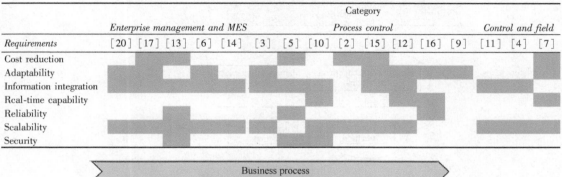

Requirements	Enterprise management and MES							Process control						Control and field		
	[20]	[17]	[13]	[6]	[14]	[3]	[5]	[10]	[2]	[15]	[12]	[16]	[9]	[11]	[4]	[7]
Cost reduction																
Adaptability																
Information integration																
Real-time capability																
Reliability																
Scalability																
Security																

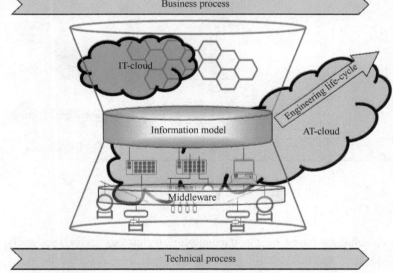

Fig. 6-2B-1 Global information architecture with use of cloud computing

As shown in Fig. 6-2B-2, the control and field levels still exist traditionally based on the well-known automation pyramid, since these levels are covering all physical devices on the factory floor. However, the upper levels have been changed. Since they are basically providing non-physical functions and services, it is possible to migrate them into the cloud. Each group of functions could be realized as individual cloud unit which addresses a particular automation level. These individual units are shown as standalone boxes for each group of functions in Fig. 6-2B-2. Consequently, these units are consisting of different applications in each level as sub-units. For instance, process control functions is including applications which deliver these kinds of functions as standard services to other units. All units are connected via a standard service bus which is offered from Service Oriented Architecture (SOA) technology to provide integration and interoperability between different objects.

The control level is divided into two parts in this architecture. The first part is the physical control level which includes common PLCs close to the technical process to be able to provide the highest performance for control loops. The second part of the control level migrated to the cloud. It offers control functions directly to field devices from the cloud with a reduced performance. This approach could be implemented as soft controllers which provide control functions via the network.

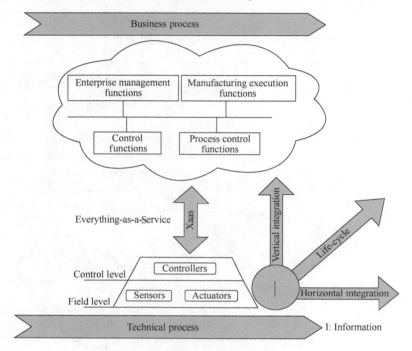

Fig. 6-2B-2 Cloud-based architecture for industrial automation

Conclusion and Outlook

In this paper, solutions based on cloud computing technology have been analyzed in order to find their potentials for a deployment in industrial automation systems. With the rise of applying information technologies such as the Internet of Things, service oriented architectures and mobile computing in industrial automation, there is a further need for a platform to provide information integration, a repository and analysis for the connected devices (things). Cloud computing could be a possible solution for this reason. On the other hand, this technology could be assumed as a solution for implementation of newly proposed architectures for automation. This would be possible with migration of current automation functions to the cloud by considering application requirements.

During the state of the art research in this paper, it was realized that there is a gap for cloud solutions in automation especially in lower levels thus far, that should be assumed as the main objectives for future work. Particularly, the control level must be further investigated in order to address reliability and real-time issues. To satisfy industrial companies for migrating to cloud-based systems, new approaches should be applied widely based on vendor independent standards to ensure their interoperability with special consideration of security issues.

WORDS AND TERMS

autonomous	adj. 自治的；自主的；自发的
enabler	n. 推动者
paradigm	n. 范例
implement	v. 实施，执行；实现，使生效
pyramid	n. 金字塔；角锥体；v. 渐增；上涨；使……渐增；使……上涨；使……成金字塔状
proactive	adj. 前摄的（前一活动中的因素对后一活动造成影响的）；有前瞻性的，先行一步的；积极主动的
scalability	n. 可扩展性；可伸缩性
cyberspace	n. 网络空间
actuator	n. 执行机构
turbine	n. ［动力］涡轮；涡轮机
interface	v. （使通过界面或接口）接合，连接；n. 界面；接口
buzzword	n. 专门术语
embed	v. 使嵌入，使插入；使深留脑中
cone	n. 圆锥体，圆锥形
encapsulate	v. 压缩；将……封进内部
interoperability	n. 互操作性，互用性
deployment	n. 调度，部署
repository	n. 知识库，智囊团
flat architecture	平面体系结构
automation architecture	自动化体系结构

NOTES

[1] Cloud computing, as a new trend from the IT area, might become an enabler for these future automation systems, because it has recently influenced many areas such as office and enterprise systems.

云计算，作为在 IT 领域的一个新趋势，可能成为未来自动化系统的推动者，因为它已经对诸如办公和企业系统等很多领域产生了影响。

[2] An information model is inserted between these two layers to standardize the information exchange in a structured way for achieving vertical and horizontal integration between devices and entities during the whole engineering life-cycle.

在这两个层之间插入一个用于标准化信息交流的信息模型。该信息交换通过在整个工程生命周期中实现设备和实体之间的垂直和水平的整合的结构化方式来实现。

C 合同与协议书常用术语和句型

合同和协议书都是两方或多方就某项事宜达成合作或约定的文件。我国的《合同法》中规定：合同是平等主体的自然人、法人、其他组织之间设立、变更、终止民事权利义务的协议。相比较而言，合同的内容通常都有详细而严格的规范，往往具有很强的法律约束力，多涉及经济领域和人事聘任等；而协议则主要涉及人员培训、技术合作等领域。

一、合同常用术语和句型

contract of employment 聘任（雇佣）合同
contract of trade 贸易合同
contract of purchase 订购合同
sales confirmation 成交确认书
name of commodity 商品名称
specification 规格

quantity 数量
country of origin 生产国别
total value 总价值
port of loading/destination 装运/目的口岸
time of shipment 装运时间
term of payment 付款条件
guarantee period 保险期
letter of credit 信用证
shipping mark 装运标记
the buyer/seller 买/卖方
the term of service 聘期，服务期
renew the contract 续约，延长合同期

unit price 单价
manufacturer 生产厂家
packing 包装
partial shipment 分装
transshipment 转船
insurance 保险
compensation allowance 补偿津贴
claim indemnity 索赔
other terms 其他条款
the engaging/engaged party 聘方/受聘方
expiration of the contract 合同到期
be covered by the seller 由卖方负责

Hereafter to be called the first/second party.
以下称甲方/乙方。

The undersigned seller and buyer have agreed to close the following transactions according to the terms and conditions stipulated below.
兹经买卖双方同意成交下列商品，特签订条款如下。

To be effected by the seller covering all risks and war risk for 2.5% of invoice value.
由卖方按发票总值的2.5%投保综合险和战争险。

To be covered by the buyer.
由买方负责。

On each package shall be stenciled conspicuously: port of destination, package number, net and gross weights, measurement and the shipping mark.
每件货物应明显标出到货口岸、件号、净重、毛重、尺码和装运标记。

Any claim shall be lodged within 120 days from the date of import.
自进口日起，索赔期为120天。

The present contract is executed in Chinese and English, both versions being equally valid.
本合同用中英文两种文字写成，两种版本具有同等效力。

Neither party shall cancel the contract without sufficient cause or reason.
如无充足理由，双方均不得解除合同。

If any other clause in this contract is conflict with the following supplementary conditions, the supplementary conditions should be taken as final and binding.
合同其他条款如与以下的附加条款相抵触时，以本附加条款为准。

二、协议书常用术语和句型

agreement on academic exchange 学术交流协议
co-operation agreement on science and technology 科技合作协议
agreement on production co-operation 生产合作协议
agreement on personnel training 人员培训协议
patterns and contents of co-operation 合作方式与内容

measures of implementation 执行措施
Sino-foreign joint venture 中外合资
duration of agreement 协议有效期
personnel matters 人员交流
the two sides 双方
technical consultation 技术咨询
financial arrangements 费用安排
conclude an agreement as follows
特签订如下协议
go into effect (be effective) from the date of signature
自签字之日起生效
The take-over will take place on...
验收将于某月某日进行
... hereby indicate the intention to enter into a program of technological co-operation on the basis of mutual benefit (to benefit both sides) ...
特此表明双方在互惠的基础上签订技术合作项目的愿望。
provide equipment, technology, engineering technicians and managerial personnel including quality inspectors
提供设备、技术、工程技术人员以及包括质量检验员在内的管理人员
establish closer co-operation in technological transfer and information exchange
在技术转让和信息交流方面建立更为密切的合作关系
must be completed 10 months after the conclusion of the present agreement
限在本协议签订后10个月内完成
This agreement is hereby made on the basis of existing contact and consultation of both sides.
根据已有的联系和协商，双方特签订此协议。
The two sides agree to co-operate with each other in research projects of common interest.
双方同意对共同感兴趣的项目进行合作。
Detailed provisions concerning the co-operation will be worked out later through consultation.
有关合作细则，由双方日后另行商定。
The provisions of this agreement may be amended at any time upon written consent of the participating co-operation.
合作双方可在任何时候经过书面协商同意后对本协议条款进行修改。
At its expiration, the agreement may be modified or the period of validity may be extended through mutual consultation.
协议期满后，可通过相互协商，对协议进行修改或延长有效期。

三、就业协议书与劳动合同

目前，我国大学毕业生与招聘单位所签订的就业协议书和劳动合同是有所区别的。

就业协议书，是由教育部高校学生司统一制定的，为高校应届毕业生在就业过程中签订就业协议的一种书面合同。其内容主要体现毕业生情况和意见、用人单位情况和意见及学校

意见，其协议约定仅指学生在毕业后到用人单位去工作的一份书面合同。毕业生就业协议书的签证部门是各级县、市人事调配部门。

劳动合同，是指劳动者同企业、事业等用人单位确立劳动关系，明确双方权利和义务的协议。劳动合同是依据《劳动法》规定以书面形式订立，其中载明合同的期限、工作内容、劳动保护和劳动条件、劳动报酬、劳动纪律、合同终止等条款，对双方当事人具有法律约束力。劳动合同的签证机关是各级劳动行政部门。

就业协议仅仅是确立了毕业生和用人单位之间的劳动关系，劳动合同更进一步确立了双方的权利和义务。因此，毕业生千万不要认为和用人单位签订了就业协议就万事大吉，应注意报到后及时和用人单位签订劳动合同。为了更加明确双方的权利和义务，毕业生可在签订就业协议时了解劳动合同的内容，尤其是工作年限和待遇等条款，毕业生也可向招聘人员索要样本或复印件。

毕业生与用人单位签订劳动合同前，也不可忽略就业协议书的签订。因为高校毕业生就业是通过各级人事调配部门在规定时间内来落实的。毕业生持学校发放的《毕业生报到证》和签订的《就业协议书》到用人单位所属县、市人事局办理报到手续，并凭其开具《行政介绍信》到具体的用人单位报到，此时的毕业生具有干部调动性。否则，毕业生就不具备干部调动性，对今后的技术职称评定、晋升、社会养老保险、退休年限等将造成一定的影响。

UNIT 3

A Smart City and the Applications

Introduction

November 6, 2008, in New York on the Foreign Relations Council, Samuel Palmisano, IBM CEO, made a speech, releasing "Smart Planet: the Agenda of the Next-Generation Leaders", and explicitly putting forward the concept of "smart planet". January 28, 2009, in the "round table", hold by the U.S. business leaders, Samuel Palmisano again putted forward the concept of "smart planet", which has received a positive response from Obama. In the same year, on February 24, during the 2009 IBM forum, D.C. Chien, the CEO of IBM in Greater China, announced "smart planet" as the latest strategy. August 7, 2009, when Premier Wen Jiabao inspected the institute of the Internet of Things in Wuxi, he proposed that in the development of sensor network, we need to plan for the future early and make breakthroughs in core technology early, reading China through "the Internet of Things".

Smart city, the important strategy of IBM, mainly focuses on applying the next-generation information technology to all walks of life, embedding sensors and equipment to hospitals, power grids, railways, bridges, tunnels, roads, buildings, water systems, dams, oil and gas pipelines and other objects in every corner of the world, and forming the "Internet of Things" via the Internet. Then we can integrate the Internet of Things through super computers and cloud computing. In this case, people can manage production and life in a more meticulous and dynamic way, achieving the state of global intelligence, and ultimately reach "Internet + Internet of Things = smart planet".

These years, with the concept of "smart planet" being putted forward, smart city, smart grid and smart enterprises have been proposed as important parts of smart planet successively. Smart city, as not only a typical application of smart planet, but also one of the most popular topics and the most cutting-edge issues, has caused widespread concern. In recent years, from London to Taipei, from New York to Singapore, one by one, the construction-project of smart city (originally known as the wireless digital city or wireless city) is the same as the spark, spreading around the world.

Smart City

"Smart city" is defined by IBM as the use of information and communication technology to sense, analyze and integrate the key information of core systems in running cities. At the same time, smart city can make intelligent response to different kinds of needs, including daily livelihood, environmental protection, public safety and city services, industrial and commercial activities.

In short, "smart city" is the actual approach of "smart planet" applying to specific region, achieving the informational and integrated management of cities. It can also be said to be an effective integration of smart planning ideas, smart construction modes, smart management methods, and smart development approaches. Through the digital grid management of urban geography, resources, environment, economic, social and other systems, as well as the digital and informational processing and application of urban infrastructure and basic environment, we can achieve intelligent urban management and services, thereby promote the more efficient, more convenient and harmonious operation of modern cities.

The structure of smart city includes perception layer, network layer and application layer, which can make the future world increasingly appreciable and measurable, increasingly interconnection and interoperability and increasingly intelligent.[1] Fig. 6-3A-1 shows the technical architecture diagram of smart city clearly.

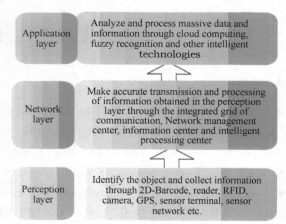

Fig. 6-3A-1 Technical architecture diagram of smart city

Relationship of Smart City and Digital City

Digital city refers to taking remote sensing (RS), global positioning system (GPS), geographic information systems (GIS) and other spatial information technologies as the main means, constructing geographic information framework of digital city, building urban geographic information platform for public service. And through the construction of infrastructure, we can complement the development and integration of all kinds of geographic information, and achieving the network, digitalization, intelligence of urban economy, social, ecology and other aspects of each operation.

Smart city is on the basis of comprehensive digital city, establishing visual and measurable urban management and operations with intelligence. The idea is that sensors are equipped to the various objects to form the Internet of Things, and achieve the integration of the Internet of Things through super computers and cloud computing. All in all, smart city is the product of digital city combined with the Internet of Things. Fig. 6-3A-2 shows the relationship of smart city and digital city.

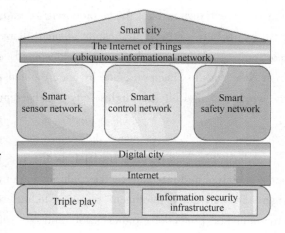

Fig. 6-3A-2 Relationship of smart city and digital city

Main Content of the Construction of Smart City Application

Smart City will be the future trend of urban development. Generally, the construction of smart city can be divided into three levels, including the construction of public infrastructure, construction of public platform for smart city, the construction of application systems. In this three-level, the construction of application systems is particularly important, and has earned great concern across the country. Currently, in addition to defense and national security applications, smart city has been typically applied to various aspects. Fig. 6-3A-3 shows some typical applications.

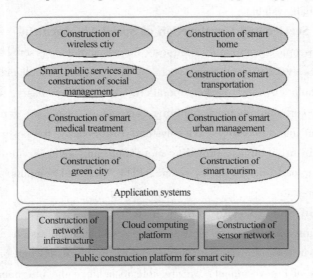

Fig. 6-3A-3 Construction frame of application systems for smart city

1. Construction of Wireless City

On the basis of powerful fiber-optic network and the technology of Wi-Fi, Mesh and WiMAX, with further extension, wireless broadband network can be built. At the same time, wireless broad base station will cover the whole city. And it can provide many functions of urban management and service systems for the public, business, foreign visitors, tourists and government agencies with its bandwidth. The functions include mobile wireless video surveillance, mobile video conferencing, mobile dispatching emergency response, and emergency telecommunications.

2. Construction of Smart Home

Sensor devices, including radio frequency identification devices, infrared sensors, global positioning system, laser scanners and so on, can be combined with the Internet to form the Internet of Things. Then all the items in life can be taken as a terminal to be brought into the network, achieving the centralized and remote control of electrical and mechanical equipment through the interaction of various networks and terminals, which can be convenient to user identification and management. For example, the realization of smart home can be convenient for us to achieve the intelligent control of lighting and electrical appliances, as well as receive the intelligent notification of home alarm messages. At the same time, whether indoors or outdoors, we can benefit from the information technological achievements of smart city.

3. Construction of Smart Transportation

According to their needs and traffic situation, every city can take good advantage of sensor network, the Internet of Things and other technical means to change the traditional transport system, and establish the smart traffic management system, including adaptive traffic signal (automatic control of traffic lights according to flow time) control system, urban traffic control system and so on. At this point, the smart traffic management system can achieve the integration of urban planning, construction, management and operations, and provide comprehensive support for other subsystems of smart urban system.

4. Smart Public Service and Construction of Social Management

In daily life, for people's complaints, requests for assistance, personal management of social affairs and other aspects, we can establish a social service system, which can cover the intelligent management of the whole city and market operation. And on this basis, we can provide basic platform services for urban comprehensive planning, emergency response, community management, and turn the government into a one-stop service system. In this case, the government can collect and analyze real-time data in urban areas, providing more rapid and agile service to the public. At this point, the public can upload information by phone, PDA, personal computer and DV, and achieve real-time query of affair-state.

5. Construction of Smart Urban Management

Based on the ubiquitous network in the future city, we can make use of 3G, wireless network, the next generation of wireless networks with broadband or the future 4G network. At the same time, through private network of e-government, we are able to achieve the interoperability of supervision center, command center and functions. In the private network, it can be possible to transfer data, work together, and form the core of the urban management system, achieving seamless management.

Smart city management can achieve the management and service of urban grid. In this case, it can bring us effective management and service of urban infrastructure, population and events through intelligent collection and analysis of data.

6. Construction of Smart Medical Treatment

With great potential to be applied into smart medical treatment, the Internet of Things can help hospitals to achieve the smart medical care and intelligent management of medical materials, and support the digital collection, processing, storage, transmission and sharing of internal medical information, equipment information, drug information, personnel information and management information. Besides, it can also meet the needs of intelligent management and supervision in medical information, medical equipment and supplies, intelligent management and supervision of public health, solving so many issues, for example the weak support of health care platform, the overall low level of medical services and the medical safety hazards.

7. Construction of Green City

Within the city limits, we can achieve the networking and interoperability of various systems posed by different devices, and make comprehensive use of various resources of monitoring and alarm to establish a new urban model and a system of green city. [2] With the technological platform, we can achieve not only the networking, interoperability and mutual control of various devices and systems, but also the collection, transmission, storage, display and control of audio, video and alarm information. At the same time, it can also achieve the linkage with the alarm system, and provide data interface to other systems.

8. Construction of Smart Tourism

Smart tourism is the only way to travel information. It should be based on the existing tourism information and infrastructure, taking good advantage of digital information and the Internet of Things to achieve the establishment of a set of solutions, which can consider and fulfill the management and tourism-related tasks, such as tourism online services, management of customer relation, management of operational area, development of domestic and overseas tourism market, intelligent management system of monitor, collection of tourism information and forecast of tourism development. Moreover, based on the integration of hardware and software platform for information and services of smart city, smart tourism can be taken good advantage of to fully integrated tourism market, tourist attractions, government departments and relevant information and services of enterprises to promote the development of tourism.

Key and Difficulty of Construction of Smart City

With the increasing needs of urban management, construction and operation in reasonable planning of urban space and function layout, detection of incident, emergency response and public information services, the construction of smart city is facing great difficulties, including the follows:

1. Management, Integration and Release of Massive Urban Spatial-Temporal Data

At present, the existing data sources of digital urban information systems are still too simple (mainly basic remote sensing, mapping data and three-dimensional model of urban street), and often appear in the style of simple query and analysis of data and the performance of three-dimensional visualization, without taking good advantage of the multi-source and multi-temporal data to make high spatial and temporal analysis to assist decisions on urban management. Particularly, in current urban information systems, the modeling of temporal data is still weak. The data structure and organization of multi-temporal data existing in the database is a quasi-static solution, and it cannot meet the practical needs of digital real-time updates, historical reconstruction and future prediction. Therefore, the key of static digital city developing to smart city is the breakthrough of integration of multi-source heterogeneous urban information, urban address code, management of urban infrastructure and components, quickly update of online spatial data, multi-dimensional visualization of spatial data, and the construction of multi-dimensional temporal data model with flexible structure and adaptability.

2. Large-Scale Space-Time Information and Efficient Services

Spatial information of smart city comes from a wide variety of sensors, controllers and computing terminals, and is maintained by computers and storage nodes of different departments, so how to manage and coordinate the equipment with various structures and wide-area distribution is a great challenge of constructing service platform. On the other hand, information on smart city contains not only a large amount of structured data, such as temperature, voltage, geographical coordinates and so on, but also a lot of unstructured data, such as pictures, audio and video files. And whether we can store and manage the huge amounts of data effectively will directly affect the performance of information services. Finally, smart city is related to intelligent analysis of urban information, decision support, public affairs and many other applications. Besides, a large amount of real-time tasks also need to respond to user requests quickly, which has higher demands for information services. For the above character of spatial information services and unsolved problems, we need to study smart urban information service system at all levels, proposing effective efficient methods, which is a wide integration of internet devices, mass data and large numbers of users.

3. Model of Heterogeneous Sensor Data and Expression of the Internet of Things

As the important basis of developing smart city, the Internet of Things is also an important part of smart city. But as the demands in sensor platform, observation mechanisms, processes, location information and technical requirements are different, how to build models describing sensor information, including location attribute, observation object, time and status is a difficult technical problem. As the urban affairs are perennial gradual change and instantaneous mutations, how to observe data and detect abnormal events effectively is also a difficulty.

4. Technology of Intelligent Analysis and Decision Support

With diverse sources and so many related departments, spatio-temporal data can achieve real-time update quickly. So how to create a unified understanding of data semantics, and extract new knowledge based on specific cycle data and real-time data is a technical difficulty in establishing knowledge base of smart city. And the specific implementation needs the effective support of municipal departments. At the same time, the extraction of knowledge should be based on manual extraction, and supplemented by automatic computer analysis. Through the accumulation of knowledge, we can analyze the law, establishing an effective intellectual model and developing tools of knowledge extraction. Besides, we can accomplish the extraction and update of real-time knowledge, which is based on automatic analysis of computer.

5. Sound Information Service and Shared Policy Mechanism and Legal Protection

As smart city involves many sectors and industries, we need to break trade barriers so as to achieve information sharing and information exchange between traffic, public security and other departments. We also need to learn from developed countries in information sharing and services, establishing coalition mechanism of national spatial information sharing and services, and accomplishing sound information services and sharing policy mechanism and legal protection to arouse the enthusiasm of departments and industries so as to try for broad social partition.

Value and Outlook of Smart City

Internationally, it is a good opportunity for China that smart planet starts with smart city. At present, China has achieved good development in information technology, and as to the technology covered in smart planet, including sensor technology, network technology, physical networking technology and intelligent information processing technology, our country have certain R&D infrastructure and industrial capacity. On this basis, we should combine our economic and social development needs to increase the investment in material, technical, and personnel infrastructure. At the same time, we should select a number of developing priorities in a planned way, for example smart transport, smart grid, deploying as soon as possible, achieving a more thorough sense, more comprehensive interconnection and more intelligence.

In terms of our country, with the development of the Internet of Things, people's daily life will be changed dramatically. At the same time, it also brings us to the development of smart city, which is based on the Internet of Things. In the encourage of global trend of smart city and national policy, a numbers of cities, for example Beijing, Shanghai, Guangdong, Wuhan, have taken smart city as an important research, and participated in the construction of "Smart City" and "reading China", trying to stand out in the future economic competition with the layout of the Internet of Things.

WORDS AND TERMS

forum *n.* 论坛，讨论
sensor *n.* 传感器
grid *n.* 输电网，网格
pipeline *n.* 管道，输油管
integrate *v.* 使成整体，使结合，使合并
meticulous *adj.* 一丝不苟的，小心翼翼的，拘泥小节的
cutting-edge *n.* 尖端，前沿；*adj.* 先进的，尖端的
wireless *adj.* 无线的，无线电的；*n.* 无线电
livelihood *n.* 生计，生活，营生
infrastructure *n.* 基础设施，公共建设
appreciable *adj.* 可评估的，可感知的

interconnection *n.* 互连，互相联络
interoperability *n.* 互操作性，互用性
spatial *adj.* 空间的
platform *n.* 平台
fiber-optic *adj.* 光学纤维的；*n.* 光纤，光纤技术
comprehensive *adj.* 综合的，广泛的，全面的
mesh *n.* 网
broadband *n.* 宽频，宽波段，宽带
bandwidth *n.* ［电子］［物］带宽
Internet of Things 物联网
spatial information technology 空间信息技术

NOTES

［1］The structure of smart city includes perception layer, network layer and application layer, which can make the future world increasingly appreciable and measurable, increasingly interconnection and interoperability and increasingly intelligent.

智慧城市的结构包括感知层、网络层和应用程序层，可以使未来世界的规模变得越来越可观并且是可以衡量出来的，同时也可使城市之间互相连接并具有互操作性，进而使城市越来越智能。

［2］Within the city limits, we can achieve the networking and interoperability of various systems posed by different devices, and make comprehensive use of various resources of monitoring and alarm to establish a new urban model and a system of green city.

在城市范围内，我们可以实现网络和不同设备构成的不同系统之间的互操作性，并综合利用各种资源的监控和报警系统，以建立一种新的城市模式——绿色城市。

B Knowledge Management System Design Model for Smart Enterprises

Introduction

The increasing number of companies and businesses every year in Indonesia based on the 2006 Indonesian Economic Census indicates that the competition between companies in Indonesia, moreover the world, has increased. Therefore, companies must have very good performance in order to increase their competitive advantage.

Smart enterprise can be assumed as a developed enterprise that has strong human capital, social capital, and information and communication technology (ICT) infrastructure. These components will increase an organization's competitive advantage if managed properly.

Malcolm Baldrige Criteria for Performance Excellence (MBCfPE) is a set of criteria that provides a system's perspective for understanding performance management in an organization.[1]

Nowadays, about 76 countries in the world have adopted MBCfPE. MBCfPE is updated annually. Therefore, performance improvement through MBCfPE is believed as the most comprehensive method in the world.

Business world today is faced with the reality that the only sustainable source of competitive advantage is knowledge. Knowledge is a frame of experiences, values, contextual information, expert insights, and intuition that provides a basic framework for evaluating and incorporating new experiences and information. In an organization, knowledge is planted not only in repositories, but also in routines, processes, practices, and norms, which as a whole is known as knowledge management (KM).

KPMG Consulting said that based on the experiences of some companies, KM has given various benefits for them. Through KM, 71% of respondent companies have been able to take better decisions, 68% have achieved faster response to key business issues, and 64% have been providing better customer service.

KM also serves as a foundation for the performance management system in MBCfPE to build a knowledge-based system in order to improve the performance and competitiveness of the organization. By looking at the needs of KM in an organization, the term Knowledge Management System (KMS) is emerging as an information system that can support the sustainability of KM.

Literature Review

1. Smart Enterprise

Smart city is a developed urban area that creates sustainable economic development and high quality of life by excelling in multiple key areas: economy, mobility, environment, people, living, and governance. Excelling in these key areas can be done through strong human capital, social capital, and/or ICT infrastructure.

Based on that definition, smart enterprise can be assumed as a developed enterprise that has strong human capital, social capital, and/or ICT infrastructure.

2. Malcolm Baldrige Criteria for Performance Excellence

MBCfPE is a framework that is used as a tool to diagnose the performance of the organization as a whole. The framework is illustrated in Fig. 6-3B-1. A feedback based on the diagnostic results is used to guide the company to improve its performance towards performance excellence.

MBCfPE criteria are embodied in seven categories, which are:

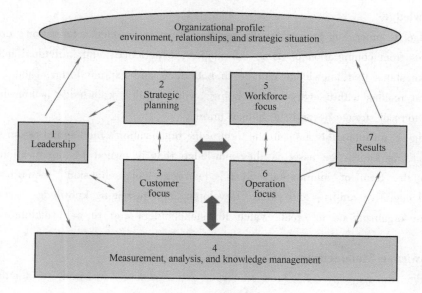

Fig. 6-3B-1 MBCfPE Framework

1) Leadership. This category essentially discusses the main aspects of the responsibilities of senior leadership and organizational governance system.

2) Strategic planning. Through this category, MBCfPE emphasizes that the long-term resilience of the organization and the competitive environment are key strategic issues that are important to complete the organization planning.

3) Customer focus. Through this category, MBCfPE emphasizes that relationship with the customer is an important end goal of learning and performance excellence strategies as a whole.

4) Measurement, analysis, and knowledge management. This category indicates that the information, analysis, and knowledge management is the main resource of competitive advantage and increased productivity.

5) Workforce focus. This category discusses key workforce practices, which directly leads to the creation and maintenance of a high-performing work environment.

6) Operation focus. This category emphasizes the importance of organizational core

competencies and the way to protect and exploit these competencies to the success and sustainability of the organization.

7) Results. This category examines the performances and the improvements in the five key outcome areas, which are Product and Process Outcomes, Customer-Focused Outcomes, Workforce-Focused Outcomes, Leadership and Governance Outcomes, and Financial and Market Outcomes.

Categories 1 through 6 are grouped into process category group and category 7 is included in result category group. In total, there are 17 items that are spread in those categories. Each item consists of one or more areas to be discussed. Organization must provide a response to the specific requirements in the form of questions for each area. These responses are taken as the basis for organizational performance excellence assessment.

3. Knowledge

Knowledge in general is human faculty resulting from interpreted information; understanding that germinates from combination of data, information, experience, and individual interpretation. Ikujiro Nonaka states that knowledge can be in both tacit and explicit form. Tacit knowledge is knowledge that resides within every human being, while explicit knowledge is knowledge that is stored in the form of storage beyond the human brain.

Knowledge is an intangible asset in the form of the organization's intellectual resources that have accumulated. This knowledge asset is the knowledge that is owned by organizations and their workforce in the form of information, ideas, learning, comprehension, memories, insights, technical and cognitive ability, and skills. Developing and managing knowledge asset is the key component for organizations to create value for stakeholders and to help maintain competitive advantage.

4. Knowledge Management

Knowledge Management (KM) is strategies and processes designed to identify, capture, structure, value, leverage, and share an organization's intellectual assets to enhance its performance and competitiveness.[2]

To achieve an effective KM, there are four stages in the process of establishing KM, illustrated in Fig. 6-3B-2, which are:

1) Knowledge creation. Ikujiro Nonaka and Hirotaka Tekuchi created a model named SECI (Socialization, Externalization, Combination, and Internalization). This model is a process model to understand the dynamic nature of knowledge creation and to manage that process effectively. This model shows that tacit and explicit knowledge interacts with each other in a continuous process until new knowledge is created. Socialization is knowledge creation process from tacit to tacit form, Externalization is from tacit to explicit form, Combination is from explicit to explicit form, and Internalization is from

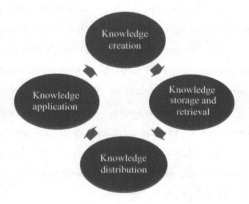

Fig. 6-3B-2 KM process

explicit to tacit form.

2) Knowledge storage and retrieval. Organization memory is an important aspect of effective KM. This memory includes knowledge in various forms, including written documentations, structured information in electronic databases, codified human knowledge in expert systems, processes, documented organization procedures, and tacit knowledge from individual and network.

3) Knowledge distribution. By considering knowledge that is distributed naturally, an important process of KM in organization arrangement is knowledge distribution to the location where that knowledge is needed and used. There are five components that enable knowledge distribution, which are the knowledge that is considered as the value of a unit source of knowledge, the nature of the motivation of the source, a transmission line, and the motivation or the absorptive capacity of the receiver.

4) Knowledge application. An important aspect of knowledge-based theory is that competitive advantage is situated in knowledge application, not in the knowledge itself. There are three key mechanisms in knowledge integration application, which are direction, routines, and independent team creation.

5. Knowledge Management System

Knowledge Management System (KMS) can be regarded as an information system that is used to run the KM.[3] Information system is a combination of hardware, software, infrastructure, and trained personnel, organized to facilitate planning, control, coordination, and decision making in an organization. Similar to information systems in general, KMS also consists of several constituent components, which are:

1) Human.
2) Process.
3) Content of knowledge.
4) Technology.

In category 4 (Measurement, Analysis, and Knowledge Management), MBCfPE gives the requirements to manage information technology and knowledge through these aspects:

1) Providing information and easy access for employees, suppliers, partners, customers, and other interested parties.

2) Ensuring that the hardware and software are reliable, safe, and easy to use (user-friendly).

3) Storing data and information, including maintaining hardware and software systems to ensure availability at all times necessary.

4) Guaranteeing the accuracy, integrity, reliability, timeliness, security, and confidentiality of organizational data, information, and knowledge.

5) Managing organizational knowledge related to the following matters:

a. Collection and transfer of employees' knowledge.

b. Transfer of relevant knowledge from and to customers, suppliers, and business partners.

c. Rapid identification, sharing, and implementation of best practices.

d. Summary and transfer of relevant knowledge for use in the company's strategic planning process.

6. Knowledge Management Road Map

According to Amrit Tiwana, KM implementation can be done by following these 10 steps:

1) Analyzing the existing infrastructure.

2) Aligning KM and business strategy.

3) Designing the KM infrastructure.

4) Auditing existing knowledge assets and systems.

5) Designing the KM team.

6) Creating the KM blueprint.

7) Developing the KMS.

8) Deploying the KMS.

9) Managing change, culture, and reward structures.

10) Evaluating performance.

Proposed KMS Design Model

This chapter contains a discussion of KMS design model that adapts the KM Road Map proposed by Amrit Tiwana. MBCfPE framework will be used in several stages in this model. In general, this model comprises ten stages that are grouped into 3 phases as illustrated in Fig. 6-3B-3.

1. KM Maturity Assessment

This stage aims to determine how mature the existing KM in the organization at the time so that the pace of change to achieve improvement can be seen. Data collection method in this stage is distributing questionnaires to members of the organization.

Categories in MBCfPE framework can be used as categories in measurement of organizational performance improvement as a result from the application of KM. Based on that relationship, the KM maturity assessment in this model can be performed using MBCfPE categories that focused on the scope of KM.

2. KMS Components Identification

Through identifying the components of KM, context for the KMS to be designed is expected to be generated.

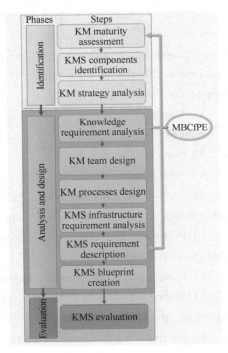

Fig. 6-3B-3 Proposed KMS design model

This is necessary in order to design effective KMS. According to the types of KMS component that have been mentioned in the literature review, the objects that will be identified at this stage include:

1) Processes involving the KMS.

2) Technology infrastructures used in the KMS.

3) Stakeholders involved in the KMS.

4) Existing content of knowledge in the KMS which is needed by stakeholders.

3. KM Strategy Analysis

KM strategy should align with the organization's business strategy. Therefore, KM strategies in organization will vary from one organization to the other. This phase is similar to the second phase of the KM Road Map proposed by Amrit Tiwana, which identifies whether the organization is likely to have a Codification or Personalization KM strategy. This identification is done by weighting the two characteristics of the strategy based on the vision, mission, and strategy of the organization's business.

4. Knowledge Requirement Analysis

After identifying the components of knowledge that already exists in the organization, the thing that will be executed at this step is the analysis of the knowledge that is supposed to be managed in the organization. Analysis can be conducted by identifying of important knowledge in each category of MBCfPE and aligning it with the KM strategy.

5. KM Team Design

In this case, KM team is a stakeholder associated with KMS. In general, the KM team serves to design, develop, implement, and deploy KMS in organizations. Roles in the KM team that are proposed in this model are Administrator, Chief Knowledge Officer, Knowledge Officer, and Knowledge Worker. After identifying stakeholder component in identification phase, this step can be conducted by analyzing whether there is a role which not be acted by stakeholder in organization or not.

6. KM Processes Design

If there is a KM process which not supported by organization activities, knowledge flow in organization cannot run well. Thus, this step proposes flow chart of complete KM processes according to literature review as solution to achieve effective and efficient KM.

7. KM Infrastructure and Method Requirement Analysis

This step is conducted by analyzing the right technology for each layer in KMS architecture proposed by Amrit Tiwana. The technology must support the goals of the KM. To ensure that the technology is able to support each KM process, each of these technologies is mapped into four KM processes that have been described in literature review. In this mapping process, it was found that not all processes could be run with only the support of technology alone. Therefore, support in the form of the method is also proposed in this model.

8. KMS Requirement Description

KMS functional and non-functional requirements will be described in this step. Functional requirement is the need to meet the goals and strategies of KM in the organization, while the non-functional requirement is the need to increase system performance. This requirement description will facilitate the implementation of later step in the design model.

9. KMS Blueprint Creation

The result of this step is a description of KMS in the form of a mock-up. Mock-up is a model or replica built full-size or to scale for testing or training purposes. Some things that have to be conducted in the creation of a blueprint are KMS global design, software design, database design, and mock-up design.

10. Evaluation

KMS is expected to be effective and is designed according to the needs of its users. This phase is conducted to evaluate whether the mock-up KMS meets the criteria or not. The questions in the questionnaire are formulated based on organization needs that are already identified in the previous phases.

Summary

The categories in MBCfPE framework can be used as categories in the measurement of organizational performance improvement resulting from the application of KM. To be able to improve organizational performance through KM, KMS design model is proposed by adapting KM road map of Amrit Tiwana and incorporating MBCfPE categories.

Although the MBCfPE framework is used in the model proposed in this paper, not all categories of MBCfPE are discussed in depth. The model is focused only on meeting some of the criteria in category 4 (Measurement, Analysis, and KM) of MBCfPE. The point is that the proposed design model can help organization in becoming a smart enterprise, thereby improving the KM.

The next step that should be done after the study is to implement the proposed KMS design model into a case study in an organization. Through this process, the factors that should be corrected from this model can be determined, so that a better design model can be implemented in the organization in order to become a smart enterprise.

WORDS AND TERMS

performance   n. 性能
sustainable   adj. 可持续的
mobility   n. 移动性，机动性
integrity   n. 完整性
retrieval   n. 检索
mechanism   n. 机制，原理
accuracy   n. 定位
timeliness   n. 快速性
confidentiality   n. 机密性
deploy   v. 配置

NOTES

[1] Malcolm Baldrige Criteria for Performance Excellence (MBCfPE) is a set of criteria that provides a system's perspective for understanding performance management in an organization.

马尔科姆·波多里奇卓越绩效标准（MBCfPE）是一组从系统的角度来理解组织中绩效管理的标准。

[2] Knowledge Management (KM) is strategies and processes designed to identify, capture, structure, value, leverage, and share an organization's intellectual assets to enhance its performance and competitiveness.

知识管理（KM）是一种战略和过程，旨在识别、捕捉、建构、评价、利用和共享组织的知识资产，以提高其性能和竞争力。

［3］Knowledge Management System（KMS）can be regarded as an information system that is used to run the KM.

知识管理系统（KMS）可以看作是用于运行知识管理的信息系统。

C 广　告

科技产品的广告（advertisement 或 ad）是对某种产品进行宣传的一种方式。用户通常在通过广告获得某生产厂家产品的有关信息后，经过进一步了解才能决定是否购买，所以说广告在当今的市场经济中占有很重要的地位。对生产厂家来说，广告应以精炼的词句、新颖的表现手法和具有吸引力的创意为主，配以相关的照片和图表，说明产品的特点、技术规格、性能用途等，吸引相关人员的兴趣，同时将购买该产品有关的价格、地址、电话等写清楚，以便于产品的销售，达到广告宣传的目的。对用户来说，广告一方面为购买产品和设备提供参考资料，另一方面也可通过广告了解相关专业的发展水平和动向。

广告的结构没有特别固定的格式，形式比较自由，大体上包括标题、主体、通信地址和电话等部分。

标题（headline）除了要有特殊醒目的字体外，其内容更要明确突出，从而达到吸引人详细阅读广告主体正文的作用。下面是一些不同类型的广告标题句：

1. 简短型

SAFE　以安全一词强调产品性能和吸引人的注意力。

2. 具体型

Aircraft Connectors, AC & DC, Power Attachable Plugs, Cable Assembly Kit, Receptacles, and more！（飞机用各种交直流电气接插件、电源插头、电缆成套组件、插座、应有尽有！）以产品的名称直接引起有关用户的兴趣和注意。

3. 叙述型

General Electric technologies let you put electricity to work at its fullest. 这是通用电气公司的广告，为叙述型。

4. 设问型

If 23 out of 30 manufactures use it, what kind of secret is it?

5. 修辞型

Spend a dime, save your time. 押韵（dime：十美分的硬币）

Airlink goes where wires won't. 拟人（Airlink：一种网络产品）

The last mile solution with the longest range & maximum data rate. 排比（三个最高级并列）

Here is the proof that something small can be powerful. 引用（Something small can be powerful 是一句英语成语）

广告的主体部分应进一步介绍产品的特点、规格、性能和用途，其形式有两种：条目式和叙述式。条目式是指使用短句或简单的名词结构分项对产品进行说明，每项可以以数字或实心圆点等方式开头；叙述式是以使用简单句为主的分段说明方式，其内容要求重点突出，简明易懂，由于复合复杂句会使文句难以理解，应避免过多使用。主体的内容可以从突出产品特点、强调公司信誉、关心用户利益等多个角度入手。

广告的最后应附上厂方的通信地址、电话、网址、E-mail 地址等信息以便客户进行咨询和购买产品。

具体的例子请参考相关的专业期刊或其他资料，本书不再一一举例。

参 考 文 献

[1] Electric power systems, Microsoft ® Encarta ® Online Encyclopedia, 2006.
[2] http://itcofe.web.cern.ch/itcofe/Services/PLC/WhatIsPLC/welcome.html, 2006.
[3] http://zone.ni.com/devzone/cda/tut/p/id/3755, 2006.
[4] WAJID ALI. Embedded systems. http://www.programmersheaven.com, 2006.
[5] http://www.chiltern.gov.uk/site/scripts/documents_info.php documentID = 357&pageNumber = 1, 2006.
[6] Electric vehicles. http://www.pge.com/about_us/environment/electric_vehicles/#topic7, 2006.
[7] BIMAL K BOSE. Modern power electronics and ac drives [M]. Upper Saddle River: Prentice-Hall Inc., 2004.
[8] EXANDER FARNSWORTH. Virtual manufacturing—a growing trend in automation [J]. Evolution (Bussiness and Technology Magazine from SKF), 2004(4).
[9] JAMES A RHG, Henry W Kraebber. Computer-integrated manufacturing [M]. 3rd ed. Upper Saddle River: Prentice-Hall, Inc., 2004.
[10] ANDREW BATEMAN, IAIN PATERSON STEPHENS. The DSP handbook—algorithms, applications and design techniques [M]. 北京: 机械工业出版社, 2003.
[11] THEODORE WILDI. Electrical machine, drive, and power system [M]. Upper Saddle River: Prentice-Hall Inc., 2002.
[12] THEODORIDIS, KOUTROUMBAS. Pattern recognition [M]. 2nd ed. 北京: 机械工业出版社, 2003.
[13] JONAS BERGE. Fieldbuses for process control: engineering operation, and maintenance [M]. North Carolina: ISA Press, 2002.
[14] ZHANG, WEI MICHAEL S BRANICKY, STEPHEN M PHILLIPS. Stability of networked control systems [J]. IEEE Control Systems Magazine, 2001.
[15] 蔡临宁. 电力专业英语阅读与翻译 [M]. 北京: 机械工业出版社, 2000.
[16] DAVID J DOLEZILEK. Power system automation [M]. Washington: Schweitzer Engineering Laboratories, Inc., 1999.
[17] DANIEL KLERFORS. Artificial neural networks [M]. St. Louis: St. Louis University, 1998.
[18] ANTSAKLIS P J. Defining intelligence control—report of task force on intelligent control [J]. IEEE Control Syst Mag, 1994, 14(3).
[19] 陈素英, 等. 土木建筑系列英语3. 计算机与自动化 [M]. 北京: 建筑工业出版社, 1987.
[20] GILBERT D P. Recent advances and future trends in electrical machine drives [J]. GEC Review, 1995, 10(2).
[21] SMITH C A. Principles and practice of automatic process control [M]. New York: John Wiley & Sons, 1985.
[22] 赵静鹏, 等. 学术英语教程 [M]. 北京: 国防工业出版社, 1995.
[23] 韩其顺, 等. 英汉科技翻译教程 [M]. 上海: 上海外语教育出版社, 1990.
[24] 尤毓国. 英语科技情报文献阅读 [M]. 北京: 新时代出版社, 1991.
[25] 周季特, 等. 写译 [M]. 哈尔滨: 哈尔滨工业大学出版社, 1994.
[26] 周虹, 等. 电脑英语快易通 [M]. 北京: 电子工业出版社, 1995.
[27] 丁往道, 等. 英语写作手册 [M]. 北京: 外语教学与研究出版社, 1994.
[28] 赵圣骅, 等. 英语面试成功之路 [M]. 上海: 上海外语教育出版社, 1995.
[29] AZEGAMI M. A systematic approach to intelligent building design [J]. IEEE Communications Magazine., 1993, 10(46).

[30] 赵链. 研究生英语写作 [M]. 北京:北京大学出版社, 1995.
[31] 冯子良, 等. 电气与电子文献检索与利用 [M]. 大连:大连理工大学出版社, 1992.
[32] 廖世翘. 英语应用文大全 [M]. 北京:机械工业出版社, 1987.
[33] VEGTE J V. Feedback control system [M]. Upper Saddle River:Prentice-Hall Inc., 1986.
[34] DISTEFANO J J. Schaum's outline of theory and problems of feedback and control systems [M]. New York: McGraw-Hill Inc., 1990.
[35] WANG FEIYUE. Scanning the Issue and Begond:Toward Knowledge Automation [J]. IEEE Transactions on Intelligent Transportation Systems, 2014, 15(1).
[36] KENWRIGHT D. Automation or Interaction:What's best for big data? [C]. Visualization '99. Proceedings. IEEE, 1999.
[37] TARIQ M, ZHOU Z, WU J, et al. Smart grid standards for home and building automation [C]. IEEE International Conference on Power System Technology(POWERCON). Auckland, 2012
[38] GIVEHCHI O, TRSEK H, JASPERNEITE J. Cloud computing for industrial automation systems-Acomprehensive overview [C]. IEEE 18th Conference on Emerging Technologies & Factory Automation. Cagliari, 2013.
[39] KEHUA SU, JIE LI, HONGBO FU. Smart city and the applications [C]. International Conference on Electronics Communications and Control (ICECC). Ningbo, 2011.
[40] RAHMATIA D. SURENDRO K. Knowledge management system design model for smart enterprise [C]. International Conference on ICT for Smart Society. Jakarta, 2013.
[41] https://en.wikipedia.org/wiki/Artificial_intelligence#History, 2018.
[42] BENGIO Y, COURVILLE A, VINCENT P. Representation Learning:A Review and New Perspectives [J]. IEEE Transactions on Pattern Analysis and Machine Intelligence, 2013, 35(8): 1798-1828.